Fields Institute Communications

Volume 89

The Communications series features conference proceedings, surveys, and lecture notes generated from the activities at the Fields Institute for Research in the Mathematical Sciences. The publications evolve from each year's main program and conferences. Many volumes are interdisciplinary in nature, covering applications of mathematics in science, engineering, medicine, industry, and finance.

Anthony Bahri • Lisa Jeffrey • Taras Panov •
Donald Stanley • Stephen Theriault
Editors

Toric Topology and Polyhedral Products

Springer

Editors
Anthony Bahri
Department of Mathematics
Rider University
Lawrence Township, NJ, USA

Lisa Jeffrey
Department of Mathematics
University of Toronto
Toronto, ON, Canada

Taras Panov
Mathematics & Mechanics
Moscow State University
Moscow, Russia

Donald Stanley
Mathematics and Statistics
University of Regina
Regina, SK, Canada

Stephen Theriault
Mathematical Sciences
University of Southampton
Southampton, UK

ISSN 1069-5265 ISSN 2194-1564 (electronic)
Fields Institute Communications
ISBN 978-3-031-57203-6 ISBN 978-3-031-57204-3 (eBook)
https://doi.org/10.1007/978-3-031-57204-3

Mathematics Subject Classification: 05E14, 13F55, 14F45, 14M15, 14M25, 14P25, 53D05, 55N31, 55N91, 55P35, 55U10, 57R19, 57S12, 57S26, 58D27

This Springer imprint is published by the registered company Springer Nature Switzerland AG
The registered company address is: Gewerbestrasse 11, 6330 Cham, Switzerland

Preface

The subject of the January–June 2020 Fields Thematic Program was toric topology and polyhedral products. Toric topology at its core is the study of torus actions on manifolds whose orbit space is a simple polytope and whose homotopy orbit space has cohomology a Stanley–Reisner ring. It lies at the intersection of a wide array of mathematical disciplines, including topology, convex geometry, combinatorics, commutative algebra, and algebraic and symplectic geometry. A key construction is a moment-angle complex, which is formed by gluing together Cartesian products of discs and spheres in a manner determined by a simplicial complex. This has a functorial generalization to Cartesian products of spaces with fixed subspaces, again with the gluing determined by a simplicial complex. Remarkably, this unifies several seemingly distinct constructions, for example, the Whitehead filtration in homotopy theory, complements of complex coordinate subspace arrangements in combinatorics, intersections of quadrics in complex-analytic geometry, and the Davis and Salvetti complexes in geometric group theory that have been used to great effect in studying right-angled Artin and Coxeter groups.

Toric topology arose relatively recently. Its genesis was a paper of Davis and Januszkiewicz from 1991 that identified an elegant family of manifolds with torus actions and described many of their geometric, topological, and combinatorial properties. Over the next several years, Buchstaber and Panov thoroughly investigated and generalized these spaces, laying the groundwork for toric topology as its own discipline. They also anticipated the functorial generalization to polyhedral products, later given full development in a seminal paper by Bahri, Bendersky, Cohen, and Gitler. Some of the great success stories in the area occurred at the intersection of different disciplines, such as the insightful use of moment-angle complexes by Bosio and Meersseman to study intersections of quadrics in complex-analytic geometry and the use of moment-angle complexes by Grbić and Theriault and later Iriye and Kishimoto to identify the homotopy types of the complements of families of complex coordinate subspace arrangements. Driving problems include describing the homology, cohomology, and homotopy types of polyhedral products and describing the subtle interplay between the algebraic and symplectic geometry and the combinatorics of toric varieties and symplectic manifolds.

The Fields Thematic Program was aimed at intensifying research into the driving problems, attracting talented young researchers to the area, and exploring potential interactions with other areas of mathematics. The program had two schools and four workshops: the first school and two workshops were focused directly on studying toric topology and polyhedral products while the second school and two workshops were aimed at crossing boundaries by investigating the interplay between toric topology, polyhedral products, geometric group theory, and applied topology. The program was also honored to have two distinguished speakers, Victor Buchstaber and Ulrike Tillmann, and two clay lecturers, Gunnar Carlsson and Shmuel Weinberger.

This volume consists of original articles that emerged from the discussions and interactions of researchers involved with the program. It spans a wide array of topics, reflecting the very wide reach of toric topology and polyhedral products.

The program organizers would like to thank the Fields Institute and its staff for all their support. Special mention should go to the Fields Directors Ian Hambleton, who patiently answered our many questions during the application process, and Kumar Murty, who encouraged us to not let the Covid outbreak and lockdown in 2020 put a stop to what was a thriving program. We would also like to thank the other workshop organizers: Peter Bubenik, Graham Denham, Matthias Franz, Jelena Grbić, Ian Leary, Vidit Nanda, and Piotr Przytyckyi, and the many speakers and participants who made the program a decisive success.

Lawrenceville, NJ, USA — Anthony Bahri
Toronto, ON, Canada — Lisa Jeffrey
Moscow, Russia — Taras Panov
Regina, SK, Canada — Donald Stanley
Southampton, UK — Stephen Theriault
October, 2022

Contents

Connected Sums of Sphere Products and Minimally Non-Golod Complexes

Steven Amelotte

1 Introduction

A central construction in toric topology functorially assigns to each finite simplicial complex K on m vertices a finite CW complex $\mathcal{Z}_K$, called the *moment-angle complex*, which comes equipped with a natural action of the m-torus $T^m = (S^1)^m$. Various homological invariants of Stanley-Reisner rings of basic importance in combinatorial commutative algebra are given geometric realizations by $\mathcal{Z}_K$ and related spaces. For example, the homotopy orbit space of $\mathcal{Z}_K$ is the *Davis-Januszkiewicz space* whose cohomology (with coefficients in a commutative ring $\mathbf{k}$) is the Stanley-Reisner ring $\mathbf{k}[K]$ itself, while the ordinary cohomology of $\mathcal{Z}_K$ recovers its Koszul homology (see [2, 8]):

$$H^*(\mathcal{Z}_K; \mathbf{k}) \cong \mathrm{Tor}^*_{\mathbf{k}[v_1,\dots,v_m]}(\mathbf{k}[K], \mathbf{k}).$$

Combinatorial properties of simplicial complexes and, in particular, homological properties of their Stanley-Reisner rings can therefore be studied by investigating the homotopy types of moment-angle complexes. This point of view has recently been useful in establishing the Golod property for $\mathbf{k}[K]$ for certain families of simplicial complexes by applying homotopy theoretic methods to show that the corresponding moment-angle complex $\mathcal{Z}_K$ is homotopy equivalent to a wedge of spheres (see, e.g. [11–14]). Here, $\mathbf{k}[K]$ is *Golod* if all products and higher Massey products vanish in $\mathrm{Tor}^*_{\mathbf{k}[v_1,\dots,v_m]}(\mathbf{k}[K], \mathbf{k})$. (Golodness for a graded or local ring implies that its Poincaré series is a rational function, and an equivalent definition can be given in

S. Amelotte (✉)
Department of Mathematics, University of Rochester, Rochester, NY, USA

Department of Mathematics, University of Western Ontario, London, ON, Canada
e-mail: samelot@uwo.ca

A. Bahri et al. (eds.), *Toric Topology and Polyhedral Products*, Fields Institute Communications 89, https://doi.org/10.1007/978-3-031-57204-3_1

terms of a certain equality of formal power series; cf. [10].) A simplicial complex K is called *Golod* if $\mathbf{k}[K]$ is Golod for every field $\mathbf{k}$.

Berglund and Jöllenbeck observed in [3] that the Golod property is stable under deletion of vertices and introduced the notion of a *minimally non-Golod* complex, that is, a non-Golod simplicial complex that becomes Golod after deleting any of its vertices. Using combinatorial arguments, they showed that the boundary complexes of stacked polytopes are minimally non-Golod. Further examples have been given by Limonchenko [15], who showed that the nerve complexes of even-dimensional dual neighbourly polytopes and certain generalized truncation polytopes are minimally non-Golod. In each of these cases, the corresponding moment-angle complex $\mathcal{Z}_K$ is well known to be a smooth manifold diffeomorphic to a connected sum of sphere products with two spheres in each product (see [4] and [9]). Moreover, in [12], it was shown that if K is a flag complex, then K being minimally non-Golod is equivalent to the condition that $\mathcal{Z}_K$ is a connected sum of twofold products of spheres. The authors raised the question of whether, more generally, K is minimally non-Golod for all simplicial complexes for which $\mathcal{Z}_K$ has such a diffeomorphism type ([12, Question 3.5]). The purpose of the present paper is to answer this question affirmatively.

Theorem 1.1 *If $\mathcal{Z}_K$ is homeomorphic to a connected sum of sphere products with two spheres in each product, then K is minimally non-Golod.*

Remark 1.2 The statement of Theorem 1.1 is not true if the "homeomorphic" condition is replaced by "homotopy equivalent". In Sect. 3, we give a counterexample in the form of a cone over a minimally non-Golod complex K for which $\mathcal{Z}_K$ is homotopy equivalent to a connected sum of sphere products (see Example 3.4), and we prove that iterated cones of this type are the only such counterexamples. We also remark that the converse of Theorem 1.1 is not true. In [16], stellar subdivisions of minimal triangulations of T^2 and $\mathbb{C}P^2$ are shown to be minimally non-Golod complexes whose corresponding moment-angle complexes are not connected sums of sphere products.

We give two proofs of Theorem 1.1. The first is a direct proof that makes crucial use of the assumption that the moment-angle complex $\mathcal{Z}_K$ has the structure of a closed manifold. The second comes as a corollary of the slightly more general Theorem 1.3 below. See Sect. 4 for an analogue for real moment-angle complexes.

For $n \geqslant -1$, let Δ^n be the standard n-simplex, where $\Delta^{-1} = \varnothing$ is the empty simplicial complex.

Theorem 1.3 *If $\mathcal{Z}_K$ is homotopy equivalent to a connected sum of sphere products with two spheres in each product, then $K = \Delta^n * L$ for some $n \geqslant -1$ where L is Gorenstein* and minimally non-Golod.*

Let K be a simplicial complex on the vertex set $[m]$. The *star* of a vertex $v \in K$ is the subcomplex

$$\mathrm{star}_K(v) = \{\sigma \in K \mid \{v\} \cup \sigma \in K\}.$$

The *core* of K is then defined to be the full subcomplex

$$\mathrm{core}(K) = K_{\{v\in[m]\,|\,\mathrm{star}_K(v)\neq K\}}$$

of K on the restricted vertex set $\{v \in [m] \mid \mathrm{star}_K(v) \neq K\}$. Note that any simplicial complex can be written as a join

$$K = \Delta^n * \mathrm{core}(K)\,, \tag{1}$$

where Δ^n is the simplex with (possibly empty) vertex set $[m]\backslash\mathrm{core}(K) = \{v \in [m] \mid \mathrm{star}_K(v) = K\}$. Since the moment-angle complex functor carries simplicial joins to Cartesian products and $\mathcal{Z}_{\Delta^n} \cong D^{2(n+1)}$ is contractible, it follows from (1) that the homotopy type of a moment-angle complex $\mathcal{Z}_K$ is determined by the core of K. In the notation of Theorem 1.3, it will be shown that $L = \mathrm{core}(K)$ and hence that any simplicial complex satisfying the hypothesis of Theorem 1.3 has a minimally non-Golod core. The Gorenstein* property implies that $\mathcal{Z}_{\mathrm{core}(K)}$ is a closed orientable manifold.

The author would like to thank the Fields Institute for Research in Mathematical Sciences and the organizers of the Thematic Program on Toric Topology and Polyhedral Products for providing both an excellent setting for research and an opportunity to present this work. In particular, comments and questions from Taras Panov, Don Stanley and Stephen Theriault led to improvements in this paper.

2 Preliminaries

Throughout this paper, K will denote a finite abstract simplicial complex on the vertex set $[m] = \{1, \ldots, m\}$. We always assume that $\varnothing \in K$ and that K has no ghost vertices, that is, $\{i\} \in K$ for all $i = 1, \ldots, m$.

Let $(\underline{X}, \underline{A}) = \{(X_i, A_i)\}_{i=1}^m$ be a sequence of pointed CW pairs. For each simplex $\sigma \in K$, define $(\underline{X}, \underline{A})^\sigma$ to be the subspace of $\prod_{i=1}^m X_i$ given by

$$(\underline{X}, \underline{A})^\sigma = \{(x_1, \ldots, x_m) \in \textstyle\prod_{i=1}^m X_i \mid x_i \in A_i \text{ for } i \notin \sigma\}.$$

The *polyhedral product* of $(\underline{X}, \underline{A})$ corresponding to K is then defined by

$$(\underline{X}, \underline{A})^K = \bigcup_{\sigma\in K} (\underline{X}, \underline{A})^\sigma \subseteq \prod_{i=1}^m X_i. \tag{2}$$

In the case where $(X_i, A_i) = (D^2, S^1)$ for each $i = 1, \ldots, m$, the polyhedral product corresponding to K is called the *moment-angle complex*, denoted $\mathcal{Z}_K$. Similarly, the *real moment-angle complex* $\mathcal{R}_K$ is defined by the polyhedral product $(\underline{X}, \underline{A})^K$ with $(X_i, A_i) = (D^1, S^0)$ for each $i = 1, \ldots, m$. Generalizing these two cases of special interest, much of the work to date on the homotopy theory of polyhedral products has focused on pairs of the form (CX_i, X_i), where CX_i is the reduced cone on X_i. For a sequence of spaces $\underline{X} = \{X_i\}_{i=1}^m$, let $C\underline{X} = \{CX_i\}_{i=1}^m$.

For $I \subseteq [m]$, the *full subcomplex* of K on the vertex set I is defined by

$$K_I = \{\sigma \in K \mid \sigma \subseteq I\}.$$

We will simply write $(\underline{X}, \underline{A})^{K_I}$ for the polyhedral product of $\{(X_i, A_i)\}_{i\in I}$ corresponding to K_I. For any vertex $i \in [m]$, we denote by $K - \{i\}$ the *deletion complex* of i defined by

$$K - \{i\} = \{\sigma \in K \mid i \notin \sigma\}.$$

Note that $K - \{i\}$ is the full subcomplex of K on the restricted vertex set $[m]\backslash\{i\}$. We will need the following basic but useful property of polyhedral products associated with full subcomplexes.

Proposition 2.1 *Let K be a simplicial complex on the vertex set $[m]$ and let $I \subseteq [m]$ be a non-empty subset. Then $(\underline{X}, \underline{A})^{K_I}$ is a retract of $(\underline{X}, \underline{A})^K$.*

Proof Let $I = \{i_1, \ldots, i_k\} \subseteq [m]$ where $1 \leqslant i_1 < \cdots < i_k \leqslant m$ and $k \geqslant 1$. The simplicial inclusion $K_I \longrightarrow K$ induces a map of polyhedral products $j_I\colon (\underline{X}, \underline{A})^{K_I} \longrightarrow (\underline{X}, \underline{A})^K$. It is straightforward to check that the projection $\prod_{j=1}^m X_j \longrightarrow \prod_{j=1}^k X_{i_j}$ restricts to a map $r\colon (\underline{X}, \underline{A})^K \longrightarrow (\underline{X}, \underline{A})^{K_I}$ such that

$$(\underline{X}, \underline{A})^{K_I} \xrightarrow{j_I} (\underline{X}, \underline{A})^K \xrightarrow{r} (\underline{X}, \underline{A})^{K_I}$$

is the identity map. □

Let $\widehat{(\underline{X}, \underline{A})}^K$ denote the image of $(\underline{X}, \underline{A})^K$ under the natural quotient map $\prod_{i=1}^m X_i \longrightarrow \bigwedge_{i=1}^m X_i$. After suspending, the retraction maps of Proposition 2.1 for all full subcomplexes of K can be added together using the co-H-space structure on $\Sigma(\underline{X}, \underline{A})^K$ to obtain the following splitting due to Bahri, Bendersky, Cohen and Gitler.

Theorem 2.2 ([1, Theorem 2.10]) *Let K be a simplicial complex on the vertex set $[m]$ and let $(\underline{X}, \underline{A}) = \{(X_i, A_i)\}_{i=1}^m$ be a sequence of pointed CW pairs. Then there is a natural homotopy equivalence*

$$\Sigma(\underline{X}, \underline{A})^K \simeq \bigvee_{I\subseteq[m]} \Sigma\widehat{(\underline{X}, \underline{A})}^{K_I}.$$

The authors of [1] go on to further identify the spaces appearing on the right-hand side of the wedge decomposition above in various cases of interest.

Theorem 2.3 ([1]) *Let K be a simplicial complex on the vertex set $[m]$ and let $\underline{X} = \{X_i\}_{i=1}^m$ be a sequence of pointed CW complexes. Then there is a homotopy equivalence*

$$\Sigma(C\underline{X}, \underline{X})^K \simeq \bigvee_{I \notin K} \Sigma^2|K_I| \wedge \widehat{X}^I$$

where $\widehat{X}^I = X_{i_1} \wedge \cdots \wedge X_{i_k}$ for $I = \{i_1, \ldots, i_k\}$.

In the special case where $X_i = S^1$ for all $i = 1, \ldots, m$, Theorem 2.3 gives the following suspension splitting for moment-angle complexes, which can be regarded as a geometric realization of the description of $\mathrm{Tor}^*_{\mathbf{k}[v_1,\ldots,v_m]}(\mathbf{k}[K], \mathbf{k})$ given by Hochster's formula.

Corollary 2.4 *There is a homotopy equivalence*

$$\Sigma \mathcal{Z}_K \simeq \bigvee_{I \notin K} \Sigma^{|I|+2}|K_I|.$$

Next, we use the splittings above to prove a lemma that will be needed in the proof of Theorem 1.3 to compare the homotopy types of moment-angle complexes associated with a simplicial complex K and its deletion complexes $K - \{i\}$.

Lemma 2.5 *Let K be a simplicial complex on the vertex set $[m]$ and let $\underline{X} = \{X_j\}_{j=1}^m$ be a sequence of pointed CW complexes, which are non-contractible. Let $i \in [m]$. Then the natural inclusion $(C\underline{X}, \underline{X})^{K-\{i\}} \longrightarrow (C\underline{X}, \underline{X})^K$ is a homotopy equivalence if and only if $K = \{i\} * (K - \{i\})$.*

Proof If $K = \{i\} * (K - \{i\})$ is the cone over the deletion complex $K - \{i\}$, then permuting coordinates defines a homeomorphism

$$(C\underline{X}, \underline{X})^K \cong CX_i \times (C\underline{X}, \underline{X})^{K-\{i\}},$$

where the sequence of pairs of spaces $(C\underline{X}, \underline{X})$ on the right-hand side is understood to be $\{(CX_j, X_j)\}_{j \in [m]\setminus\{i\}}$. The natural inclusion $(C\underline{X}, \underline{X})^{K-\{i\}} \longrightarrow (C\underline{X}, \underline{X})^K$ composed with the homeomorphism above is the inclusion of the right-hand factor in the product $CX_i \times (C\underline{X}, \underline{X})^{K-\{i\}}$, which is a homotopy equivalence since CX_i is contractible.

Conversely, suppose $(C\underline{X}, \underline{X})^{K-\{i\}} \longrightarrow (C\underline{X}, \underline{X})^K$ is a homotopy equivalence. By Theorem 2.3, there is a suspension splitting

$$\begin{aligned}\Sigma(C\underline{X}, \underline{X})^K &\simeq \bigvee_{I \notin K} \Sigma^2|K_I| \wedge \widehat{X}^I \\ &\simeq \Big(\bigvee_{\substack{I \notin K \\ i \notin I}} \Sigma^2|K_I| \wedge \widehat{X}^I\Big) \vee \Big(\bigvee_{\substack{I \notin K \\ i \in I}} \Sigma^2|K_I| \wedge \widehat{X}^I\Big)\end{aligned}$$

$$\simeq \quad \Sigma(C\underline{X}, \underline{X})^{K-\{i\}} \vee \Big(\bigvee_{\substack{I \notin K \\ i \in I}} \Sigma^2 |K_I| \wedge \widehat{X}^I \Big),$$

and, up to homotopy, the suspended inclusion $\Sigma(C\underline{X}, \underline{X})^{K-\{i\}} \longrightarrow \Sigma(C\underline{X}, \underline{X})^K$ is given by the inclusion of the first wedge summand. Since this is a homotopy equivalence by assumption, it follows that $\Sigma^2|K_I| \wedge \widehat{X}^I$ must be contractible for every non-face $I \notin K$ containing the vertex i. As the CW complexes $X_1, \ldots, X_m$ are all non-contractible, so are their smash products $\widehat{X}^I = \bigwedge_{j \in I} X_j$, so this implies in particular that $\Sigma^2|K_I|$ is contractible for every non-face $I \notin K$ containing i.

To show that $K = \{i\} * (K - \{i\})$, it suffices to show that $\{i\} \cup \sigma \in K$ whenever $\sigma \in K$. First observe that $\{i, j\} \in K$ for all $j \in [m]$, since otherwise we would have $\Sigma^2|K_{\{i,j\}}| = \Sigma^2 S^0 \simeq S^2 \not\simeq *$, contradicting the conclusion of the previous paragraph. Next, assume inductively that $\{i\} \cup \sigma \in K$ for every simplex $\sigma \in K$ with $|\sigma| = n$. Let $\tau = \{j_1, \ldots, j_{n+1}\} \in K$. Then for each $1 \leqslant k \leqslant n+1$, we have $\{j_1, \ldots, \widehat{j_k}, \ldots, j_{n+1}\} \in K$, which implies $\{i, j_1, \ldots, \widehat{j_k}, \ldots, j_{n+1}\} \in K$. Now every proper subset of $\{i, j_1, \ldots, j_{n+1}\}$ is a simplex of K, so if $\{i, j_1, \ldots, j_{n+1}\} \notin K$, then

$$\Sigma^2|K_{\{i,j_1,\ldots,j_{n+1}\}}| = \Sigma^2 \partial \Delta^{n+1} \simeq S^{n+2} \not\simeq *,$$

which is a contradiction. Therefore, $\{i\} \cup \tau \in K$, which completes the induction. □

By iterating Lemma 2.5, we obtain the following simple combinatorial characterization of when the inclusion of a full subcomplex induces a homotopy equivalence of polyhedral products.

Proposition 2.6 *Let K be a simplicial complex on the vertex set $[m]$ and let $\underline{X} = \{X_j\}_{j=1}^m$ be a sequence of pointed CW complexes, which are non-contractible. For $I \subseteq [m]$, let $j_I \colon (C\underline{X}, \underline{X})^{K_I} \longrightarrow (C\underline{X}, \underline{X})^K$ be the natural map induced by the inclusion $K_I \subseteq K$. The following conditions are equivalent:*

(a) $j_I \colon (C\underline{X}, \underline{X})^{K_I} \longrightarrow (C\underline{X}, \underline{X})^K$ *is a homotopy equivalence;*
(b) $\mathrm{core}(K) \subseteq K_I$;
(c) $\mathrm{star}_K(v) = K$ *for all* $v \in [m] \backslash I$;
(d) $\mathrm{link}_K(v) = K - \{v\}$ *for all* $v \in [m] \backslash I$;
(e) $K = \Delta^{m-|I|-1} * K_I$.

Proof The equivalence of conditions (b), (c), (d) and (e) follows immediately from the definitions.

(e) ⇒ (a): This can be proved exactly as in the proof of Lemma 2.5.

(a) ⇒ (e): Write $[m] \backslash I = \{j_1, \ldots, j_p\}$ and note that the map j_I factors as a composite of inclusions

$$(C\underline{X}, \underline{X})^{K-\{j_1,\ldots,j_p\}} \longrightarrow \cdots \longrightarrow (C\underline{X}, \underline{X})^{K-\{j_1\}} \longrightarrow (C\underline{X}, \underline{X})^K$$

where each map above has a left inverse by Proposition 2.1. Therefore, if j_I is a homotopy equivalence, then so is each map in the composite, so by Lemma 2.5, we obtain

$$\begin{aligned} K &= \{j_1\} * (K - \{j_1\}) \\ &\doteq \{j_1\} * \{j_2\} * (K - \{j_1, j_2\}) \\ &\doteq \{j_1\} * \cdots * \{j_p\} * (K - \{j_1, \ldots, j_p\}) \\ &= \Delta^{m-|I|-1} * K_I , \end{aligned}$$

as desired. □

3 Proofs of Theorems 1.1 and 1.3

In this section, we restate and prove the main results and discuss some consequences. We begin with a lemma well known to homotopy theorists and include a short proof for completeness.

Lemma 3.1 *If a space Y is a homotopy retract of a simply connected wedge of spheres $\bigvee_{\alpha \in \mathcal{I}} S^{n_\alpha}$, then Y has the homotopy type of a wedge of spheres.*

Proof Suppose Y is a homotopy retract of $\bigvee_{\alpha \in \mathcal{I}} S^{n_\alpha}$ where $n_\alpha \geqslant 2$ for all $\alpha \in \mathcal{I}$. Then there is a map $r\colon \bigvee_{\alpha \in \mathcal{I}} S^{n_\alpha} \longrightarrow Y$ inducing a split epimorphism in integral homology, and the Hurewicz natural transformation gives a commutative diagram

$$\begin{array}{ccc} \pi_*(\bigvee_{\alpha \in \mathcal{I}} S^{n_\alpha}) & \longrightarrow & H_*(\bigvee_{\alpha \in \mathcal{I}} S^{n_\alpha}) \\ \downarrow r_* & & \downarrow r_* \\ \pi_*(Y) & \longrightarrow & H_*(Y). \end{array}$$

The bottom horizontal arrow is an epimorphism since the top horizontal and right vertical ones are. By hypothesis, $H_*(Y)$ is a graded free abelian group, so by choosing Hurewicz pre-images of the elements of a basis for $H_*(Y)$ and taking their wedge sum, we obtain a map from a wedge of spheres into Y inducing an isomorphism in homology. This map is therefore a homotopy equivalence by Whitehead's theorem since it also follows from the hypothesis that Y has the homotopy type of a simply connected CW complex. □

Theorem 3.2 *If $\mathcal{Z}_K$ is homeomorphic to a connected sum of sphere products with two spheres in each product, then K is minimally non-Golod.*

Proof Suppose there is a homeomorphism

$$\mathcal{Z}_K \cong \mathop{\#}_{k=1}^{\ell} (S^{n_k} \times S^{n-n_k})$$

where ℓ is finite and $3 \leqslant n_k \leqslant n-3$ for each $k = 1, \ldots, \ell$ since every moment-angle complex is a finite *two*-connected CW complex. Note that $H^*(\mathcal{Z}_K)$ has a non-trivial cup product, so K is not Golod. Let $i \in [m]$ be a vertex of K and let $j : \mathcal{Z}_{K-\{i\}} \longrightarrow \mathcal{Z}_K$ be the map induced by the inclusion $K - \{i\} \subseteq K$. It follows from the definition of a polyhedral product (2) that any point $(z_1, \ldots, z_m) \in (D^2)^m$ with $|z_i| < 1$ lies in $\mathcal{Z}_K$ outside the image of j, and hence j is not surjective. Since $\mathcal{Z}_K$ is a closed manifold by assumption, the complement of a point in $\mathcal{Z}_K$ deformation retracts onto the $(n-1)$ skeleton of $\mathcal{Z}_K$. Therefore, up to homotopy, j lifts through the $(n-1)$ skeleton of $\#_{k=1}^{\ell}(S^{n_k} \times S^{n-n_k})$, which is $\bigvee_{k=1}^{\ell}(S^{n_k} \vee S^{n-n_k})$ since the connected sum of sphere products has the homotopy type of a wedge of spheres with a single top cell attached by a sum of Whitehead products of the form $w_k : S^{n-1} \longrightarrow S^{n_k} \vee S^{n-n_k}$.

Combining the above observation with the fact that j admits a retraction $r : \mathcal{Z}_K \longrightarrow \mathcal{Z}_{K-\{i\}}$ by Proposition 2.1, we obtain a diagram

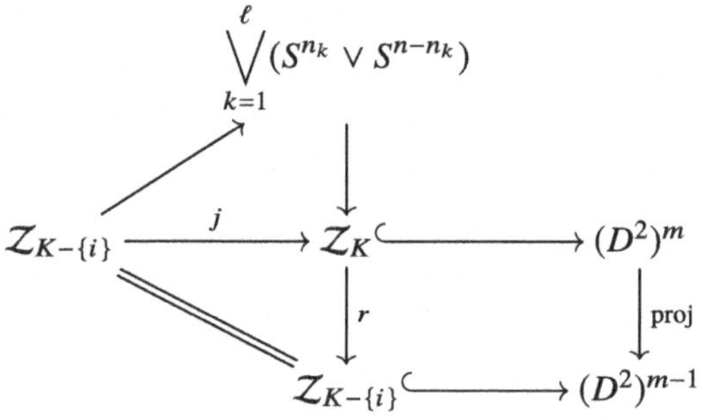

where the bottom triangle and square commute and the top triangle commutes up to homotopy. It follows that $\mathcal{Z}_{K-\{i\}}$ is a homotopy retract of $\bigvee_{k=1}^{\ell}(S^{n_k} \vee S^{n-n_k})$ and hence is homotopy equivalent to a wedge of spheres by Lemma 3.1. Consequently, $K - \{i\}$ is Golod, which implies K is minimally non-Golod as this holds for every vertex i of K. □

Let $\mathcal{K}$ denote the collection of simplicial complexes whose corresponding moment-angle complexes are homeomorphic to connected sums of sphere products with two spheres in each product. Then $\mathcal{K}$ includes the nerve complexes of all simple polytopes obtained by vertex truncations of one or a product of two simplices and all even-dimensional dual neighbourly polytopes, as well as all simplicial complexes obtained from these by applying the simplicial wedge construction or vertex truncation operations in any order (see [9] and [7]).

Corollary 3.3 *Let $K \in \mathcal{K}$. Then every proper full subcomplex K_I of K has the property that $\mathcal{Z}_{K_I}$ is homotopy equivalent to a wedge of spheres. In particular, the Stanley-Reisner ring* $\mathbf{k}[K_I]$ *is Golod over any ring* $\mathbf{k}$.

It is not true that K is minimally non-Golod whenever $\mathcal{Z}_K$ has the *homotopy type* of a connected sum of twofold products of spheres. We describe the smallest possible counterexample below before turning to the proof of Theorem 1.3.

Example 3.4 Consider the simplicial complex K on $five$ vertices with facets $\{1, 2, 5\}$, $\{2, 3, 5\}$, $\{3, 4, 5\}$ and $\{1, 4, 5\}$. Observe that K is the cone over the boundary of a square and can be written as the join $K = K_4 * \{5\}$. It is easy to see that $\mathcal{Z}_{K_4} \cong S^3 \times S^3$. (More generally, if K_m is the boundary of an m-gon with $m \geqslant 4$, then $\mathcal{Z}_{K_m}$ is homeomorphic to a connected sum of sphere products by [4].) It follows that $\mathcal{Z}_K$ has the homotopy type of a connected sum of sphere products since

$$\mathcal{Z}_K \cong \mathcal{Z}_{K_4} \times \mathcal{Z}_{\{5\}} \cong S^3 \times S^3 \times D^2 \simeq S^3 \times S^3,$$

but K is not minimally non-Golod since its deletion complex $K - \{5\} = K_4$ is not Golod.

Theorem 3.5 *If $\mathcal{Z}_K$ is homotopy equivalent to a connected sum of sphere products with two spheres in each product, then $K = \Delta^d * L$ for some $d \geqslant -1$ where L is Gorenstein* and minimally non-Golod.*

Proof Suppose there is a homotopy equivalence

$$\mathcal{Z}_K \simeq \mathop{\#}_{k=1}^{\ell} (S^{n_k} \times S^{n-n_k})$$

for some $\ell \geqslant 1$ and $3 \leqslant n_k \leqslant n-3$ for each $k = 1, \ldots, \ell$. For each vertex $i \in [m]$, consider the natural inclusion $j \colon \mathcal{Z}_{K-\{i\}} \longrightarrow \mathcal{Z}_K$ and the induced homomorphism

$$j^* \colon \mathbb{Z} \cong H^n(\mathcal{Z}_K) \longrightarrow H^n(\mathcal{Z}_{K-\{i\}}).$$

By Proposition 2.1, j^* has a right inverse, so either j^* is an isomorphism or else $H^n(\mathcal{Z}_{K-\{i\}}) = 0$. If j^* is an isomorphism, then the Poincaré duality of $H^*(\mathcal{Z}_K)$ implies that j induces an isomorphism in cohomology in all dimensions and is thus a homotopy equivalence. In this case, we obtain that $K = \{i\} * (K - \{i\})$ by Lemma 2.5. It follows that the set of all vertices $i \in [m]$ for which the map j^* above is an isomorphism span a simplex Δ^d in K and that $K = \Delta^d * L$, where L is the full subcomplex of K on the set of vertices $i \in [m]$ for which $j \colon \mathcal{Z}_{K-\{i\}} \longrightarrow \mathcal{Z}_K$ is not a homotopy equivalence. (Note that $L = \mathrm{core}(K)$ by Proposition 2.6 and that $-1 \leqslant d \leqslant m-5$ since L is a simplicial complex on $m-d-1$ vertices and $\mathcal{Z}_L \simeq \mathcal{Z}_K$ cannot have the homotopy type of a connected sum of twofold products of spheres if L has less than $four$ vertices.)

For each vertex i of L, we have that $H^n(\mathcal{Z}_{L-\{i\}}) = 0$. Since $\mathcal{Z}_{L-\{i\}}$ is a retract of $\mathcal{Z}_L \simeq \mathcal{Z}_K \simeq \#_{k=1}^{\ell}(S^{n_k} \times S^{n-n_k})$, it follows that $\mathcal{Z}_{L-\{i\}}$ has the homotopy type of a simply connected CW complex of dimension less than n. Therefore, the map $\mathcal{Z}_{L-\{i\}} \longrightarrow \mathcal{Z}_L$ lifts up to homotopy through the $(n-1)$ skeleton of $\mathcal{Z}_L \simeq \#_{k=1}^{\ell}(S^{n_k} \times S^{n-n_k})$. The same argument as in the proof of Theorem 1.1 now shows that $\mathcal{Z}_{L-\{i\}}$ is homotopy equivalent to a wedge of spheres and hence that $L - \{i\}$ is Golod.

Finally, it follows from the Poincaré duality of $H^*(\mathcal{Z}_K) \cong \mathrm{Tor}^*_{\mathbb{Z}[v_1,\ldots,v_m]}(\mathbb{Z}[K], \mathbb{Z})$ that K is a Gorenstein complex (see [5, Theorem 4.6.8]). Thus $L = \mathrm{core}(K)$ is a Gorenstein* complex. □

Remark 3.6 A combinatorial-topological characterization due to Stanley [18] states that a simplicial complex K is Gorenstein* if and only if K is a generalized homology sphere. In [6], it was shown that a moment-angle complex $\mathcal{Z}_K$ is a closed topological manifold of dimension $m+n$ if and only if K is a generalized homology $(n-1)$ sphere. In particular, any moment-angle complex satisfying the hypothesis of Theorem 1.3 is in fact homeomorphic to a product of disks and a closed orientable manifold with the homotopy type of a connected sum of sphere products.

4 An Analogue for Real Moment-Angle Complexes

In this section, we prove an analogue of Theorem 1.3 for real moment-angle complexes. Recall that the real moment-angle complex corresponding to K is defined by the polyhedral product $\mathcal{R}_K = (C\underline{X}, \underline{X})^K$ for the sequence $\underline{X} = \{X_i\}_{i=1}^m$ with $X_i = S^0$ for each $i = 1, \ldots, m$.

Example 4.1 Let K_m be the boundary of an m-gon. If $m \geqslant 4$, then the corresponding real moment-angle complex $\mathcal{R}_{K_m} \cong \#_{k=1}^{g}(S^1 \times S^1)$ is an orientable surface of genus $g = 1 + (m-4)2^{m-3}$ by a result attributed to Coxeter (see [5, Proposition 4.1.8]). In this case, each deletion complex $K_m - \{i\}$ is a path graph, which is Golod, and $\mathcal{R}_{K_m-\{i\}}$ is homotopy equivalent to a wedge of circles.

As the example above illustrates, real moment-angle complexes need not be simply connected. For this reason, we will need a stronger version of Lemma 3.1. A proof that the statement of Lemma 3.1 still holds without the simply connectedness hypothesis, provided that the index set $\mathcal{I}$ is finite, is given in [17, Theorem 3.3].

A further modification to the proof of Theorem 1.3 is required to relate the homotopy type of $\mathcal{R}_K$ to the homotopy type of $\mathcal{Z}_K$ and hence to the Golodness of K. For this, we refer to the work of Iriye and Kishimoto [14] on the fat wedge filtration of $\mathcal{R}_K$ and its relation to the homotopy type of polyhedral products of the form $(C\underline{X}, \underline{X})^K$.

Theorem 4.2 *If $\mathcal{R}_K$ is homotopy equivalent to a connected sum of sphere products with two spheres in each product, then $K = \Delta^d * L$ for some $d \geqslant -1$ where L is minimally non-Golod.*

Proof Suppose $\mathcal{R}_K$ is homotopy equivalent to a connected sum of sphere products $\#_{k=1}^{\ell}(S^{n_k} \times S^{n-n_k})$ with $\ell \geqslant 1$ and $1 \leqslant n_k \leqslant n-1$ for each $k = 1, \ldots, \ell$. As in the proof of Theorem 1.3, $K = \Delta^d * L$ where $L = \mathrm{core}(K)$ has the property that $j\colon \mathcal{R}_{L-\{i\}} \longrightarrow \mathcal{R}_L$ is not a homotopy equivalence for any vertex i of L by Proposition 2.6. A priori, this does not immediately imply that j does not induce an isomorphism in cohomology since $\mathcal{R}_L$ and its retract $\mathcal{R}_{L-\{i\}}$ are not necessarily simply connected. However, the proof of the forward implication in Lemma 2.5 shows that if $\Sigma j: \Sigma\mathcal{R}_{L-\{i\}} \longrightarrow \Sigma\mathcal{R}_L$ is a homotopy equivalence, then $L = \{i\} * (L - \{i\})$, contradicting that $\{i\} \in L = \mathrm{core}(L)$. Thus, Σj is not a homotopy equivalence, which implies that Σj does not induce an isomorphism in cohomology since the suspensions $\Sigma\mathcal{R}_{L-\{i\}}$ and $\Sigma\mathcal{R}_L$ are simply connected. It follows from the Poincaré duality of $H^*(\mathcal{R}_L) \cong H^*(\#_{k=1}^{\ell}(S^{n_k} \times S^{n-n_k}))$ that

$$j^*\colon \mathbb{Z} \cong H^n(\mathcal{R}_L) \longrightarrow H^n(\mathcal{R}_{L-\{i\}})$$

is not an isomorphism, and hence the retract $\mathcal{R}_{L-\{i\}}$ does not contain the top cell of $\mathcal{R}_L$. Since $\mathcal{R}_{L-\{i\}}$ is then a homotopy retract of $\bigvee_{k=1}^{\ell}(S^{n_k} \times S^{n-n_k})$, we conclude that $\mathcal{R}_{L-\{i\}}$ is homotopy equivalent to a wedge of spheres by Mohareri et al. [17, Theorem 3.3].

In [14], the fat wedge filtration of a real moment-angle complex is shown to be a cone decomposition. Since, for each vertex i of L, $\mathcal{R}_{L-\{i\}}$ is homotopy equivalent to a wedge of spheres, it follows that the attaching maps in this cone decomposition for $\mathcal{R}_{L-\{i\}}$ are null homotopic, and by [14, Theorem 1.2], the decomposition of $\Sigma(C\underline{X}, \underline{X})^{L-\{i\}}$ in Theorem 2.3 desuspends for any $\underline{X}$. In particular, $\mathcal{Z}_{L-\{i\}}$ is a suspension, which implies that $L - \{i\}$ is Golod (see [14, Proposition 6.5]). □

Corollary 4.3 *If $\mathcal{R}_K$ is homeomorphic to a connected sum of sphere products with two spheres in each product, then K is minimally non-Golod.*

Proof Under the given assumption, $K = \Delta^d * L$ for some minimally non-Golod complex L by Theorem 4.2. Therefore, $\mathcal{R}_K \cong \mathcal{R}_{\Delta^d} \times \mathcal{R}_L \cong D^{d+1} \times \mathcal{R}_L$. But since $\mathcal{R}_K$ is a manifold without boundary by assumption, the disk D^{d+1} must have dimension 0, so $d = -1$ and $K = L$ is minimally non-Golod. □

References

1. A. Bahri, M. Bendersky, F.R. Cohen, S. Gitler, The polyhedral product functor: a method of decomposition for moment-angle complexes, arrangements and related spaces. Adv. Math. **225**, 1634–1668 (2010)
2. I. Baskakov, V. Buchstaber, T. Panov, Cellular cochain complexes and torus actions. Uspekhi Mat. Nauk **59**(3), 159–160 (2004) (Russian); Russian Math. Surveys **59**(3), 562–563 (2004) (English translation)
3. A. Berglund, M. Jöllenbeck, On the Golod property of Stanley–Reisner rings. J. Algebra **315**(1), 249–273 (2007)
4. F. Bosio, L. Meersseman, Real quadrics in $\mathbb{C}^n$, complex manifolds and convex polytopes. Acta Math. **197**, 53–127 (2006)
5. V. Buchstaber, T. Panov, *Toric Topology*. Mathematical Surveys and Monographs, vol. 204 (American Mathematical Society, Providence, 2015)
6. L. Cai, On products in a real moment-angle manifold. J. Math. Soc. Japan **69**(2), 503–528 (2017)
7. L. Chen, F. Fan, X. Wang, The topology of moment-angle manifolds–on a conjecture of S. Gitler and S. López de Medrano. Sci. China Math. **63**, 2079–2088 (2020)

8. M. Franz, On the integral cohomology of smooth toric varieties. Proc. Steklov Inst. Math. **252**, 53–62 (2006)
9. S. Gitler, S. López de Medrano, Intersections of quadrics, moment-angle manifolds and connected sums. Geom. Topol. **17**(3), 1497–1534 (2013)
10. E.S. Golod, On the cohomology of some local rings. Soviet Math. Dokl. **3**, 745–749 (1962)
11. J. Grbić, S. Theriault, The homotopy type of the polyhedral product for shifted complexes. Adv. Math. **245**, 690–715 (2013)
12. J. Grbić, T. Panov, S. Theriault, J. Wu, The homotopy types of moment-angle complexes for flag complexes. Trans. Am. Math. Soc. **368**, 6663–6682 (2016)
13. K. Iriye, D. Kishimoto, Decompositions of polyhedral products for shifted complexes. Adv. Math. **245**, 716–736 (2013)
14. K. Iriye, D. Kishimoto, Fat wedge filtration and decomposition of polyhedral products. Kyoto J. Math. **59**(1), 1–51 (2019)
15. I. Limonchenko, Stanley–Reisner rings of generalized truncation polytopes and their moment-angle manifolds. Proc. Steklov Inst. Math. **286**, 188–197 (2014)
16. I. Limonchenko, Families of minimally non-Golod complexes and their polyhedral products. Far East. Math. J. **15**(2), 222–237 (2015)
17. M. Mohareri, B. Mashayekhy, H. Mirebrahimi, The capacity of wedge sum of spheres of different dimensions. Topol. Appl. **246**, 1–8 (2018)
18. R. Stanley, *Combinatorics and Commutative Algebra*, 2nd edn. Progress in Mathematics, vol. 41 (Birkhäuser, Boston, 1996)

Toric Manifolds over 3-Polytopes

Anton Ayzenberg

1 Introduction

For a three-dimensional simple polytope P, one can construct a six-dimensional manifold with the action of the compact torus T^3, whose orbit space is P. The topology of this manifold tells a lot about the combinatorics of the polytope. There exist several constructions of such manifolds arising in different areas of mathematics: toric varieties in algebraic geometry and singularity theory, symplectic toric manifolds in symplectic geometry, and quasitoric manifolds in algebraic topology. Each construction requires certain properties from the polytope, and these properties affect the geometrical structure of the resulting manifold. For example, the construction of a quasitoric manifold as an identification space [1] requires only the combinatorial type of a polytope, and the resulting manifold is just a topological manifold. However, if we fix the affine realization of a polytope, the resulting quasitoric manifold attains smooth structure; see [2]. The construction of a symplectic toric manifold requires a polytope to be Delzant [3]. There also exist certain conditions on the polytope, which imply the existence of almost complex or algebraical structure on the corresponding manifold.

There are several natural questions. What does the existence of certain geometrical structure on a toric space say about the combinatorics of the polytope? The constructions of toric topology and toric geometry allow to construct a lot of examples of six-dimensional manifolds. But how large is the set of examples having certain geometrical structure? In this paper, we gather and review some known results.

This work is supported by the Russian Science Foundation under grant 20-11-19998

A. Ayzenberg (✉)
Steklov Mathematical Institute of Russian Academy of Sciences, Moscow, Russia

A. Bahri et al. (eds.), *Toric Topology and Polyhedral Products*, Fields Institute Communications 89, https://doi.org/10.1007/978-3-031-57204-3_2

There is a well-known correspondence between toric varieties and rational fans. Projective toric varieties correspond to normal fans of convex polytopes, and smooth projective toric varieties correspond to unimodular polytopal fans (i.e., normal fans of Delzant polytopes). Quasitoric manifolds are the algebro-topological generalization of smooth projective toric varieties. A smooth compact manifold M of real dimension $2n$ with the action of half-dimensional compact torus T^n is called *quasitoric* if:

1. The action is locally standard (i.e., locally modeled by the standard action of T^n on $\mathbb{C}^n$ by coordinate-wise rotations).
2. The orbit space is diffeomorphic to some simple polytope P as a manifold with corners.

In this case, we say that M is a quasitoric manifold over P. Recall that n-dimensional convex polytope is called *simple* if each of its vertices lies in exactly n facets (equivalently, each vertex lies in exactly n edges). Among all convex polytopes, only simple polytopes are manifolds with corners.

Every smooth projective toric variety X is a quasitoric manifold: we can restrict the action of an algebraic torus $(\mathbb{C}^{\times})^n$ on X to its compact subtorus $T^n \subset (\mathbb{C}^{\times})^n$; and the orbit space of this action can be identified with the image of the moment map, which is a simple polytope. However, there exist many quasitoric manifolds, which are not toric varieties. The simplest example is $\mathbb{C}P^2\#\mathbb{C}P^2$: this is a quasitoric manifold, which is not even a complex algebraic variety [4, Sect.7.6]. On the other hand, there also exist smooth non-projective toric varieties, which are not quasitoric [5].

In this paper, we discuss two basic theorems:

Theorem 1.1 ([1]) *There exists a quasitoric manifold over any three-dimensional simple polytope.*

Theorem 1.2 ([6]) *If there exists a smooth projective toric variety over a simple three-dimensional polytope P, then P has at least one triangular or quadrangular face.*

The recent interest to these results arose in connection with fullerenes. Mathematically, a fullerene is a simple three-dimensional polytope having only pentagonal and hexagonal faces. Buchstaber [7] suggested to study fullerenes from the perspective of toric topology. The celebrated paper [8] provides links between geometry and combinatorics of three-dimensional polytopes, the cohomological rigidity of the related six-dimensional quasitoric manifolds, and their real three-dimensional hyperbolic counterparts.

Theorems 1.1 and 1.2 above show that (1) there exist quasitoric manifolds over fullerenes and (2) there are no smooth projective toric varieties over fullerenes. Therefore, due to their rigid geometrical nature, smooth projective toric varieties are not suited for the study of fullerenes.

We make a remark that Theorems 1.1 and 1.2 are based on completely different methods. Theorem 1.1 essentially relies on the four colors theorem: there are no known proofs, which avoid this result. This argument is well-known in toric topology: it is included in the paper for completeness. Theorem 1.2 was formulated and proved by Delaunay in [6], and its proof is based on the work of Reid [9] concerning Mori's minimality theory for toric varieties. We restate the proof in more combinatorial topological terms, without referring to this algebro-geometrical theory, to make the difference between toric and quasitoric cases more transparent.

Notice that the analogue of Theorem 1.2 holds in any dimension $\geqslant 3$. We even get a stronger statement.

Corollary 1.3 *If there exists a smooth projective toric variety over a simple n-dimensional polytope P, $n \geqslant 3$, then any three-dimensional face of P has at least one triangular or quadrangular face.*

If P is Delzant, so any of its faces is Delzant as well, so this is a direct corollary of Theorem 1.2.

2 Quasitoric Manifolds

Let M be a quasitoric manifold of dimension $2n$. Its orbit space under the action of T^n is a simple polytope P. Let $\mathcal{F}_1, \dots, \mathcal{F}_m$ be the facets of the polytope P. Any point x in the interior of a facet $\mathcal{F}_i$ represents an $(n-1)$-dimensional orbit of the action. The stabilizer of this orbit is a one-dimensional toric subgroup $G_i \subset T^n$. We may assume that $G_i = \exp(\lambda_{i,1}, \dots, \lambda_{i,n})$, where $(\lambda_{i,1}, \dots, \lambda_{i,n}) \in \mathbb{Z}^n$ is a primitive integral vector determined uniquely up to sign. One-dimensional stabilizer subgroups define the so-called characteristic function. Let $[m] = \{1, \dots, m\}$ be the index set of facets of P. Consider the function $\lambda \colon [m] \to \mathbb{Z}^n$, $\lambda \colon i \mapsto (\lambda_{i,1}, \dots, \lambda_{i,n})$. Actually, the value is determined uniquely up to sign; however, we make a choice of this sign arbitrarily (this corresponds to the choice of orientation of each stabilizer G_i). Since the action of the torus is locally standard, the characteristic function satisfies the condition:

$$\text{if facets } \mathcal{F}_{i_1}, \dots, \mathcal{F}_{i_n} \text{ intersect in a vertex, then } \{\lambda(i_1), \dots, \lambda(i_n)\} \text{ is a basis of } \mathbb{Z}^n. \tag{1}$$

Therefore, for any quasitoric manifold, there is an associated characteristic pair (P, λ), where P is the simple polytope representing the orbit space and λ is the characteristic function.

Construction 2.1 The construction above can be reverted [1]. Given any simple polytope P with facets $\mathcal{F}_1, \dots, \mathcal{F}_m$ and a function $\lambda \colon [m] \to \mathbb{Z}^n$ satisfying condition (1), we can construct a quasitoric manifold $M_{(P,\lambda)}$ as follows. For each $i \in [m]$, let $G_i = \exp(\lambda(i)) \subset T^n$ be the corresponding circle subgroup. Take any point $x \in P$; it lies in the interior of some face $F \in P$. We have $F = \mathcal{F}_{i_1} \cap \dots \cap \mathcal{F}_{i_k}$. Let G_x denote the toric subgroup $G_{i_1} \times \dots \times G_{i_k} \subset T^n$. Consider the identification space

$$M_{(P,\lambda)} = (P \times T^n)/\sim$$

where $(x, t) \sim (x', t')$ whenever the points x, x' coincide and $t't^{-1} \in G_x$. One can check that $M_{(P,\lambda)}$ is a topological manifold, and there is a locally standard action of T^n, which rotates the second coordinate. Naturally, the orbit space of this action is P itself. The canonical smooth structure on $M_{(P,\lambda)}$ was constructed in [2], so that $M_{(P,\lambda)}$ becomes a quasitoric manifold.

To prove Theorem 1.1, one needs to show that every simple 3-polytope admits a function λ, satisfying condition (1). This is done by the four colors theorem.

Proof (Proof of Theorem 1.1) Let $c\colon \{\mathcal{F}_1, \ldots, \mathcal{F}_m\} \to \{a, b, c, d\}$ be the coloring of facets of P by four colors such that the adjacent facets have distinct colors. Let e_1, e_2, e_3 be the basis of the lattice $\mathbb{Z}^3$. Replace colors by the vectors as follows: $a \mapsto e_1, b \mapsto e_2, c \mapsto e_3, d \mapsto e_1 + e_2 + e_3$. This gives a characteristic function, since every three vectors among $(e_1, e_2, e_3, e_1 + e_2 + e_3)$ form a basis of the lattice. □

3 Toric Varieties

Let $V \cong \mathbb{R}^n$ be an oriented real vector space with the fixed lattice $\mathbb{Z}^n \cong N \subset V$. Recall that *a fan* in $\mathbb{R}^n$ is a collection of convex cones with apex at the origin such that the intersection of each two cones of the collection is a face of both and belongs to the collection. The fan is called *complete* if the union of all cones is the whole space V. The fan is called *rational* if all cones are generated by rational vectors. The cone is called simplicial (resp. unimodular) if it is generated by linearly independent vectors of V (resp. part of a basis of the lattice N). The fan is called *simplicial* (resp. *unimodular*) if all its cones are simplicial. Every unimodular fan is simplicial.

Let P be a convex polytope in the dual space V^*. With any such polytope, one associates *the normal fan*: for each face $F \subset P$, take the cone spanned by outward normal vectors to the facets of P containing F, and take the collection of these cones. Normal fan is complete. Normal fan of a simple polytope is simplicial. Normal fans of polytopes are called *polytopal fans*. Note that there exist non-polytopal complete fans [10].

Definition 3.1 A polytope P is called *Delzant* if its normal fan is unimodular.

It follows that every Delzant polytope is simple.

Toric varieties are classified by rational fans. Compact toric varieties correspond to complete fans. Smooth toric varieties correspond to unimodular fans. Projective toric varieties correspond to polytopal fans. Therefore, smooth projective toric varieties correspond to normal fans of Delzant polytopes (i.e., polytopal unimodular fans). Theorem 1.2 can be restated as follows.

Proposition 3.2 ([6]) *Let P be a three-dimensional Delzant polytope. Then P has at least one triangular or quadrangular face.*

Let P be an n-dimensional Delzant polytope with facets $\mathcal{F}_1, \ldots, \mathcal{F}_m$, and let $\lambda(i) \in \mathbb{Z}^n$ be the primitive outward normal vector to $\mathcal{F}_i$. The unimodularity property of the normal fan of P implies that the function $\lambda\colon [m] \to \mathbb{Z}^n$ satisfies condition (1). Therefore, each Delzant polytope determines the characteristic pair (P, λ) in a natural way. Smooth projective toric variety corresponding to the normal fan of P is equivariantly diffeomorphic to the quasitoric manifold determined by the pair (P, λ); see [1]. Due to this observation, smooth projective toric varieties are particular cases of quasitoric manifolds (from topological point of view).

Let us introduce a notation to be used in the following. Let Δ denote a complete unimodular fan in $V \cong \mathbb{R}^n$ and m be the number of rays in Δ. Let X_Δ be the smooth compact toric variety corresponding to this fan. The underlying simplicial sphere K of the fan Δ has m vertices and $\dim K = n - 1$. Let $\lambda\colon [m] \to N$ be the characteristic function, that is, $\lambda(i) \in N$ is the primitive generator of ith ray of the fan Δ.

Cohomology

Theorem 3.3 (Danilov–Jurkiewicz) *$H^*(X_\Delta; \mathbb{Z}) \cong \mathbb{Z}[K]/\Theta$, where*

$$\mathbb{Z}[K] = \mathbb{Z}[v_1, \ldots, v_m]/(v_{i_1} \cdots v_{i_s} \mid \{i_1, \ldots, i_s\} \notin K), \qquad |v|_i = 2$$

is the Stanley–Reisner ring of the sphere K, and ideal Θ is generated by linear forms $\sum_{i\in[m]} \langle \mu, \lambda(i)\rangle v_i$, for each linear functional $\mu\colon N \to \mathbb{Z}$.

A similar theorem was proved by Davis and Januszkiewicz for quasitoric manifolds: in this case, K is a simplicial sphere dual to a polytope, and λ is a general characteristic function. Similar theorems hold for coefficients in $\mathbb{R}$ (or any other field).

Let $\int_{X_\Delta}\colon H^{2n}(X_\Delta; \mathbb{Z}) \to \mathbb{Z}$ denote the pairing with the fundamental class of a toric variety X_Δ. Consider a subset $I = \{i_1, \ldots, i_n\} \subset [m]$. We have

$$\int_{X_\Delta} v_{i_1} \cdots v_{i_n} = \begin{cases} 1, & \text{if } I \in K \\ 0, & \text{otherwise.} \end{cases} \tag{2}$$

Indeed, the class $v_i \in H^2(X_\Delta; \mathbb{Z})$ is Poincare dual to the preimage of the facet F_i under the momentum map. Therefore, the class $v_{i_1} \cdots v_{i_n} \in H^{2n}(X_\Delta; \mathbb{Z})$ vanishes whenever the corresponding facets have empty intersection and is Poincare dual to the class of a point in $H_0(X_\Delta; \mathbb{Z})$ taken with plus sign (the latter is due to the existence of complex structure on X_Δ).

In the following, we also need the description of tangent Chern classes of X_Δ.

Theorem 3.4 ([11]) *Under the isomorphism of Theorem 3.3, the jth Chern class of the tangent bundle of the manifold X_Δ is the elementary symmetric polynomial in the variables v_i:*

$$c_j(X_\Delta) = \sigma_j(v_1, \ldots, v_m) = \sum_{I\in K, |I|=j} \prod_{i\in I} v_i \in H^{2j}(X_\Delta; \mathbb{Z}).$$

A completely similar theorem was proved in [2] for a quasitoric manifold after introducing the canonical stably complex structure on it.

Effective Cone

The notion of effective cone is one of the essential points in the proof of Theorem 1.2. This notion is defined in algebraic geometry for arbitrary projective varieties; however, we restrict to the smooth case, where it has a clear geometrical meaning. This subsection is needed only for the completeness of the exposition: for toric varieties, all necessary notions will be defined in combinatorial–geometrical manner below.

Let X be an arbitrary smooth Kähler manifold. Each compact complex curve $C \subset X$ determines a homology class $[C] \in H_2(X; \mathbb{R})$, which is called *effective*. The set of all nonnegative linear combinations of effective classes in $H_2(X; \mathbb{R})$ is called *the effective cone* of the manifold X:

$$\mathrm{NE}(X) = \left\{ \sum r_i[C_i] \in H_2(X; \mathbb{R}) \mid r_i \geqslant 0 \right\}.$$

Proposition 3.5 NE(X) *is a strictly convex cone in* $H_2(X; \mathbb{R})$.

Proof We need to prove that all nonzero effective classes lie in some open half-space of $H_2(X; \mathbb{R})$. Consider the class of a Kähler form $\omega \in H^2(X; \mathbb{R})$. For each complex curve C, we have

$$\langle \omega, [C] \rangle = \int_C \omega|_C = \mathrm{Vol}(C) > 0.$$

This means that all effective classes lie in the half-space

$$\{\alpha \in H_2(X; \mathbb{R}) \mid \langle \omega, \alpha \rangle > 0\},$$

which implies the statement. □

Proposition 3.6 ([9]) *Let X be a smooth projective toric variety. Then its effective cone* NE(X) *is polyhedral and is generated by the fundamental classes of torus-invariant 2-spheres (preimages of edges of the polytope under the projection to the orbit space).*

The generators of the effective cone are called *extremal cycles*. Note that in general not all edges of the polytope define extremal cycles: some of them may lie in the cone generated by others.

Effective Cone in Toric Case: Combinatorial–Geometrical Approach

Here we introduce all the necessary notions from the previous paragraph in combinatorial manner. Algebraic details can be found in [12, Sections 6.3 and 6.4]. Let X_Δ be the smooth projective variety corresponding to a polytopal fan Δ.

The simplices of K of codimension 1 as well as the corresponding cones of Δ will be called *the walls*. For each wall $J = \{i_1, \dots, i_{n-1}\} \in K$, consider the class $v_J = v_{i_1} \cdots v_{i_{n-1}} \in H^{2n-2}(X; \mathbb{R})$. Note that $v_J \neq 0$, as follows, for example, from (2). Consider the cone in $H^{2n-2}(X_\Delta; \mathbb{R})$ generated by the classes v_J for all walls $J \in K$:

$$\mathrm{NE}(X_\Delta) = \left\{ \sum r_J v_J \in H^{2n-2}(X_\Delta; \mathbb{R}) \mid r_J \geqslant 0 \right\}$$

Proposition 3.7 *For each smooth projective toric variety, the effective cone* $\mathrm{NE}(X_\Delta)$ *is a strictly convex polyhedral cone in* $H^{2n-2}(X_\Delta; \mathbb{R})$.

Proof Let $V_\Delta \in \mathbb{R}[c_1, \dots, c_m]$ be the volume polynomial of the fan Δ. By definition,

$$V_\Delta(c_1, \dots, c_m) = \frac{1}{n!} \int_{X_\Delta} (c_1 v_1 + \ldots + c_m v_m)^n.$$

It is known (see [13]) that the values of this polynomial are the volumes of simple polytopes with the normal fan Δ. More precisely, let $P = \{x \in V^* \mid \langle x, \lambda(i) \rangle \leqslant \widetilde{c}_i\}$ be a simple convex polytope with the normal fan Δ (since X_Δ is projective, at least one such polytope exists). The numbers $\widetilde{c}_i$ are called the support parameters of P. Then we have $\mathrm{Vol}(P) = V_\Delta(\widetilde{c}_1, \dots, \widetilde{c}_m)$. To avoid the mess, we denote the formal variables of the volume polynomial by c_i, while concrete real numbers substituted in this polynomial are denoted by $\widetilde{c}_i$.

Let $\partial_i = \frac{\partial}{\partial c_i}$ be the differential operators, acting on $\mathbb{R}[c_1, \dots, c_m]$. Let $\mathcal{D} = \mathbb{R}[\partial_1, \dots, \partial_m]$ be the commutative algebra of differential operators with constant coefficients and $\mathrm{Ann} V_\Delta = \{D \in \mathcal{D} \mid D V_\Delta = 0\}$ be the annihilating ideal of the polynomial V_Δ. According to [14, 15], we have

$$\mathcal{D}/\mathrm{Ann} V_\Delta \cong H^*(X_\Delta; \mathbb{R}), \qquad \partial_i \leftrightarrow v_i.$$

Moreover, the integration map $\int_{X_\Delta} : H^{2n}(X_\Delta; \mathbb{R}) \to \mathbb{R}$ coincides with the natural map $(\mathcal{D}/\mathrm{Ann} V_\Delta)_n \to \mathbb{R}$, $D \mapsto D V_\Delta$. To prove the proposition, it suffices to show that the classes

$$\{\partial_J = \partial_{i_1} \cdots \partial_{i_{n-1}} \in (\mathcal{D}/\mathrm{Ann} V_\Delta)_{n-1} \mid J = \{i_1, \dots, i_{n-1}\} \in K\}$$

lie in one open half-space.

Let P be a convex polytope with the normal fan Δ and support parameters $\widetilde{c}_i$. Consider the element $\partial_c = \sum_{i \in [m]} \widetilde{c}_i \partial_i \in (\mathcal{D}/\mathrm{Ann} V_\Delta)_1$. Recall a simple fact: for each homogeneous polynomial $\Psi \in \mathbb{R}[c_1, \dots, c_m]$ of degree k, there holds $\frac{1}{k!} \partial_c^k \Psi = \Psi(\widetilde{c}_1, \dots, \widetilde{c}_m)$ (this is an instance of Euler's theorem on homogeneous functions). □

Lemma 3.8 *Let* $J \in K$ *be a wall. Then* $\partial_c \partial_J V_\Delta > 0$.

Proof Note that $\partial_J V_\Delta$ is a linear polynomial in variables c_i. Therefore, the number $\partial_c \partial_J V_\Delta$ coincides with the value of the polynomial $\partial_J V_\Delta$ at the point $\widetilde{c} = (\widetilde{c}_1, \ldots, \widetilde{c}_m)$ by the preceding remark. It is known that the value of the polynomial $\partial_J V_\Delta$ at the point $\widetilde{c}$ coincides, up to a positive factor, with the length of the edge $F_J \subset P$ dual to the wall $J \in K$ (this was noted by Timorin in [15], and in [16], we proved that the factor is the volume of the parallelepiped spanned by $\lambda(i_1), \ldots, \lambda(i_{n-1})$). Thus, $\partial_c \partial_J V_\Delta > 0$. □

According to the lemma, all classes $\partial_J \in \mathcal{D}/\mathrm{Ann}V_\Delta$ lie in the half-space $\{D \mid D\partial_c V_\Delta > 0\}$, which implies the statement. □

Definition 3.9 Let $J = \{i_1, \ldots, i_{n-1}\} \in K$ be a wall such that $v_J \in H^{2n-2}(X_\Delta; \mathbb{R})$ is a generating element of the effective cone $\mathrm{NE}(X_\Delta)$. Then J is called *an extremal simplex* and v_J is called *an extremal class.*

The condition of being extremal can be written as follows. Suppose an extremal class v_J is expressed as a sum $\nu_1 + \nu_2$, where $\nu_1, \nu_2 \in \mathrm{NE}(X_\Delta)$. Then both ν_1 and ν_2 are proportional to v_J.

Remark 3.10 This definition agrees with the general theory. The vector spaces $H^{2n-2}(X_\Delta; \mathbb{R})$ and $H_2(X_\Delta; \mathbb{R})$ can be identified by Poincare duality, and under this identification, the class $v_J = v_{i_1} \cdots v_{i_{n-1}}$ corresponds to the fundamental class of torus-invariant 2-sphere obtained as a transversal intersection of characteristic submanifolds $X_{i_1}, \ldots, X_{i_{n-1}}$ (preimages of facets $\mathcal{F}_{i_1}, \ldots, \mathcal{F}_{i_{n-1}}$ under the projection to the orbit space).

4 Unimodular Geometry of Fans

An arbitrary wall $J = \{i_1, \ldots, i_{n-1}\} \in K$ is contained in exactly two maximal simplices: $I = \{i_1, \ldots, i_{n-1}, i\}$ and $I' = \{i_1, \ldots, i_{n-1}, i'\}$. Both sets of vectors

$$\{\lambda(i_1), \ldots, \lambda(i_{n-1}), \lambda(i)\}, \qquad \{\lambda(i_1), \ldots, \lambda(i_{n-1}), \lambda(i')\}$$

are the bases of the lattice. Write $\lambda(i')$ in the first basis:

$$\lambda(i') = a_1\lambda(i_1) + \ldots + a_{n-1}\lambda(i_{n-1}) - \lambda(i).$$

(Unimodularity condition of the set $\lambda(I')$ guarantees that the coefficient at $\lambda(i)$ is ± 1. The fact that the cones at I and I' lie on the opposite sides of the wall J guarantees that the coefficient at $\lambda(i)$ is exactly -1.) In what follows, we assume that the vertices $i_1, \ldots, i_{n-1}$ are ordered such that $\{\lambda(i_1), \ldots, \lambda(i_{n-1}), \lambda(i)\}$ is a positive basis of the lattice, while, respectively, $\{\lambda(i_1), \ldots, \lambda(i_{n-1}), \lambda(i')\}$ is a negative basis.

Definition 4.1 The number

$$\operatorname{curv}(J) = 2 - a_1 - \ldots - a_{n-1} \in \mathbb{Z}$$

is called *the unimodular curvature* of the wall J.

The underlying simplicial complex K of a fan Δ may be realized in $V \cong \mathbb{R}^n$ as a star-shaped sphere as follows: let us send the vertex i to the point $\lambda(i) \in V$ and continue the map on each simplex by linearity. We denote the image of this map by $\operatorname{st}(K)$; it is a piecewise linear sphere in V winding around the origin one time.

We say that $\operatorname{st}(K)$ is concave (resp. convex, resp. flat) at the wall J if the affine hyperplane through the points $\lambda(i_1), \ldots, \lambda(i_{n-1}), \lambda(i)$ separates $\lambda(i')$ from the origin (resp. does not separate, resp. contains $\lambda(i')$).

Lemma 4.2 *A unimodular curvature and parameters $a_1, \ldots, a_{n-1}$ defined above satisfy the following properties:*

1. $a_s = \det(\lambda(i_1), \ldots, \lambda(i_{s-1}), \lambda(i'), \lambda(i_{s+1}), \ldots, \lambda(i_{n-1}), \lambda(i))$.
2. *The star-shaped sphere* $\operatorname{st}(K)$ *is convex (resp. flat, resp. concave) at a wall J if and only if* $\operatorname{curv}(J) > 0$ *(resp.* $\operatorname{curv}(J) = 0$, *resp.* $\operatorname{curv}(J) < 0$*).*
3. *There exists a wall of positive curvature in a complete simplicial fan.*
4. $\int_{X_\Delta} v_J v_{i_s} = -a_s, \quad \operatorname{curv}(J) = \int_{X_\Delta} v_J (\sum_{t\in[m]} v_t)$.

Proof (1) Take an exterior product of the relation

$$\lambda(i) + \lambda(i') = \sum_{t=1}^{n-1} a_t \lambda(i_t) \tag{3}$$

with the exterior form $\lambda(i_1) \wedge \cdots \wedge \widehat{\lambda(i_s)} \wedge \cdots \wedge \lambda(i_{n-1}) \wedge \lambda(i)$. The result is the desired relation. Relation (3) is known in toric geometry as a wall relation; see [12, Formula 6.4.4]. It traces back to Reid [9]. (2) The convexity of the star-shaped sphere $\operatorname{st}(K)$ at a wall J depends on spatial relationship between the affine line through the points $\lambda(i), \lambda(i')$ and the codimension 2 affine subspace through the points $\lambda(i_1), \ldots, \lambda(i_{n-1})$, which is on the sign of the determinant

$$\det(\lambda(i_1) - \lambda(i'), \ldots, \lambda(i_{n-1}) - \lambda(i'), \lambda(i) - \lambda(i')) =$$
$$= \det(\lambda(i_1), \ldots, \lambda(i_{n-1}), \lambda(i)) - \det(\lambda(i_1), \ldots, \lambda(i_{n-1}), \lambda(i')) -$$
$$- \sum_{s=1}^{n-1} \det(\lambda(i_1), \ldots, \overset{[}{s}]\lambda(i'), \ldots \lambda(i_{n-1}), \lambda(i)) = 1 - (-1) - \sum_{s=1}^{n-1} a_s = \operatorname{curv}(J).$$

(3) If the curvature of any wall is non-positive, then the star-shaped sphere $\operatorname{st}(K)$ could not wind around the origin.

(4) Let us write the class v_{i_s} as a linear combination of v_j, $j \notin J$, using linear relations in the cohomology ring. Consider the linear functional μ on the space V, such that $\langle \mu, \lambda(i) \rangle = 0$ and

$$\langle \mu, \lambda(i_t) \rangle = \begin{cases} 0, & \text{if } t \neq s, \\ 1, & \text{if } t = s. \end{cases}$$

Applying μ to relation (3), we get $\langle \mu, \lambda(i') \rangle = a_s$. It follows that there is a linear relation $v_{i_s} + a_s v_{i'} + \sum_{j \notin \{i_1, \dots, i_{n-1}, i, i'\}} C_j v_j$ in the cohomology ring. Let us multiply this relation by v_J. Since J forms a simplex only with vertices i, i', Stanley–Reisner relations imply

$$\int_{X_\Delta} v_J v_{i_s} = \int_{X_\Delta} -a_s v_J v_{i'} = -a_s \int_{X_\Delta} v_{J \sqcup \{i'\}} = -a_s$$

according to (2). The formula for the curvature easily follows

$$\int_{X_\Delta} \left(v_J \cdot \sum\nolimits_{t \in [m]} v_t \right) = \int_{X_\Delta} v_J v_i + \int_{X_\Delta} v_J v_{i'} + \sum_{s=1}^{n-1} \int_{X_\Delta} v_J v_{i_s}$$
$$= 2 - \sum_{s=1}^{n-1} a_s = \operatorname{curv}(J).$$

□

One corollary of Lemma 4.2(4) is worth mentioning (this also gives an alternative proof of pt.3 in the above lemma in dimension 3).

Proposition 4.3 (Unimodular Gauss–Bonnet Theorem) *Let Δ be a unimodular simplicial fan of dimension* 3. *Then the sum of curvatures of all its walls equals* 24.

Proof It follows from the previous lemma that

$$\sum_{J \in K, |J|=2} \operatorname{curv}(J) = \int_{X_\Delta} \Big(\sum_{J \in K, |J|=2} v_J \Big) \Big(\sum_{t \in [m]} v_t \Big) = \int_{X_\Delta} c_2(X_\Delta) c_1(X_\Delta) = c_{1,2}(X_\Delta).$$

It is known that for stably complex manifolds of real dimension 6, the Chern number $c_{1,2}(X_\Delta)$ coincides with $24\operatorname{Td}(X_\Delta)$. The Todd genus of a smooth compact toric variety equals 1, and the statement follows. □

Now we prove Theorem 1.2. Let Δ be the normal fan of a Delzant polytope P. The walls of this fan are simply the edges of the two-dimensional triangulated sphere K.

Proof (Proof of Theorem 1.2) According to Lemma 4.2(3), there exists a wall $\tilde{J} \in K$ of positive curvature. On the other hand, Lemma 4.2(4) implies that the curvature of the wall $\tilde{J}$ coincides with the value of the linear functional $H^4(X_\Delta; \mathbb{R}) \to \mathbb{R}$, $u \mapsto \int_{X_\Delta} (u \cdot c_1(X_\Delta))$ on the effective class $v_{\tilde{J}}$. Since a linear functional takes positive value on some element of the effective cone, this functional should take positive value on some generator of this cone. Therefore, there exists an extremal wall $J = \{i_1, i_2\} \in K$ of positive curvature.

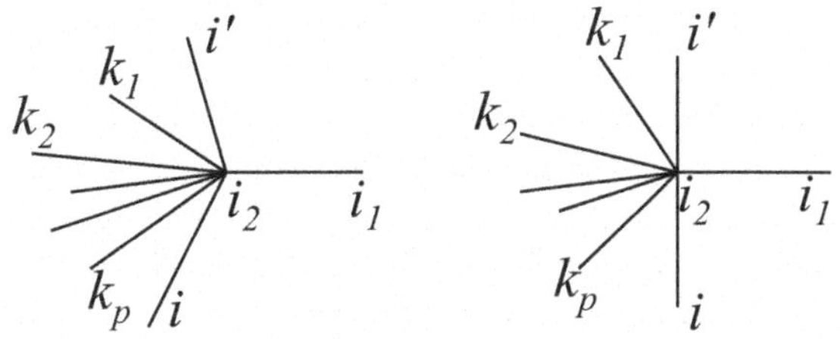

Fig. 1 The vicinity of the ray $\mathbb{R}_{\geqslant 0}\lambda(i_2)$ in the first and the second cases. The ray $\mathbb{R}_{\geqslant 0}\lambda(i_2)$ points to the reader

Let a_1, a_2 be the parameters of the wall J, defined earlier. Since $\operatorname{curv}(J) = 2 - a_1 - a_2 > 0$ and the numbers a_1, a_2 are integers, we have either $a_1 \leqslant 0$ or $a_2 \leqslant 0$. Assume $a_1 \leqslant 0$. Consider two cases:

(1) $a_1 < 0$. Let us prove that in this case, i_2 is contained in exactly three maximal cones. As before, let $I = \{i, i_1, i_2\}$, $I' = \{i', i_1, i_2\}$ be the maximal simplices containing the wall J. Suppose that apart from the vertices i_1, i, i', the vertex i_2 is connected to the vertices $k_1, \ldots, k_p$, $p \geqslant 1$ (we assume that the neighbors of the vertex i_2 are cyclically ordered as $i_1, i', k_1, \ldots, k_p, i$; see Fig. 1).

According to Lemma 4.2(1), $a_1 = \det(\lambda(i'), \lambda(i_2), \lambda(i)) < 0$. This means that the sum of dihedral angles of the cones $C(I) = \operatorname{cone}(\lambda(i_1), \lambda(i_2), \lambda(i))$ and $C(I') = \operatorname{cone}(\lambda(i_1), \lambda(i_2), \lambda(i'))$ at the edge $\mathbb{R}_{\geqslant 0}\lambda(i_2)$ exceeds one straight angle (see left part of Fig. 1). There exists a two-plane Π, which contains the ray $\mathbb{R}_{\geqslant 0}\lambda(i_2)$ and separates $\lambda(i_1)$ from the vectors

$$\{\lambda(i), \lambda(i'), \lambda(k_1), \ldots, \lambda(k_p)\}. \tag{4}$$

Let μ be the linear functional on $\mathbb{R}^3$, annihilating the plane Π and taking value 1 on the vector $\lambda(i_1)$. By construction, μ takes negative values on all vectors from the list (4). Using μ, we obtain a linear relation

$$v_{i_1} = \sum_{t \in \{i, i', k_1, \ldots, k_p\}} C_t v_t + \sum_{t \notin \{i_1, i_2, i, i', k_1, \ldots, k_p\}} D_t v_t,$$

in $H^*(X_\Delta; \mathbb{Z})$, where all coefficients C_t are positive. Multiplying this relation by v_{i_2}, we get

$$v_J = v_{i_1} v_{i_2} = \sum_{t \in \{i, i', k_1, \ldots, k_p\}} C_t v_t v_{i_2}$$

(the part of expression, having coefficients D_t, vanishes due to Stanley–Reisner relations). Therefore, the class v_J is expressed as a positive linear combination of the classes $v_t v_{i_2}$, $t \in \{i, i', k_1, \ldots, k_p\}$. Since v_J was chosen to be extremal, each of the classes $v_t v_{i_2}$ is proportional to the class v_J. Since all these classes are nonzero, they are all proportional to each other. This leads to contradiction. Indeed, according to relation (2), we have $(v_{i'} v_{i_2}) v_{k_1} \neq 0$ since $\{i', i_2, k_1\} \in K$, but $(v_{i_1} v_{i_2}) v_{k_1} = 0$ since $\{i_1, i_2, k_1\} \notin K$.

(2) $a_1 = 0$. We prove that in this case, the vertex i_2 is contained in four maximal cones. The proof is similar to the previous case. Assume the contrary: let the vertex i_2 have the neighbors $i, i_1, i', k_1, \dots, k_p$, $p \geqslant 2$, written in the cyclic order.

According to Lemma 4.2(1), the condition $a_1 = 0$ implies that the vectors $\lambda(i_2), \lambda(i), \lambda(i')$ belong to a single 2-plane, say Π. Let μ be the linear functional annihilating Π and taking value 1 on the vector $\lambda(i_1)$. Consequently, μ takes strictly negative values on the vectors $\lambda(k_1), \dots, \lambda(k_p)$. By the same arguments as before, the class $v_J = v_{i_1} v_{i_2}$ is written as a positive linear combination of the classes $v_t v_{i_2}$, $t \in \{k_1, \dots, k_p\}$. The extremality of the wall J implies that all these classes (there are at least two of them by assumption) are proportional to the class v_J. Again, this leads to contradiction: $v_{k_1} v_{i_2} v_i = 0$ since $\{k_1, i_2, i\} \notin K$, but $v_{i_1} v_{i_2} v_i \neq 0$ since $\{i_1, i_2, i\} \in K$.

We proved that there are no more than four maximal cones containing $\lambda(i_2)$. There cannot be three maximal cones by obvious geometrical reasons: the vectors $\lambda(i_2), \lambda(i), \lambda(i')$ belong to a 2-plane and therefore do not span a maximal cone.

It was shown that in the 2-sphere K, there exists a vertex having either three or four neighbors. This means that in the dual 3-polytope P, there exists either a triangular or quadrangular face. □

Remark 4.4 The existence of a strictly convex effective cone, and as a corollary, extremal classes, is the fact, which marks out projective smooth toric varieties among all quasitoric manifolds. For general quasitoric manifolds, we may still define the cohomology classes $v_J \in H^{2n-2}(X; \mathbb{R})$ corresponding to the walls; however, their nonnegative linear combinations may span the whole space $H^{2n-2}(X; \mathbb{R})$ rather than a strictly convex cone. This is why it is impossible to find "extremal" classes with nice properties.

References

1. M. Davis, T. Januszkiewicz, Convex polytopes, Coxeter orbifolds and torus actions. Duke Math. J. **62**(2), 417–451 (1991)
2. V.M. Buchstaber, T.E. Panov, N. Ray, Spaces of polytopes and cobordism of quasitoric manifolds. Mosc. Math. J. **7**(2), 219–242 (2007)
3. T. Delzant, Hamiltoniens periodiques et image convex de l'application moment. Bull. Soc. Math. France **116**, 315–339 (1988)
4. V. Buchstaber, T. Panov. *Toric Topology*. Mathematical Surveys and Monographs, vol. 204 (AMS, Providence, 2015)
5. Y. Suyama, Examples of toric manifolds which are not quasitoric manifolds. Algebr. Geom. Topol. **14**, 3097–3106 (2014). Preprint. arXiv:1312.5973
6. C. Delaunay, *On Hyperbolicity of Toric Real Threefolds*. International Mathematics Research Notices (2005), No. 51
7. V.M. Buchstaber, Toric topology and applications, in *International Conference "Combinatorial and Toric Homotopy"*, 1–31 August, 2015 (IMS, Singapore, 2015)
8. V.M. Buchstaber, N.Yu. Erokhovets, M. Masuda, T.E. Panov, S. Park, Cohomological rigidity of manifolds defined by 3-dimensional polytopes. Russian Math. Surveys **72**(2), 199–256 (2017)

9. M. Reid, *Decomposition of Toric Morphisms*. Arithmetic and Geometry, vol. II, ed. by M. Artin, J. Tate. Progress in Mathematics, vol. 36 (1983), pp. 395–418
10. T. Oda, *Convex Bodies and Algebraic Geometry: An Introduction to the Theory of Toric Varieties* (Springer, Berlin, 1988)
11. F. Ehlers, Eine Klasse komplexer Mannigfaltigkeiten und die Auflösung einiger isolierter Singularitäten. Math. Ann. **218**, 127–156 (1975)
12. D.A. Cox, J.B. Little, H. Schenck, Toric varieties. Graduate Studies in Mathematics, Am. Math. Soc. **124**, 731 (2011)
13. A.G. Khovanskii, A.V. Pukhlikov, Finitely additive measures of virtual polyhedra. Algebra i Analiz **4**(2), 161–185 (1992); English transl., St. Petersburg Math. J. **4**, 337–356 (1992)
14. A.G. Khovanskii, A.V. Pukhlikov, The Riemann–Roch theorem for integrals and sums of quasipolynomials on virtual polytopes. Algebra i Analiz **4**(4), 188–216 (1992); English transl., St. Petersburg Math. J. **4**(4), 789–812 (1993)
15. V.A. Timorin, An analogue of the Hodge–Riemann relations for simple convex polytopes. Russian Math. Surveys **54**(2), 381–426 (1999)
16. A. Ayzenberg, M. Masuda, Volume polynomials and duality algebras of multi-fans. Arnold Math J. **2**(3), 329–381 (2016)

Symmetric Products and a Cartan-Type Formula for Polyhedral Products

A. Bahri, M. Bendersky, F. R. Cohen, and S. Gitler

1 Introduction

Our purpose is to recall some standard properties of infinite symmetric products, known also as the Dold-Thom construction [10], and to develop some related maps, which are defined for polyhedral products. The main feature is that topological maps on the level of infinite symmetric products applied to polyhedral products can be defined directly from homological information.

Polyhedral products $Z\big(K; (\underline{X}, \underline{A})\big)$ [1] are defined for a simplicial complex K on the vertex set $[m] = \{1, 2, \ldots, m\}$ and a family of pointed CW pairs

$$(\underline{X}, \underline{A}) = \big\{(X_i, A_i) : i = 1, 2, \ldots, m\big\}.$$

They are natural subspaces of the Cartesian product $X_1 \times X_2 \times \cdots \times X_m$, in such a way that if $K = \Delta^{m-1}$, the $(m-1)$-simplex, then

$$Z\big(K; (\underline{X}, \underline{A})\big) = X_1 \times X_2 \times \cdots \times X_m.$$

More specifically, we consider K to be a category where the objects are the simplices of K and the morphisms $d_{\sigma,\tau}$ are the inclusions $\sigma \subset \tau$. A polyhedral

A. Bahri (✉)
Department of Mathematics, Rider University, Lawrenceville, NJ, USA
e-mail: bahri@rider.edu

M. Bendersky
Department of Mathematics CUNY, New York, NY, USA
e-mail: mbenders@hunter.cuny.edu

F. R. Cohen (deceased) · S. Gitler (deceased)

A. Bahri et al. (eds.), *Toric Topology and Polyhedral Products*, Fields Institute Communications 89, https://doi.org/10.1007/978-3-031-57204-3_3

product is given as the colimit of a diagram $D_{(\underline{X},\underline{A})} : K \to CW_*$, where at each $\sigma \in K$, we set

$$D_{(\underline{X},\underline{A})}(\sigma) = \prod_{i=1}^{m} W_i, \quad \text{where} \quad W_i = \begin{cases} X_i \text{ if } i \in \sigma \\ A_i \text{ if } i \in [m] - \sigma. \end{cases} \tag{1}$$

Here, the colimit is a union given by

$$Z(K; (\underline{X}, \underline{A})) = \bigcup_{\sigma \in K} D_{(\underline{X},\underline{A})}(\sigma),$$

but the full colimit structure is used heavily in the development of the elementary theory. Notice that when $\sigma \subset \tau$, then $D_{(\underline{X},\underline{A})}(\sigma) \subseteq D_{(\underline{X},\underline{A})}(\tau)$. In the case that K itself is a simplex,

$$Z(K; (\underline{X}, \underline{A})) = \prod_{i=1}^{m} X_i.$$

Polyhedral products were formulated first for the case $(X_i, A_i) = (D^2, S^1)$ by V. Buchstaber and T. Panov in [6]; they called their spaces *moment-angle complexes*.

In a way entirely similar to that above, a related space $\widehat{Z}(K; (\underline{X}, \underline{A}))$, called the *polyhedral smash product*, is defined by replacing the Cartesian product everywhere above by the smash product. That is,

$$\widehat{D}_{(\underline{X},\underline{A})}(\sigma) = \bigwedge_{i=1}^{m} W_i \quad \text{and} \quad \widehat{Z}(K; (\underline{X}, \underline{A})) = \bigcup_{\sigma \in K} \widehat{D}_{(\underline{X},\underline{A})}(\sigma)$$

with

$$\widehat{Z}(K; (\underline{X}, \underline{A})) \subseteq \bigwedge_{i=1}^{m} X_i.$$

The polyhedral smash product is related to the polyhedral product by the stable decomposition discussed in [1] and [2]. We denote by $(\underline{X}, \underline{A})_J$ the restricted family of CW pairs $\{(X_j, A_j)\}_{j \in J}$ and by K_J the full subcomplex on $J \subset [m]$.

Theorem 1.1 ([2, Theorem 2.10]) *Let K be an abstract simplicial complex on vertices $[m]$. Given a family $\{(X_j, A_j)\}_{j=1}^{m}$ of pointed pairs of CW complexes, there is a natural pointed homotopy equivalence*

$$H\colon \Sigma\big(Z(K; (\underline{X}, \underline{A}))\big) \longrightarrow \Sigma\big(\bigvee_{J \subseteq [m]} \widehat{Z}(K_J; (\underline{X}, \underline{A})_J)\big). \tag{2}$$

In many of the most important cases, the spaces $\widehat{Z}\big(K_J; (\underline{X}, \underline{A})_J\big)$ can be identified explicitly [2]. Aside from the various unstable and stable splitting theorems

[1, 12–15], there is an extensive history of computations of the cohomology groups and rings of various families of polyhedral products [4, Sections 5, 8 and 11]; see also [3, 5, 7, 8, 11, 16, 18].

Some very early calculations of the cohomology of certain moment-angle complexes (the case $(X_i, A_i) = (D^2, S^1)$ for all $i = 1, 2, \ldots, m$) appeared in the work of Santiago López de Medrano [16], though at that time, the spaces he studied were not recognized to have the structure of a moment-angle complex. The cohomology algebras of all moment-angle complexes were computed first by M. Franz [11] and by I. Baskakov, V. Buchstaber, and T. Panov in [5].

The cohomology of the polyhedral product $Z\big(K; (\underline{X}, \underline{A})\big)$, for $(\underline{X}, \underline{A})$, satisfying certain freeness conditions (e.g., coefficients in a field k), was computed using a spectral sequence by the authors in [3]. A computation using different methods by Q. Zheng can be found in [18].

The special family of CW pairs $(\underline{U}, \underline{V}) = (\underline{B \vee C}, \underline{B \vee E})$ satisfying the condition that for all i, $(U_i, V_i) = (B_i \vee C_i, B_i \vee E_i)$, where $E_i \hookrightarrow C_i$ is a null-homotopic inclusion, is called *wedge decomposable.* As announced in [4, Section 12], one goal of the current paper is to show that for wedge decomposable pairs $(\underline{U}, \underline{V})$, the algebraic decomposition given by the spectral sequence calculation [3, Theorem 5.4] is a consequence of an underlying geometric splitting. *Moreover, the results of this observation extend to general based CW pairs of finite type.*

This paper is partly a revised version of the authors' unpublished preprint from 2014, which in turn originated from an earlier preprint from 2010. In addition, the results of this paper have been extended to describe the product structure in the cohomology, and these will appear separately.

We begin in Sect. 2 by deriving for wedge decomposable pairs $(\underline{U}, \underline{V})$ an explicit decomposition of the polyhedral product into a wedge of much simpler spaces (Theorem 2.2 and Corollary 2.4). In particular, this allows us to identify explicit additive generators for $H^*\big(Z\big(K; (\underline{U}, \underline{V})\big)$. The proof in Sect. 4 is an induction based on a filtration of the polyhedral product, which is introduced in Sect. 3.

These decompositions give a direct framework for deducing (Theorem 7.1), an analogous homological Cartan formula for the additive structure of the homology of $Z\big(K; (\underline{X}, \underline{A})\big)$ for *any* family of pairs of finite, pointed, path-connected CW complexes $(\underline{X}, \underline{A})$. This is done by applying properties of the infinite symmetric product $SP(-)$ and the polyhedral product. Namely, given pointed pairs of finite, path-connected CW complexes $(\underline{X}, \underline{A})$, there exist pointed pairs of path-connected CW complexes

$$(\underline{U}, \underline{V}) = (\underline{B \vee C}, \underline{B \vee E})$$

together with a homotopy equivalence

$$SP\big(\widehat{Z}(K; (\underline{U}, \underline{V}))\big) \longrightarrow SP\big(\widehat{Z}(K; (\underline{X}, \underline{A}))\big)$$

Applications of the additive results comprise Sects. 9 and 10.

2 The Polyhedral Product of Wedge Decomposable Pairs

We begin with a definition.

Definition 2.1 The special family of CW pairs $(\underline{U}, \underline{V}) = (\underline{B \vee C}, \underline{B \vee E})$ satisfying $(U_i, V_i) = (B_i \vee C_i, B_i \vee E_i)$ for all i, where $E_i \hookrightarrow C_i$ is a null-homotopic inclusion, is called *wedge decomposable.*

The fact that the smash product distributes over wedges of spaces leads to the characterization of the smash polyhedral product in a way that resembles a *Cartan formula.*

Theorem 2.2 (Cartan Formula) *Let* $(\underline{U}, \underline{V}) = (\underline{B \vee C}, \underline{B \vee E})$ *be a wedge decomposable pair, and then there is a homotopy equivalence*

$$\widehat{Z}\big(K; (\underline{U}, \underline{V})\big) \longrightarrow \bigvee_{I \leq [m]} \Big(\widehat{Z}\big(K_I; (\underline{C}, \underline{E})_I\big) \wedge \widehat{Z}\big(K_{[m]-I}; (\underline{B}, \underline{B})_{[m]-I}\big)\Big)$$

which is natural with respect to maps of decomposable pairs. Of course,

$$\widehat{Z}\big(K_{[m]-I}; (\underline{B}, \underline{B})_{[m]-I}\big) \;=\; \bigwedge_{j \,\in\, [m]-I} B_j$$

with the convention that

$$\widehat{Z}\big(K_{\varnothing}; (\underline{B}, \underline{B})_{\varnothing}\big),\ \widehat{Z}\big(K_{\varnothing}; (\underline{C}, \underline{E})_{\varnothing}\big) \text{ and } \widehat{Z}\big(K_I; (\varnothing, \varnothing)_I\big) \;=\; S^0.$$

We can decompose $\widehat{Z}\big(K; (\underline{U}, \underline{V})\big)$ further by applying (a generalization of) the Wedge Lemma. We recall first the definition of a link. For σ, a simplex in a simplicial complex $\mathcal{K}$, $\mathrm{lk}_\sigma(\mathcal{K})$, *the link of* σ *in* $\mathcal{K}$, is defined to be the simplicial complex for which

$$\tau \in \mathrm{lk}_\sigma(K) \quad \text{if and only if} \quad \tau \cup \sigma \in \mathcal{K}.$$

Theorem 2.3 ([1, Theorem 2.12], [19, Lemma 1.8]) *Let* $\mathcal{K}$ *be a simplicial complex on* $[m]$ *and* $(\underline{C}, \underline{E})$ *a family of CW pairs satisfying* $E_i \hookrightarrow C_i$ *is null-homotopic for all* i*; then*

$$\widehat{Z}(\mathcal{K}; (\underline{C}, \underline{E})) \simeq \bigvee_{\sigma \in \mathcal{K}} |\Delta(\overline{\mathcal{K}})_{<\sigma}| * \widehat{D}^{[m]}_{\underline{C},\underline{E}}(\sigma)$$

where $|\Delta(\overline{\mathcal{K}})_{<\sigma}| \cong |lk_\sigma(\mathcal{K})|$*, the realization of the link of* σ *in the simplicial complex* $\mathcal{K}$*, and*

$$\widehat{D}^{[m]}_{\underline{C},\underline{E}}(\sigma) = \bigwedge_{j=1}^{m} W_{i_j}, \quad \text{with} \quad W_{i_j} = \begin{cases} C_{i_j} & \text{if } i_j \in \sigma \\ E_{i_j} & \text{if } i_j \in [m] - \sigma. \end{cases} \tag{3}$$

□

Applying this to the decomposition of Theorem 2.2, we get

Corollary 2.4 *There is a homotopy equivalence*

$$\widehat{Z}\big(K;(\underline{U},\underline{V})\big)\longrightarrow \bigvee_{I\leq[m]}\Big(\Big(\bigvee_{\sigma\in K_I}|lk_\sigma(K_I)|*\widehat{D}^I_{\underline{C},\underline{E}}(\sigma)\Big)\wedge\widehat{Z}\big(K_{[m]-I};(\underline{B},\underline{B})_{[m]-I}\big)\Big).$$

where $\widehat{D}^I_{\underline{C},\underline{E}}(\sigma)$ *is as in* (3) *with* I *replacing* $[m]$.

Combined with Theorem 1.1, this gives a complete description of the topological spaces $Z\big(K;(\underline{U},\underline{V})\big)$ for wedge decomposable pairs $(\underline{U},\underline{V})$.

The case $E_i \simeq *$ simplifies further by [2, Theorem 2.15] to give the next corollary.

Corollary 2.5 *For wedge decomposable pairs of the form* $(\underline{B\vee C},\underline{B})$, *corresponding to* $E_i\simeq *$ *for all* $i=1,2,\ldots,m$, *there are homotopy equivalences*

$$\widehat{Z}\big(K_I;(\underline{C},\underline{E})_I\big)\simeq\widehat{Z}\big(K_I;(\underline{C},*)_I\big)\simeq\widehat{C}^I,$$

and so Theorem 2.2 gives $\widehat{Z}\big(K;(\underline{B\vee C},\underline{B})\big)\simeq\bigvee_{I\leq[m]}\big(\widehat{C}^I\wedge\widehat{B}^{([m]-I)}\big).$ □

Notice here that the Poincaré series for the space $\widehat{Z}\big(K;(\underline{B\vee C},\underline{B})\big)$ follows easily from Corollary 2.5.

Remark 2.6 In comparing these observations with [3, Theorem 5.4], notice that the links appear in the terms $\widehat{Z}\big(K_I;(\underline{C},\underline{E})_I\big)$. Also, while Theorem 2.2 and Corollary 2.4 give a geometric underpinning for the cohomology calculation in [3, Theorem 5.4] for wedge decomposable pairs, the geometric splitting does not require that E, B or C have torsion-free cohomology.

3 A Filtration

We begin by reviewing the filtration on polyhedral products used for the spectral sequence calculation in [3], following [3, Section 2], where more details can be found. The length-lexicographical ordering on the faces of the $(m-1)$-simplex $\Delta[m-1]$ is induced by an ordering on the vertices. This is the left lexicographical ordering on strings of varying lengths with shorter strings taking precedence. The ordering gives a filtration on $\Delta[m-1]$ by

$$F_t(\Delta[m-1])=\bigcup_{s\leq t}\sigma_s.$$

In turn, this gives a total ordering on the simplices of a simplicial K on m vertices

$$\sigma_0=\varnothing<\sigma_1<\sigma_2<\ldots<\sigma_t<\ldots<\sigma_s \tag{4}$$

via the natural inclusion

$$K \subset \Delta[m-1].$$

This is filtration preserving in the sense that $F_t K = K \cap F_t \Delta[m-1]$.

Example 3.1 Consider $[m] = [3]$ and

$$K = \big\{\phi, \{v_1\}, \{v_2\}, \{v_3\}, \big\{\{v_1\}, \{v_3\}\big\}, \big\{\{v_2\}, \{v_3\}\big\}\big\}$$

with the realization consisting of two edges with a common vertex. Here the length-lexicographical ordering on the two-simplex $\Delta[2]$ is

$$\phi < v_1 < v_2 < v_3 < v_1v_2 < v_1v_3 < v_2v_3 < v_1v_2v_3$$

and so the induced ordering on K is

$$\phi < v_1 < v_2 < v_3 < v_1v_3 < v_2v_3.$$

Remark 3.2 Notice that if $t < m$, then $F_t K$ will contain *ghost* vertices, that is, vertices that are in $[m]$ but are not considered simplices. They do, however, label Cartesian product factors in the polyhedral product.

As described in [3, Section 2], this induces a natural filtration on the polyhedral product $Z\big(K; (\underline{X}, \underline{A})\big)$ and the smash polyhedral product $\widehat{Z}\big(K; (\underline{X}, \underline{A})\big)$ as follows:

$$F_t Z(K; (\underline{X}, \underline{A})) = \bigcup_{k \leq t} D_{(\underline{X}, \underline{A})}(\sigma_k) \quad \text{and} \quad F_t \widehat{Z}(K; (\underline{X}, \underline{A})) = \bigcup_{k \leq t} \widehat{D}_{\underline{X}, \underline{A}}(\sigma_k).$$

Notice also that the filtration satisfies

$$F_t \widehat{Z}(K; (\underline{X}, \underline{A})) = \widehat{Z}(F_t K; (\underline{X}, \underline{A})). \tag{5}$$

4 The Proof of Theorem 2.2

Let the family of CW pairs $(\underline{U}, \underline{V})$ be wedge decomposable as in Definition 2.1. We begin by checking that Theorem 2.2 holds for $F_0 \widehat{Z}(K; (\underline{U}, \underline{V}))$. In this case, $F_0 K$ consists of the empty simplex (the boundary of a point) and $m-1$ ghost vertices. So,

$$\begin{aligned} \widehat{Z}(F_0 K; (\underline{U}, \underline{V})) &= V_1 \wedge V_2 \cdots \wedge V_m \\ &= (B_1 \vee E_1) \wedge (B_2 \vee E_2) \wedge \cdots \wedge (B_m \vee E_m). \end{aligned} \tag{6}$$

Next, fix $I = (i_1, i_2, \ldots, i_k) \subset [m]$ and set $[m] - I = (j_i, j_2, \ldots j_{m-k})$. Then

$$\begin{aligned}\widehat{Z}\big(F_0 K_I; (\underline{C}, \underline{E})_I\big) &\wedge \widehat{Z}\big(K_{[m]-I}; (\underline{B}, \underline{B})_{[m]-I}\big)\\ &= (E_{i_1} \wedge E_{i_2} \wedge \cdots E_{i_k}) \wedge (B_{j_1} \wedge B_{j_2} \wedge \cdots B_{j_{m-k}}).\end{aligned}$$

is the Ith wedge term in the expansion of the right-hand side of (6). This confirms Theorem 2.2 for $t = 1$.

We suppose next the induction hypothesis that

$$F_{t-1}\widehat{Z}(K; (\underline{U}, \underline{V})) \simeq \bigvee_{I \leq [m]} \widehat{Z}\big(F_{t-1}K_I; (\underline{C}, \underline{E})_I\big) \wedge \widehat{Z}\big(K_{[m]-I}; (\underline{B}, \underline{B})_{[m]-I}\big),$$

with a view to verifying it for F_t. The definition of the filtration gives

$$\begin{aligned}F_t\widehat{Z}(K; (\underline{U}, \underline{V})) &= \widehat{D}_{\underline{U},\underline{V}}(\sigma_t) \cup F_{t-1}\widehat{Z}(K; (\underline{U}, \underline{V})) \qquad (7)\\ &\simeq \widehat{D}_{\underline{U},\underline{V}}(\sigma_t) \cup \bigvee_{I \leq [m]} \widehat{Z}\big(F_{t-1}K_I; (\underline{C}, \underline{E})_I\big)\\ &\qquad \wedge \widehat{Z}\big(K_{[m]-I}; (\underline{B}, \underline{B})_{[m]-I}\big).\end{aligned}$$

The space $\widehat{D}_{\underline{U},\underline{V}}(\sigma_t)$ is the smash product

$$\bigwedge_{j=1}^{m} B_j \vee Y_j, \quad \text{with} \quad Y_j = \begin{cases} C_j \text{ if } j \in \sigma_t \\ E_j \text{ if } j \notin \sigma_t. \end{cases} \qquad (8)$$

After a shuffle of wedge factors, the space $\widehat{D}_{\underline{U},\underline{V}}(\sigma_t)$ becomes

$$\begin{aligned}&\bigvee_{I \leq [m],\ \sigma_t \in I} \widehat{D}^I_{\underline{C},\underline{E}}(\sigma_t) \wedge \widehat{Z}\big(K_{[m]-I}; (\underline{B}, \underline{B})_{[m]-I}\big) \vee\\ &\qquad \bigvee_{I \leq [m],\ \sigma_t \notin I} \widehat{Z}\big(K_I; (\underline{C}, \underline{E})_I\big) \wedge \widehat{Z}\big(K_{[m]-I}; (\underline{B}, \underline{B})_{[m]-I}\big) \qquad (9)\end{aligned}$$

where the space $\widehat{D}^I_{\underline{C},\underline{E}}(\sigma_t)$ is defined by (3).

Remark 4.1 Notice here the relevant fact that the number of subsets $I \leq [m]$ is the same as the number of wedge summands in the expansion of (8), namely, 2^m.

The right-hand wedge summand in (9) is a subset of

$$\bigvee_{I \leq [m]} \widehat{Z}\big(F_{t-1}K_I; (\underline{C}, \underline{E})_I\big) \wedge \widehat{Z}\big(K_{[m]-I}; (\underline{B}, \underline{B})_{[m]-I}\big)$$

and so,

$$\bigvee_{\substack{I \leq [m] \\ \sigma_t \notin I}} \widehat{Z}\big(K_I; (\underline{C}, \underline{E})_I\big) \wedge \widehat{Z}\big(K_{[m]-I}; (\underline{B}, \underline{B})_{[m]-I}\big)$$

$$\bigcup \bigvee_{I \leq [m]} \widehat{Z}\big(F_{t-1}K_I; (\underline{C}, \underline{E})_I\big) \wedge \widehat{Z}\big(K_{[m]-I}; (\underline{B}, \underline{B})_{[m]-I}\big)$$

$$= \bigvee_{I \leq [m]} \widehat{Z}\big(F_{t-1}K_I; (\underline{C}, \underline{E})_I\big) \wedge \widehat{Z}\big(K_{[m]-I}; (\underline{B}, \underline{B})_{[m]-I}\big).$$

Finally, for each $I \leq [m]$ with $\sigma_t \in I$, we have

$$\widehat{D}^I_{\underline{C},\underline{E}}(\sigma_t) \cup \widehat{Z}\big(F_{t-1}K_I; (\underline{C}, \underline{E})_I\big) = \widehat{Z}\big(F_t K_I; (\underline{C}, \underline{E})_I\big). \tag{10}$$

This concludes the inductive step to give

$$F_t\widehat{Z}(K; (\underline{U}, \underline{V})) \simeq \bigvee_{I \leq [m]} \widehat{Z}\big(F_t K_I; (\underline{C}, \underline{E})_I\big) \wedge \widehat{Z}\big(K_{[m]-I}; (\underline{B}, \underline{B})_{[m]-I}\big). \tag{11}$$

It is straightforward to explicitly check the steps above in the case of F_0 and F_1. This completes the proof. □

5 Symmetric Products

We begin with a definition.

Definition 5.1 Let $(X, *)$ denote a pointed topological space. The *m-fold symmetric product* for $(X, *)$ is the orbit space

$$SP^m(X) = X^m \big/ \Sigma_m$$

where the symmetric group on m-letters Σ_m acts on the left by permutation of coordinates. There are natural maps

$$\begin{array}{cccc} \boldsymbol{e}\colon & SP^m(X) & \longrightarrow & SP^{m+1}(X) \\ & [x_1, x_2, \ldots, x_m] & \mapsto & [x_1, x_2, \ldots, x_m, *] \end{array} \tag{12}$$

which allow for the definition of the *infinite symmetric product* as a colimit

$$SP(X) = \operatorname*{colim}_{1 \leq m} SP^m(X).$$

The colimit becomes a filtered unital commutative monoid under concatenation, with $*$ as the unit. Furthermore, there is a natural inclusion

$$\begin{array}{rccc} E_X\colon & X & \longrightarrow & SP(X) \\ & p & \mapsto & [p] \end{array} . \tag{13}$$

One version of the classical Dold-Thom theorem is as follows.

Theorem 5.2 ([10]) *Given a pointed, path-connected pair of finite CW complexes $(X, A, *)$ (where A is a closed subcomplex of X), then the following hold.*

1. *$SP(X)$ is homotopy equivalent to a product of Eilenberg-MacLane spaces*

$$\prod_{1 \leq q \leq \infty} K(H_q(X), q)$$

2. *The natural map*

$$SP(X) \longrightarrow SP(X/A)$$

is a quasi-fibration with quasi-fiber $SP(A)$.

A natural map,

$$\widehat{\theta}\colon SP^{q_1}(X_1) \wedge SP^{q_2}(X_2) \wedge \cdots \wedge SP^{q_m}(X_m) \longrightarrow SP^q(X_1 \wedge X_2 \wedge \cdots \wedge X_m),$$

for $q = q_1 q_2 \cdots q_m$, is constructed next by setting

$$\widehat{\theta}\big([[x_{11}, x_{12}, \ldots, x_{1q_1}], [x_{21}, x_{22}, \ldots, x_{2q_2}], \ldots, [x_{m1}, x_{m2}, \ldots, x_{mq_m}]]^{\wedge}\big)$$

$$= \Big[\prod_{\substack{1 \leq j_t \leq q_t \\ 1 \leq t \leq m}} [x_{1j_1}, x_{2j_2}, \ldots, x_{mj_m}]^{\wedge} \Big] \tag{14}$$

where, here, square brackets [ ] are used to denote equivalence classes in the symmetric product and $[\ ]^{\wedge}$ for the smash products. Next we introduce an extension of $\widehat{\theta}$, which we shall use throughout to deduce the main results.

Theorem 5.3 *The construction of the map $\widehat{\theta}$ extends in a natural way to give*

$$\widehat{\theta}\colon SP(X_1) \wedge SP(X_2) \wedge \cdots \wedge SP(X_m) \longrightarrow SP(X_1 \wedge X_2 \wedge \cdots \wedge X_m)$$

a map of colimits.

Proof It suffices to check that the diagram below commutes

$$\begin{array}{ccc} SP^{q_1}(X_1)\wedge\cdots\wedge SP^{q_k}(X_k)\wedge\cdots\wedge SP^{q_m}(X_m) & \xrightarrow{\widehat{\theta}} & SP^{q}(X_1\wedge X_2\wedge\cdots\wedge X_m) \\ \downarrow e & & \downarrow e \\ SP^{q_1}(X_1)\wedge\cdots\wedge SP^{q_k+1}(X_k)\wedge\cdots\wedge SP^{q_m}(X_m) & \xrightarrow{\widehat{\theta}} & SP^{q'}(X_1\wedge X_2\wedge\cdots\wedge X_m) \end{array}$$

where, here, $q = q_1q_2\cdots q_m$, $q' = q_1q_2\cdots q_{k-1}(q_k+1)q_{k+1}\cdots q_m$, and the map e is as in (12). Consider then,

$$\begin{aligned} & e\circ\theta\big([[x_{11},x_{12},\ldots,x_{1q_1}],[x_{21},x_{22},\ldots,x_{2q_2}],\ldots,[x_{m1},x_{m2},\ldots,x_{mq_m}]]^{\wedge}\big) \\ &\quad = e\Big(\Big[\prod_{\substack{1\le j_t\le q_t\\ 1\le t\le m}}[x_{1j_1},x_{2j_2},\ldots,x_{mj_m}]^{\wedge}\Big]\Big) \\ &\quad = \Big[\prod_{\substack{1\le j_t\le q_t\\ 1\le t\le m}}[x_{1j_1},x_{2j_2},\ldots,x_{mj_m}]^{\wedge},\ \underbrace{[*]^{\wedge},\ldots,[*]^{\wedge}}_{q'-q}\Big]. \end{aligned}$$

On the other hand,

$$\begin{aligned} & \theta\circ e\big([[x_{11},x_{12},\ldots,x_{1q_1}],[x_{21},x_{22},\ldots,x_{2q_2}],\ldots,[x_{m1},x_{m2},\ldots,x_{mq_m}]]^{\wedge}\big) \\ &= \theta\big([[x_{11},x_{12},\ldots,x_{1q_1}],\ldots,[x_{21},x_{22},\ldots,x_{2q_k},*],\ldots,[x_{m1},x_{m2},\ldots,x_{mq_m}]]^{\wedge}\big) \\ &= \Big[\prod_{\substack{1\le j_t\le q_t\\ 1\le t\le m}}[x_{1j_1},x_{2j_2},\ldots,x_{mj_m}]^{\wedge}\prod_{\substack{1\le j_t\le q_t\\ 1\le t\le m,\ t\ne k}}[x_{1j_1},x_{2j_2},\ldots,x_{mj_m},*]^{\wedge}\Big] \\ &= \Big[\prod_{\substack{1\le j_t\le q_t\\ 1\le t\le m}}[x_{1j_1},x_{2j_2},\ldots,x_{mj_m}]^{\wedge},\ \underbrace{*,\ldots,*}_{q'-q}\Big] \end{aligned}$$

A simple example illustrates the proof of Theorem 5.3 for $m = 2$, $q = 2$, and $q' = 4$.

$$\begin{array}{ccc} SP^1(X_1)\wedge SP^2(X_2) & \xrightarrow{\widehat{\theta}} & SP^2(X_1\wedge X_2) \\ \downarrow e\times 1 & & \downarrow e \\ SP^2(X_1)\wedge SP^2(X_2) & \xrightarrow{\widehat{\theta}} & SP^4(X_1\wedge X_2) \end{array} \tag{15}$$

Here,

$$
\begin{aligned}
(e \circ \widehat{\theta})\big([[x_{11}], [x_{21}, x_{22}]]^{\wedge}\big) &= e\big[[x_{11}, x_{21}]^{\wedge}, [x_{11}, x_{22}]^{\wedge}\big] \\
&= \big[[x_{11}, x_{21}]^{\wedge}, [x_{11}, x_{22}]^{\wedge}, [*]^{\wedge}, [*]^{\wedge}\big],
\end{aligned}
$$

whereas

$$
\begin{aligned}
\big(\widehat{\theta} \circ (e \times 1)\big)\big([[x_{11}], [x_{21}, x_{22}]]^{\wedge}\big) &= \widehat{\theta}\big([[x_{11}, *], [x_{21}, x_{22}]]^{\wedge}\big) \\
&= \big[[x_{11}, x_{21}]^{\wedge}, [x_{11}, x_{22}]^{\wedge}, [*, x_{21}]^{\wedge}, [*, x_{22}]^{\wedge}\big].
\end{aligned}
$$

The diagram commutes because both $[*, x_{21}]^{\wedge}$ and $[*, x_{22}]^{\wedge}$ equal $[*]^{\wedge}$.

6 Connections Between Symmetric Products and Polyhedral Products

Consider a simplicial complex K on $[m]$ and a family of pointed CW pairs $(\underline{X}, \underline{A})$. We adopt the notation

$$
\big(\underline{SP^*(X)}, \underline{SP^*(A)}\big) \;=\; \big\{\big(SP^{q_i}(X_i), SP^{q_i}(A_i)\big)\big\}_{i=1}^{m}, \tag{16}
$$

and construct a structure map

$$
\zeta\colon \widehat{Z}\big(K; \big(\underline{SP(X)}, \underline{SP(A)}\big)\big) \longrightarrow SP\big(\widehat{Z}(K; (\underline{X}, \underline{A}))\big). \tag{17}
$$

by considering first the composite map

$$
\begin{aligned}
\widehat{Z}\big(K; \big(\underline{SP^*(X)}, \underline{SP^*(A)}\big)\big) &\xrightarrow{\ \iota\ } SP^{q_1}(X_1) \wedge SP^{q_2}(X_2) \wedge \cdots \wedge SP^{q_m}(X_m) \\
&\xrightarrow{\ \widehat{\theta}\ } SP^{q}(X_1 \wedge X_2 \wedge \cdots \wedge X_m)
\end{aligned} \tag{18}
$$

Lemma 6.1 *A map ζ exists in the diagram below making the diagram commute*

$$
\begin{array}{ccc}
\widehat{Z}\big(K; \big(\underline{SP^*(X)}, \underline{SP^*(A)}\big)\big) & \xrightarrow{\ \widehat{\Theta} \circ \iota\ } & SP^{q}(X_1 \wedge X_2 \wedge \cdots \wedge X_m) \\
 & \overset{\zeta}{\searrow} & \Big\uparrow {\scriptstyle SP^{q}(\iota)} \\
 & & SP^{q}\big(\widehat{Z}(K; (\underline{X}, \underline{A}))\big)
\end{array} \tag{19}
$$

where $\iota\colon \widehat{Z}(K; (\underline{X}, \underline{A})) \hookrightarrow X_1 \wedge X_2 \wedge \cdots \wedge X_m$ is the inclusion. Moreover, the map ζ extends to a map at level of infinite symmetric products

$$
\zeta\colon \widehat{Z}\big(K; \big(\underline{SP(X)}, \underline{SP(A)}\big)\big) \longrightarrow SP\big(\widehat{Z}(K; (\underline{X}, \underline{A}))\big). \tag{20}
$$

Proof We use the indexing from (16) and begin by defining

$$\zeta\colon \widehat{D}_{(\underline{SP^*(X)},\underline{SP^*(A)})}(\sigma) \longrightarrow SP^q\big(\widehat{D}_{(\underline{X},\underline{A})}(\sigma)\big)$$

for $\sigma \in K$, where

$$\widehat{D}_{(\underline{SP^*(X)},\underline{SP^*(A)})}(\sigma) = \bigwedge_{i=1}^{m} Y_i, \quad \text{where} \quad Y_i = \begin{cases} SP^{q_i}(X_i) \text{ if } i \in \sigma \\ SP^{q_i}(A_i) \text{ if } i \in [m]-\sigma, \end{cases} \tag{21}$$

and

$$\widehat{D}_{(\underline{X},\underline{A})}(\sigma) = \bigwedge_{i=1}^{m} W_i, \quad \text{where} \quad W_i = \begin{cases} X_i \text{ if } i \in \sigma \\ A_i \text{ if } i \in [m]-\sigma, \end{cases} \tag{22}$$

by

$$\begin{aligned} &\zeta\big([[x_{11}, x_{12}, \ldots, x_{1q_1}], [x_{21}, x_{22}, \ldots, x_{2q_2}], \ldots, [x_{m1}, x_{m2}, \ldots, x_{mq_m}]]^{\wedge}\big) \\ &\qquad = \Big[\prod_{\substack{1\le j_t \le q_t \\ 1 \le t \le m}} \big[x_{1j_1}, x_{2j_2}, \ldots, x_{mj_m}\big]^{\wedge}\Big] \end{aligned} \tag{23}$$

The key point that makes the target of ζ equal to $SP^q\big(\widehat{D}_{(\underline{X},\underline{A})}(\sigma)\big)$ is the observation that if a point $[x_{r1}, x_{r2}, \ldots, x_{rq_r}]^{\wedge}$ is in $\widehat{A}_r^{q_r}$, then

$$\big[x_{1j_1}, x_{2j_2}, \ldots, x_{rj_r}, \ldots, x_{mj_m}\big]^{\wedge} \in X_1 \wedge \cdots \wedge X_{r-1} \wedge A_r \wedge X_{r+1} \wedge \cdots \wedge X_m.$$

Next, we need to check that the diagram following commutes.

$$\begin{array}{ccc} \widehat{D}_{(\underline{SP^*(X)},\underline{SP^*(A)})}(\sigma) & \xrightarrow{\quad\zeta\quad} & SP^q\big(\widehat{D}_{(\underline{X},\underline{A})}(\sigma)\big) \\ \Big\downarrow{\scriptstyle \ell} & & \Big\downarrow{\scriptstyle SP^q(\ell)} \\ \widehat{D}_{(\underline{SP^*(X)},\underline{SP^*(A)})}(\tau) & \xrightarrow{\quad\zeta\quad} & SP^q\big(\widehat{D}_{(\underline{X},\underline{A})}(\tau)\big) \end{array} \tag{24}$$

whenever $\sigma \overset{\ell}{\hookrightarrow} \tau$ in K, but this is immediate from the definitions (22) and (21). Taking colimits with respect to the diagram of K, we get the dashed arrow of (19):

$$\zeta\colon \widehat{Z}\big(K; (\underline{SP^*(X)}, \underline{SP^*(A)})\big) \longrightarrow SP^q\big(\widehat{Z}(K; (\underline{X}, \underline{A}))\big) \tag{25}$$

It remains to check the commutativity of diagram (19). Let

$$\big[[x_{11}, x_{12}, \ldots, x_{1q_1}], [x_{21}, x_{22}, \ldots, x_{2q_2}], \ldots, [x_{m1}, x_{m2}, \ldots, x_{mq_m}]\big]^{\wedge} \in D_{(\underline{SP^*(X)}, \underline{SP^*(A)})}(\sigma)\,,$$

and we have

$$\begin{aligned}
&(\widehat{\theta} \circ \iota)\Big(\big[[x_{11}, x_{12}, \ldots, x_{1q_1}], [x_{21}, x_{22}, \ldots, x_{2q_2}], \ldots, [x_{m1}, x_{m2}, \ldots, x_{mq_m}]\big]^{\wedge}\Big)\\
&\quad = \widehat{\theta}\Big(\big[[x_{11}, x_{12}, \ldots, x_{1q_1}], [x_{21}, x_{22}, \ldots, x_{2q_2}], \ldots, [x_{m1}, x_{m2}, \ldots, x_{mq_m}]\big]^{\wedge}\Big)\\
&\quad = \Big[\prod_{\substack{1 \le j_t \le q_t\\ 1 \le t \le m}} [x_{1j_1}, x_{2j_2}, \ldots, x_{mj_m}]^{\wedge}\Big]
\end{aligned}$$

Also,

$$\begin{aligned}
&(SP^q(\iota) \circ \zeta)\Big(\big[[x_{11}, x_{12}, \ldots, x_{1q_1}], [x_{21}, x_{22}, \ldots, x_{2q_2}], \ldots, [x_{m1}, x_{m2}, \ldots, x_{mq_m}]\big]^{\wedge}\Big)\\
&\quad = SP^q(\iota)\Big(\Big[\prod_{\substack{1 \le j_t \le q_t\\ 1 \le t \le m}} [x_{1j_1}, x_{2j_2}, \ldots, x_{mj_m}]^{\wedge}\Big]\Big)\\
&\quad = \Big[\prod_{\substack{1 \le j_t \le q_t\\ 1 \le t \le m}} [x_{1j_1}, x_{2j_2}, \ldots, x_{mj_m}]^{\wedge}\Big]
\end{aligned}$$

Finally, we need to check that the map ζ extends to a map of the colimits defining the infinite symmetric products, as in (20). To this end, we fix $k \in [m]$ and modify (16) by setting

$$\big(\underline{SP^{*_k}(X)}, \underline{SP^{*_k}(A)}\big) = \big\{\big(SP^{q_i'}(X_i), SP^{q_i'}(A_i)\big)\big\}_{i=1}^{m}, \tag{26}$$

where

$$q_i' = \begin{cases} q_i & \text{if } i \neq k \\ q_k + 1 & \text{if } i = k \end{cases}$$

and $q' = q_1' q_2' \cdots q_m'$. It suffices now to check the commutativity of the diagram

$$\begin{array}{ccc}
\widehat{Z}(K; (\underline{SP^*(X)}, \underline{SP^*(A)})) & \xrightarrow{\ \zeta\ } & SP^q(\widehat{Z}(K; (\underline{X}, \underline{A})))\\
\big\downarrow{\scriptstyle \widehat{Z}(K;\, e)} & & \big\downarrow{\scriptstyle e}\\
\widehat{Z}(K; (\underline{SP^{*_k}(X)}, \underline{SP^{*_k}(A)})) & \xrightarrow{\ \zeta\ } & SP^{q'}(\widehat{Z}(K; (\underline{X}, \underline{A})))
\end{array} \tag{27}$$

In the notation of (23), we have

$$(e \circ \zeta)\Big(\big[[x_{11}, x_{12}, \dots, x_{1q_1}], [x_{21}, x_{22}, \dots, x_{2q_2}], \dots, [x_{m1}, x_{m2}, \dots, x_{mq_m}]\big]^{\wedge}\Big)$$

$$= e\Big(\Big[\prod_{\substack{1\le j_t\le q_t\\ 1\le t\le m}} [x_{1j_1}, x_{2j_2}, \dots, x_{mj_m}]^{\wedge}\Big]\Big)$$

$$= \Big[\prod_{\substack{1\le j_t\le q_t\\ 1\le t\le m}} [x_{1j_1}, x_{2j_2}, \dots, x_{mj_m}]^{\wedge}, \underbrace{*, \dots, *}_{q'-q}\Big].$$

On the other hand,

$$(\zeta \circ e)\Big(\big[[x_{11}, x_{12}, \dots, x_{1q_1}], [x_{21}, x_{22}, \dots, x_{2q_2}], \dots, [x_{m1}, x_{m2}, \dots, x_{mq_m}]\big]^{\wedge}\Big)$$

$$= \zeta\Big(\big[[x_{11}, x_{12}, \dots, x_{1q_1}], \dots, [x_{21}, x_{22}, \dots, x_{2q_k}, *], \dots, [x_{m1}, x_{m2}, \dots, x_{mq_m}]\big]^{\wedge}\Big)$$

$$= \Big[\prod_{\substack{1\le j_t\le q_t\\ 1\le t\le m}} [x_{1j_1}, x_{2j_2}, \dots, x_{mj_m}]^{\wedge} \prod_{\substack{1\le j_t\le q_t\\ 1\le t\le m,\ t\neq k}} [x_{1j_1}, x_{2j_2}, \dots, x_{mj_m}, *]^{\wedge}\Big]$$

$$= \Big[\prod_{\substack{1\le j_t\le q_t\\ 1\le t\le m}} [x_{1j_1}, x_{2j_2}, \dots, x_{mj_m}]^{\wedge}, \underbrace{*, \dots, *}_{q'-q}\Big]$$

The next construction is standard.

Lemma 6.2 *A map of the form* $\phi\colon X \longrightarrow SP^k(Y)$*, which induces a map*

$$\phi\colon X \longrightarrow SP(Y)$$

admits a canonical multiplicative extension $\psi\colon SP(X) \longrightarrow SP(Y)$*. This extension is the identity map if* $X = Y$ *and the map* ϕ *is the inclusion* E_X*.*

Proof The map ψ is defined by the map ϕ as a composite

$$SP^q(X) \xrightarrow{SP^q(\phi)} SP^q\big(SP^k(Y)\big) \xrightarrow{\eta} SP^{qk}(Y) \tag{28}$$

where the map η is given by

$$\big[[y_{11}, y_{12}, \dots, y_{1k}], [y_{21}, y_{22}, \dots, y_{2k}], \ \dots\ , [y_{q1}, x_{q2}, \dots, y_{qk}]\big]$$

$$\longmapsto [y_{11}, y_{12}, \dots, y_{1k}, y_{21}, x_{22}, \dots, y_{2k}, \ \dots\ , y_{q1}, y_{q2}, \dots, y_{qk}]$$

So, writing $\phi(x_i) = [\phi(x_i)_1, \phi(x_i)_2, \dots, \phi(x_i)_k] \in SP^k(Y)$, we have

$$\begin{aligned}\psi([x_1, x_2, \ldots, x_q]) &= \eta([\phi(x_1), \phi(x_2), \ldots, \phi(x_q)]) \\ &= [\phi(x_1)_1, \phi(x_1)_2, \ldots, \phi(x_1)_k, \phi(x_2)_1, \phi(x_2)_2, \ldots, \\ &\qquad \phi(x_2)_k, \ldots, \phi(x_q)_1, \phi(x_q)_2, \ldots, \phi(x_q)_k]\end{aligned}$$

The map ψ fits into a commutative diagram as follows:

$$\begin{array}{ccc} SP^q(X) & \xrightarrow{\psi} & SP^{qk}(Y) \\ \downarrow{\scriptstyle e} & & \downarrow{\scriptstyle e} \\ SP^{q+1}(X) & \xrightarrow{\psi} & SP^{(q+1)k}(Y). \end{array} \tag{29}$$

More specifically, let $[x_1, x_2, \ldots, x_q] \in SP^q(X)$, and then

$$\begin{aligned}&(e \circ \psi)([x_1, x_2, \ldots, x_q]) \\ &\quad = e(\eta([\phi(x_1), \phi(x_2), \ldots, \phi(x_q)])) \\ &\quad = [\phi(x_1)_1, \phi(x_1)_2, \ldots, \phi(x_1)_k, \ldots, \phi(x_q)_1, \phi(x_q)_2, \ldots, \phi(x_q)_k, \underbrace{*, \ldots, *}_{k}]\end{aligned}$$

On the other hand,

$$\begin{aligned}&(\psi \circ e)([x_1, x_2, \ldots, x_q]) \\ &\quad = \psi(([x_1, x_2, \ldots, x_q, *]) \\ &\quad = \eta([\phi(x_1), \phi(x_2), \ldots, \phi(x_q), \phi(*)]) \\ &\quad = [\phi(x_1), \phi(x_2), \ldots, \phi(x_q), \underbrace{*, \ldots, *}_{k}] \\ &\quad = [\phi(x_1)_1, \phi(x_1)_2, \ldots, \phi(x_1)_k, \ldots, \phi(x_q)_1, \phi(x_q)_2, \ldots, \phi(x_q)_k, \underbrace{*, \ldots, *}_{k}]\end{aligned}$$

The second statement of the lemma follows from the definition of the map E_X. Notice further that, for the map E_X of (13), we see that

$$(\psi \circ E_X)\colon X \longrightarrow SP(Y) \tag{30}$$

coincides with $\phi\colon X \longrightarrow SP(Y)$.

Applying Lemma 6.2 to the map ζ of (20), we get its multiplicative extension

$$\psi\colon SP\big(\widehat{Z}\big(K; \big(\underline{SP(X)}, \underline{SP(A)}\big)\big)\big) \longrightarrow SP\big(\widehat{Z}(K; (\underline{X}, \underline{A}))\big). \tag{31}$$

More properties of the map ψ are given next.

Lemma 6.3 *Let K be a simplicial complex on $[m]$ and $(\underline{X}, \underline{A})$ a family of finite pointed CW pairs.*

1. *The maps* $(E_{X_i}, E_{A_i})\colon (X_i, A_i) \longrightarrow \big(SP(X_i), SP(A_i)\big)$ *induce a morphism of polyhedral smash products*

$$E^K_{(\underline{X},\underline{A})}\colon\ \widehat{Z}(K; (\underline{X}, \underline{A})) \longrightarrow \widehat{Z}\big(K; (\underline{SP(X)}, \underline{SP(A)})\big).$$

2. *There is a commutative diagram*

$$\begin{array}{ccc}
\widehat{Z}(K;(\underline{X},\underline{A})) & \xrightarrow{\ E^K_{(\underline{X},\underline{A})}\ } & \widehat{Z}(K;(\underline{SP(X)},\underline{SP(A)})) \\
 & \searrow^{E_{\widehat{Z}(K;(\underline{X},\underline{A}))}} & \downarrow \zeta \\
 & & SP\big(\widehat{Z}(K;(\underline{X},\underline{A}))\big)
\end{array}$$

3. *There is a strictly commutative multiplicative diagram*

$$\begin{array}{ccc}
SP\big(\widehat{Z}(K;(\underline{X},\underline{A}))\big) & \xrightarrow{\ SP\left(E^K_{(\underline{X},\underline{A})}\right)\ } & SP\big(\widehat{Z}(K;(\underline{SP(X)},\underline{SP(A)})\big) \\
 & \searrow^{1} & \downarrow \psi \\
 & & SP\big(\widehat{Z}(K;(\underline{X},\underline{A}))\big)
\end{array}$$

where

$$SP\big(\widehat{Z}(K;(\underline{X},\underline{A}))\big) \xrightarrow{\ SP\left(E^K_{(\underline{X},\underline{A})}\right)\ } SP\big(\widehat{Z}(K;(\underline{SP(X)},\underline{SP(A)})\big)$$

is multiplicative.

Proof Part 1 is a consequence of the functoriality of the polyhedral smash product, and part 2 follows from the definition of ζ (23). Next, applying (30) from Lemma 6.2, we see that the map ψ ((31)), of part 3, restricted to $\widehat{Z}\big(K; (\underline{SP(X)}, \underline{SP(A)})\big)$, coincides with the map ζ. The diagram of part 3 follows by applying Lemma 6.2 to the maps $E^K_{(\underline{X},\underline{A})}$ and ζ in the diagram of part 2. □

This lemma admits further extensions for subspaces of polyhedral smash products.

Lemma 6.4 *Let* K *be as in Lemma 6.3 and* $(\underline{X}, \underline{A})$ *and* $(\underline{U}, \underline{V})$ *be families of pairs of pointed, finite connected CW complexes. Assume further that there are maps of pointed pairs*

$$g_i\colon (U_i, V_i) \longrightarrow \big(SP(X_i), SP(A_i)\big).$$

Then:

1. *There are induced maps*

$$\underline{g}\colon \underline{U}^{[m]} \longrightarrow SP(X_1) \wedge SP(X_2) \wedge \cdots \wedge SP(X_m)$$

and

$$\widehat{D}_{(\underline{U},\underline{V})}(\omega) \xrightarrow{\ \underline{g}_\omega\ } \widehat{D}_{(\underline{SP(X)},\underline{SP(A)})}(\omega)$$

for $\omega \in K$.

2. *For* $\sigma \subset \tau \in K$, *there is a strictly commutative diagram, obtained by restriction, as follows:*

$$\begin{array}{ccc} \widehat{D}_{(\underline{U},\underline{V})}(\sigma) & \xrightarrow{\lambda_\sigma} & SP\big(\widehat{D}_{(\underline{X},\underline{A})}(\sigma)\big) \\ \big\downarrow{\scriptstyle \ell} & & \big\downarrow{\scriptstyle SP(\ell)} \\ \widehat{D}_{(\underline{U},\underline{V})}(\tau) & \xrightarrow{\lambda_\tau} & SP\big(\widehat{D}_{(\underline{X},\underline{A})}(\tau)\big) \end{array} \qquad (32)$$

where $i\colon \tau \hookrightarrow \sigma$ *denotes the inclusion of faces and, for* $\omega \in K$, *the map* λ_ω *is the composite*

$$\widehat{D}_{(\underline{U},\underline{V})}(\omega) \xrightarrow{\ \underline{g}_\omega\ } \widehat{D}_{(\underline{SP(X)},\underline{SP(A)})}(\omega) \xrightarrow{\ \zeta\ } SP\big(\widehat{D}_{(\underline{X},\underline{A})}(\omega)\big) \qquad (33)$$

and ζ *is constructed in Lemma 6.1.*

Proof The map $\underline{g}$ arises from the m-fold smash product of maps g_i in a natural way. The commutativity of the diagram (32) is obtained by splitting it into two diagrams corresponding to the factorization of the map λ_w as (33). The commutativity of the first diagram, corresponding to $\underline{g}_\omega$, is straightforward, and the second, corresponding to the map ζ, is (24). □

7 An Extension from Wedge Decomposable Pairs to the General Case

The purpose of this section is to begin the task of extending Theorem 2.2, the Cartan formula for wedge decomposable pairs, to a homological Cartan formula for arbitrary pointed path-connected pairs of finite CW complexes $(\underline{X}, \underline{A})$.

Theorem 7.1 (Homological Cartan Formula) *Let* K *be an abstract simplicial complex with* m *vertices. Assume that* $(\underline{X}, \underline{A})$ *are pointed, path-connected pairs of finite CW complexes for all* i. *There exist spaces* B_j, C_j, E_j, $1 \leq j \leq m$, *which* *are finite wedges of spheres and mod-n Moore spaces together with a homotopy equivalence*

$$SP(\widehat{Z}(K; (\underline{B \vee C}, \underline{B \vee E}))) \to SP(\widehat{Z}(K; (\underline{X}, \underline{A}))).$$

Thus, $H_*(\widehat{Z}(K; (\underline{X}, \underline{A})))$ *is isomorphic to* $H_*(\widehat{Z}(K; (\underline{B \vee C}, \underline{B \vee E})))$ *over the integers. This allows for a description of* $H_*(\widehat{Z}(K; (\underline{X}, \underline{A})))$ *in terms of the decompositions of Theorem 2.2 and Corollary 2.4.*

Remark 7.2 In a forthcoming paper, the authors use this and Corollary 2.4 to describe products in the cohomology of a polyhedral product.

Preliminary results required for the proof of Theorem 7.1 will occupy the remainder of this section. We begin with a definition.

Definition 7.3 The pairs $(\underline{U}, \underline{V})$ and $(\underline{X}, \underline{A})$ are said to have *strongly isomorphic homology* provided:

1. There are isomorphisms of singular homology groups

$$\alpha_j : H_*(U_j) \to H_*(X_j),$$

and

$$\beta_j : H_*(V_j) \to H_*(A_j).$$

2. There is a commutative diagram

$$\begin{array}{ccc} \bar{H}_i(V_j) & \xrightarrow{\lambda_{j*}} & \bar{H}_i(U_j) \\ {\scriptstyle\beta_j}\downarrow & & \downarrow{\scriptstyle\alpha_j} \\ \bar{H}_i(A_j) & \xrightarrow{\iota_{j*}} & \bar{H}_i(X_j), \end{array}$$

where $\lambda_j : V_j \subset U_j$ and $\iota_j : A_j \subset X_j$ are the natural inclusions.

3. There is an induced morphism of exact sequences for which all vertical arrows are isomorphisms:

$$\begin{array}{ccccccccc} 0 & \longrightarrow & ker(\lambda_{j*}) & \longrightarrow & \bar{H}_i(V_j) & \xrightarrow{\lambda_{j*}} & \bar{H}_i(U_j) & \longrightarrow & coker(\lambda_{j*}) & \longrightarrow & 0 \\ \downarrow & & {\scriptstyle\beta_j}\downarrow & & {\scriptstyle\beta_{j*}}\downarrow & & \downarrow{\scriptstyle\alpha_{j*}} & & \downarrow{\scriptstyle\bar{\alpha}_j} & & \downarrow \\ 0 & \longrightarrow & ker(\iota_{j*}) & \longrightarrow & \bar{H}_i(A_j) & \xrightarrow{\iota_{j*}} & \bar{H}_i(X_j) & \longrightarrow & coker(\iota_{j*}) & \longrightarrow & 0 \end{array}$$

where $\bar{\alpha}_j$ is induced by α_j.

4. The maps of pairs (α_j, β_j): $(H_*(U_j), H_*(V_j)) \to (H_*(X_j), H_*(A_j))$, which satisfy conditions 1–3, are said to *induce a strong homology isomorphism*.

Remark 7.4 The feature of the pairs $(\underline{U}, \underline{V})$ and $(\underline{X}, \underline{A})$ having strongly isomorphic homology groups implies, via the Künneth theorem, that the spaces $\widehat{D}_{(\underline{U},\underline{V})}(\sigma)$ and $\widehat{D}_{(\underline{X},\underline{A})}(\sigma)$ have isomorphic homology groups.

Lemma 7.5 *Given pointed, path-connected pairs of finite CW complexes $(\underline{X}, \underline{A})$ and $(\underline{U}, \underline{V})$ with strongly isomorphic homology groups, and let $\sigma \in K$ be any face of the simplicial complex K, then there is an isomorphism of singular homology groups*

$$\bar{H}_*\big(\widehat{D}_{(\underline{X},\underline{A})}(\sigma)\big) \longrightarrow \bar{H}_*\big(\widehat{D}_{(\underline{U},\underline{V})}(\sigma)\big).$$

The rest of the section is devoted to showing that isomorphisms can be chosen to be suitably compatible to pass to isomorphisms on homology for the full polyhedral product.

Lemma 7.6 *Given pointed, path-connected pairs of finite CW complexes $(\underline{X}, \underline{A})$, there exist wedges of spheres and mod-p^r Moore spaces $(\underline{B \vee C}, \underline{B \vee E})$ together with isomorphisms of singular homology groups*

$$\alpha_j \colon H_*(B_j \vee C_j) \to H_*(X_j)$$

and

$$\beta_j \colon H_*(B_j \vee E_j) \to H_*(A_j),$$

which give strong homology isomorphisms, and the pairs $(\underline{B \vee C}, \underline{B \vee E})$ satisfy the condition of wedge decomposability in Definition 2.1 that the inclusion $E_j \to C_j$ is null-homotopic.

Proof The proof of this lemma follows from the fact that X_j and A_j are finite, path-connected CW complexes, and so all homology groups as well as kernels and cokernels are finite direct sums of cyclic abelian groups. Thus, B_j, C_j, E_j may be chosen to be wedges of spheres and mod-p^r Moore spaces. Some details are given for completeness.

Consider a pair of pointed path-connected CW complexes (X, A) together with the induced map in homology $H_*A \to H_*X$. Then for any fixed $j \geq 1$, both H_jA and H_jX are finite direct sums of abelian groups. Any such finitely generated abelian group is a direct sum of cyclic abelian groups either of the form $\mathbb{Z}$ or $\mathbb{Z}/p^r\mathbb{Z}$ for some choice of n. In the case of $\mathbb{Z}$, then $H_j(S^j) = \mathbb{Z}$. In the case of $\mathbb{Z}/p^r\mathbb{Z}$, then the mod-p^r Moore space given by $P^{j+1}(\mathbb{Z}/p^r\mathbb{Z})$ satisfies $H_j\big(P^{j+1}(\mathbb{Z}/p^r\mathbb{Z})\big) = \mathbb{Z}/p^r\mathbb{Z}$. At this level, it is direct to realize maps on homology. The hypothesis of *strongly isomorphic homology* gives the naturality required. □

Our next goal is to establish a standard property of the infinite symmetric product.

Lemma 7.7 *Let U and X be finite, pointed, path-connected CW complexes and*

$$\alpha\colon \widetilde{H}_*(U) \longrightarrow \widetilde{H}_*(X)$$

a homomorphism in singular homology. Then there is a multiplicative map

$$g\colon SP(U) \longrightarrow SP(X)$$

which satisfies $\pi_*(g) = \alpha$ *and is a homotopy equivalence if the map* α *is an isomorphism.*

Proof Recall that the reduced singular homology of $SP(X)$ for any pointed, path-connected CW complex X is isomorphic to

$$\bigoplus_{1\leq m\leq\infty} \widetilde{H}_*\widehat{SP}^{(m)}(X)$$

where $\widehat{SP}^{(m)}(X)$ denotes the m-fold symmetric smash product given by

$$\widehat{SP}^{(m)}(X) = (X \wedge X \wedge \cdots \wedge X)\big/\Sigma_m.$$

This result appears in the thesis of R. J. Milgram, implicitly in [10] and in [9, Corollary 4.8]. Also, N. Steenrod proved, in unpublished notes, that there is a homotopy equivalence

$$SP\big(SP(X)\big) \to SP\Big(\bigvee_{1\leq m<\infty} \widehat{SP}^{(m)}(X)\Big).$$

Since X and U are finite, path-connected, pointed CW complexes with isomorphic singular homology groups, there is a homotopy equivalence

$$\phi_n : SP(\Sigma^n(U)) \longrightarrow SP(\Sigma^n(X))$$

for every natural number n. Next, consider the composite

$$\Sigma^n(U) \xrightarrow{\ E\ } SP(\Sigma^n(U)) \xrightarrow{\ \phi_n\ } SP(\Sigma^n(X))$$

and denote the canonical multiplicative extension (Lemma 6.2) of $\phi_n \circ E$ by

$$\psi_n : SP(\Sigma^n(U)) \to SP(\Sigma^n(X)).$$

Remark 0.1 This map might not be homotopic to the map ϕ_n. □

The next step in the proof of Theorem 7.1 is a lemma that allows a direct translation of algebraic properties concerning homology groups to geometric maps on the level of symmetric products of polyhedral products. It follows from the Dold-Thom theorem that the map ψ_n satisfies the formula

$$\psi_*(u) = \alpha(u) + \Delta_u \quad \text{where} \quad \Delta_u \in \bigoplus_{2\leq m} H_*\big(\widehat{SP}^m(\Sigma^n(X))\big).$$

Here the class Δ_u projects to zero in $H_*(\Sigma^n X) = H_*(\widehat{SP^1}(\Sigma^n X))$ on identifying the reduced homology of a space $\Sigma^n(X)$ with that of X. Thus, the canonical multiplicative extension $\psi_n : SP(\Sigma^n(U)) \to SP(\Sigma^n(X))$ induces a surjection on homology.

It follows that ψ_n induces a homology isomorphism since the homology groups of source and target are finitely generated abelian groups in each degree, which are abstractly isomorphic where the map is a surjection. □

Lemma 7.8 *Suppose that $(\underline{U}, \underline{V})$ and $(\underline{X}, \underline{A})$ are pointed, connected, pairs of finite CW complexes, with strongly isomorphic homology groups, and then the map of pairs*

$$g : \big(SP(U), SP(V)\big) \longrightarrow \big(SP(X), SP(A)\big)$$

arising from Lemma 7.7 induces strongly isomorphic homology groups. That is, the induced map on homology gives a commutative diagram

$$\begin{array}{ccc} \bar{H}_i(SP(V_j)) & \xrightarrow{\lambda_*} & \bar{H}_i(SP(U_j)) \\ {\scriptstyle g_*}\downarrow & & \downarrow{\scriptstyle g_*} \\ \bar{H}_i(SP(A_j)) & \xrightarrow{\iota_*} & \bar{H}_i(SP(X_j)) \end{array}$$

together with a second commutative diagram for which all vertical arrows are isomorphisms:

$$\begin{array}{ccccccc} ker(\lambda_*) & \longrightarrow & \bar{H}_i(SP(V_j)) & \xrightarrow{\lambda_*} & \bar{H}_i(SP(U_j)) & \longrightarrow & coker(\lambda_*) \\ {\scriptstyle g_*}\downarrow & & {\scriptstyle g_*}\downarrow & & \downarrow{\scriptstyle g_*} & & \downarrow{\scriptstyle \bar{g}_*} \\ ker(\iota_*) & \longrightarrow & \bar{H}_i(SP(A_j)) & \xrightarrow{\iota_*} & \bar{H}_i(SP(X_j)) & \longrightarrow & coker(\iota_*) \end{array}$$

where $\bar{g}_*$ is induced by g_*.

The proof of this lemma follows from the fact that X_j, A_j, U_j, and V_j are finite connected CW complexes, and so all homology groups as well as kernels and cokernels are finite direct sums of cyclic abelian groups.

Finally, we shall need a version of the Projection Lemma due to D. Quillen.

Lemma 7.9 ([19, Projection Lemma 1.6]) *Let $\mathcal{D}$ and $\mathcal{E}$ denote finite diagrams of finite CW complexes over the same finite category $\mathfrak{C}$ for which all inclusions in the intersection poset are closed cofibrations. Furthermore, assume that*

$$U = \bigcup_{\alpha \in \mathfrak{C}} D_\alpha \quad \textit{and} \quad X = \bigcup_{\alpha \in C} E_\alpha$$

and that there is a map

$$\mu \colon SP(U) \longrightarrow SP(X)$$

which restricts to homotopy equivalences on

$$\mu|_{SP(D_\alpha)} : SP(D_\alpha) \longrightarrow SP(E_\alpha)$$

for all $\alpha \in \mathfrak{C}$*. Then* μ *is a homotopy equivalence.*

8 The Proof of Theorem 7.1 Completed

The proof of Theorem 7.1 uses Theorem 2.2 and the interplay between the Dold-Thom construction and polyhedral products given in Sect. 6.

Let K be an abstract simplicial complex with m vertices. Assume that $(\underline{X}, \underline{A})$ are pointed, path-connected pairs of finite CW complexes for all i. Then by Lemma 7.6, we have wedges of spheres and mod-p^r Moore spaces $(\underline{B \vee C}, \underline{B \vee E})$ together with isomorphisms of singular homology groups

$$\alpha_j \colon H_*(B_j \vee C_j) \to H_*(X_j) \quad \text{and} \quad \beta_j \colon H_*(B_j \vee E_j) \to H_*(A_j),$$

which give strong homology isomorphisms, and the pairs $(\underline{B \vee C}, \underline{B \vee E})$ satisfy the condition of wedge decomposability in Definition 2.1 that the inclusion $E_j \to C_j$ is null-homotopic.

Next, Lemma 7.8 gives a map of pointed pairs

$$g \colon \big(\underline{SP(B \vee C)}, \underline{SP(B \vee E)}\big) \longrightarrow (\underline{SP(X)}, \underline{SP(A)})$$

which induces a strong isomorphism in homology. Applying the functor $\widehat{D}_{(-,-)}(\sigma)$ to this map, we get a map

$$\widehat{D}(\sigma; g) \colon \widehat{D}_{(\underline{SP(B \vee C)},\, \underline{SP(B \vee E)})}(\sigma) \longrightarrow \widehat{D}_{(\underline{SP(X)}, \underline{SP(A)})}(\sigma)$$

and, for each $\tau \subset \sigma$, a commutative diagram

$$\begin{array}{ccc}
\widehat{D}_{(\underline{SP(B\vee C)},\ \underline{SP(B\vee E)})}(\tau) & \xrightarrow{\widehat{D}(\tau;g)} & \widehat{D}_{(\underline{SP(X)},\underline{SP(A)})}(\tau) \\
\downarrow \beta & & \beta \downarrow \\
\widehat{D}_{(\underline{SP(B\vee C)},\ \underline{SP(B\vee E)})}(\sigma) & \xrightarrow{\widehat{D}(\sigma;g)} & \widehat{D}_{(\underline{SP(X)},\underline{SP(A)})}(\sigma)
\end{array}$$

where each horizontal arrow is a homotopy equivalence.

Further, there are induced morphisms of commutative diagrams via the structure map ζ of Lemma 6.1.

$$\begin{array}{ccc}
\widehat{D}_{(\underline{SP(B\vee C)},\ \underline{SP(B\vee E)})}(\sigma) & \longrightarrow & \widehat{D}_{(\underline{SP(X)},\underline{SP(A)})}(\sigma) \\
\zeta \downarrow & & \zeta \downarrow \\
SP\big(\widehat{D}_{(\underline{B\vee C},\ \underline{B\vee E})}(\sigma)\big) & \longrightarrow & SP\big(\widehat{D}_{(\underline{X},\ \underline{A})}(\sigma)\big)
\end{array}$$

Here, the horizontal arrows are homotopy equivalences by the Dold-Thom theorem and Lemma 7.5. Thus, there is a map

$$SP\big(\bigcup_{\sigma\in K} \widehat{D}_{\underline{B\vee C},\ \underline{B\vee E})}(\sigma)\big) \longrightarrow SP\big(\bigcup_{\sigma\in K} \widehat{D}_{(\underline{X},\ \underline{A})}(\sigma)\big)$$

which is a homotopy equivalence by Lemma 7.9. Theorem 7.1 follows from this. □

9 The Hilbert-Poincaré Series for $Z(K; (X, A))$

We begin by reviewing some of the elementary properties of Hilbert-Poincaré series. Assume now that homology is taken with coefficients in a field k and all spaces are pointed, path connected with the homotopy type of CW complexes. The Hilbert-Poincaré series

$$P(X,t) = \sum_n \big(\dim_k H_n(X;k)\big)t^n$$

and the reduced Hilbert-Poincaré series

$$\overline{P}(X,t) = -1 + P(X,t)$$

satisfy the following properties:

1. $P(X,t)P(Y,t) = P(X \times Y), t)$.
2. $\overline{P}(X,t)\overline{P}(Y,t) = P(X \wedge Y, t)$.

For a pair (X, A) satisfying the conditions of Theorem 7.1, we have

$$\overline{P}\big(\widehat{Z}(K;(\underline{X},\underline{A})),t\big)=\overline{P}\big(\widehat{Z}(K;(\underline{U},\underline{V})),t\big) \tag{34}$$

where the pair $(\underline{U},\underline{V}) = \underline{B\vee C},\ \underline{(B\vee E)})$ is the pair determined by $(\underline{X},\underline{A})$ and given by Lemma 7.6. Next, Theorem 2.2 gives

$$\overline{P}\big(\widehat{Z}(K;(\underline{U},\underline{V})),t\big)=\sum_{I\leq[m]}\Big[\overline{P}\big(\widehat{Z}(K_I;(\underline{C},\underline{E})_I),t)\big)\cdot\prod_{j\in[m]-I}\overline{P}(B_j,t)\Big] \tag{35}$$

We apply now Corollary 2.4 to refine this further and obtain the next theorem.

Theorem 9.1 *The reduced Hilbert-Poincaré series for $\widehat{Z}(K;(\underline{U},\underline{V}))$ and hence for $\widehat{Z}(K;(\underline{X},\underline{A}))$ is given as follows:*

$$\overline{P}\big(\widehat{Z}(K;(\underline{U},\underline{V})),t\big)$$
$$=\sum_{I\leq[m]}\Big[\sum_{\sigma\in K_I}\big[t\cdot\overline{P}(|lk_\sigma(K_I)|,t)\cdot\overline{P}\big(\widehat{D}^I_{\underline{C},\underline{E}}(\sigma),t\big)\big]\cdot\prod_{j\in[m]-I}\overline{P}(B_j,t)\Big]$$

where here we use the convention that $t\cdot\overline{P}(|\varnothing|,t)=1$ and $\overline{P}\big(\widehat{D}^I_{\underline{C},\underline{E}}(\sigma),t\big)$ can be read off from (3).

Finally, Theorem 1.1 gives now the Hilbert-Poincaré series for $Z(L;(U,V))$ and for $Z(L;(X,A)$, by applying (9.1) for each $K=L_J,\ J\subseteq[m]$.

10 Applications

We begin by illustrating the computation of a Poincaré series.

Example 10.1 Consider the composite

$$f\colon \mathbb{C}P^2\hookrightarrow\mathbb{C}P^3\to\mathbb{C}P^3/\mathbb{C}P^1. \tag{36}$$

and denote the mapping cylinder of (36) by M_f. We consider $\widetilde{H}^*\big(\widehat{Z}\big(K;(M_f,\mathbb{C}P^3)\big)\big)$ and $\widetilde{H}^*\big(Z(K;(M_f,\mathbb{C}P^3))\big)$ for any simplicial complex K on vertices $[m]$ and describe the Poincaré series for the special case

$$K=\big\{\{v_1\},,\{v_2\},\{v_3\},\{v_1,v_2\},\{v_1,v_3\}\big\}. \tag{37}$$

For $(X,A)=(M_f,\mathbb{C}P^2)$, we have

$$(U,V)=\Big(S^4\vee S^6,\ S^4\vee S^2\Big). \tag{38}$$

so that

$$B = S^4,\ C = S^6 \text{ and } E = S^2$$

Theorem 7.1 gives now

$$\widetilde{H}^*\big(\widehat{Z}(K; (M_f, \mathbb{C}P^3))\big) \cong \widetilde{H}^*\big(\widehat{Z}(K; (B \vee C, B \vee E))\big).$$

Applying Theorem 2.2, we get

$$\begin{aligned}\widehat{Z}(K; (B \vee C, B \vee E)) &\xrightarrow{\simeq} \bigvee_{I \leq [m]} \widehat{Z}(K_I; (S^6,\ S^2)) \wedge \widehat{Z}\big(K_{[m]-I}; (S^4,\ S^4)\big)\\ &= \bigvee_{I \leq [m]} \widehat{Z}\big(K_I; (S^6,\ S^2)\big) \wedge (S^4)^{\wedge |[m]-I|}\end{aligned}$$

where the last term represents the $(|[m] - I)|$-fold smash product. Finally, Corollary 2.4 determines completely

$$\widehat{Z}\big(K_I; (S^6,\ S^2)\big)$$

by enumerating all the links $|lk_\sigma(K_I)|$.

Next, we describe the Poincaré series for K as in (37). According to (38), the cohomology of $(M_f, \mathbb{C}P^2)$ satisfies

$$H^*(M_f) = \mathbb{Z}\{b_4, c_6\} \quad \text{and} \quad H^*(\mathbb{C}P^2)) = \mathbb{Z}\{e_2, b_4\} \tag{39}$$

where the dimensions of the classes are given by the subscripts. We denote the classes $\{e_2, b_4, c_6\}$ supported on the vertex i by $\{e_2^i, b_4^i, c_6^i\}$ and illustrate computation using Theorem 9.1 by determining the summand corresponding to

$$I = \{2, 3\} \quad \text{and} \quad \sigma = \varnothing. \tag{40}$$

(i) In this case, we have

$$\widehat{D}^I_{\underline{C},\underline{E}}(\sigma) = E_2 \wedge E_3 = S^2 \wedge S^2 \quad \text{and} \quad \widetilde{H}\big(\widehat{D}^I_{\underline{C},\underline{E}}(\sigma)\big) = k\{e_2^2 \otimes e_2^3\}$$

and so we get $\overline{P}\big(\widehat{D}^I_{\underline{C},\underline{E}}(\sigma), t\big) = t^4$.

(ii) Next, since $[m] - I = \{1\}$, we have

$$\prod_{j \in \{1\}} \overline{P}(B_j, t) = \overline{P}(B_1, t) \implies \overline{P}(b_4^1, t) = t^4.$$

(iii) Turning to the links, we have

$$|lk_\varnothing(K_I)| = |\{\{2\}, \{3\}\}| = S^0$$

so that $t \cdot \overline{P}\big(|lk_{\varnothing}(K_I), t|\big) \;=\; t.$

Finally, for the case at hand (40), Theorem 9.1 contributes t^9 to the Poincaré series for $H^*\big(\widehat{Z}(K; (\underline{X}, \underline{A}))\big)$. Continuing in this way, we arrive at the (reduced) Poincaré series

$$\overline{P}\big(H^*(\widehat{Z}(K; (M_f, \mathbb{C}P^2)), t\big) = t^9 + t^{11} + 3t^{12} + 5t^{14} + 2t^{16}.$$

Theorem 2.2 applies particularly well in cases where spaces have unstable attaching maps.

Example 10.2 The homotopy equivalence $S^1 \wedge Y \simeq \Sigma(Y)$ implies homotopy equivalences

$$\Sigma^{mq}\big(\widehat{Z}(K; (\underline{X}, \underline{A}))\big) \longrightarrow \widehat{Z}\big(K; \big(\underline{\Sigma^q(X)}, \underline{\Sigma^q(A)}\big)\big) \tag{41}$$

where, as usual, m is the number of vertices of K. Recall now that $SO(3) \cong \mathbb{R}\mathrm{P}^3$ and consider the pair

$$(X, A) = \big(SO(3), \mathbb{R}\mathrm{P}^2\big),$$

for which there is a well-known homotopy equivalence of pairs [17, Section 1]

$$\big(\Sigma^2\big(SO(3)\big),\ \Sigma^2\big(\mathbb{R}\mathrm{P}^2\big)\big) \longrightarrow \big(\Sigma^2\big(\mathbb{R}\mathrm{P}^2\big) \vee \Sigma^2(S^3),\ \Sigma^2\big(\mathbb{R}\mathrm{P}^2\big)\big), \tag{42}$$

which makes the pair $\big(SO(3), \mathbb{R}\mathrm{P}^2\big)$ *stably wedge decomposable*. Next, combining (41) and (42), we get a homotopy equivalence

$$\Sigma^{2m}\big(\widehat{Z}(K; (SO(3), \mathbb{R}\mathrm{P}^2))\big) \longrightarrow \widehat{Z}\big(K; \big(\Sigma^2(\mathbb{R}\mathrm{P}^2) \vee \Sigma^2(S^3),\ \Sigma^2(\mathbb{R}\mathrm{P}^2)\big)\big). \tag{43}$$

Finally, Theorem 7.1 allows us to conclude that $\widehat{Z}(K; (SO(3), \mathbb{R}\mathrm{P}^2))$, and hence the polyhedral product $Z(K; (SO(3), \mathbb{R}\mathrm{P}^2))$, is stably a wedge of smash products of S^3 and $\mathbb{R}\mathrm{P}^2$.

Similar splitting results exist for the polyhedral product whenever the spaces X and A split after finitely many suspensions. In particular, the fact that $\Omega^2 S^3$ splits stably into a wedge of Brown-Gitler spectra implies that the polyhedral product $Z\big(K; (\Omega^2 S^3, *)\big)$ splits stably into a wedge of smash products of Brown-Gitler spectra.

Acknowledgments This work was supported in part by grant 426160 from Simons Foundation. The authors are grateful to the Fields Institute for a conducive environment during the Thematic Program: Toric Topology and Polyhedral Products. The first author acknowledges Rider University for a Spring 2020 research leave.

References

1. A. Bahri, M. Bendersky, F.R. Cohen, S. Gitler, Decompositions of the polyhedral product functor with applications to moment-angle complexes and related spaces. PNAS **106**, 12241–12244 (2009)
2. A. Bahri, M. Bendersky, F.R. Cohen, S. Gitler, The polyhedral product functor: a method of computation for moment-angle complexes, arrangements and related spaces. Adv. Math. **225**, 1634–1668 (2010)
3. A. Bahri, M. Bendersky, F.R. Cohen, S. Gitler, A spectral sequence for polyhedral products. Adv. Math. **308**, 767–814 (2017)
4. A. Bahri, M. Bendersky, F.R. Cohen, *Polyhedral Products and Features of Their Homotopy Theory*. Handbook of Homotopy Theory, ed. by H. Miller (Chapman and Hall/CRC, Boca Raton, 2019)
5. I. Baskakov, V. Buchstaber, T. Panov, Cellular cochain complexes and torus actions. Uspeckhi. Mat. Nauk **59**(3), 159–160 (2004) (Russian); Russian Math. Surveys **89**(3), 562–563 (2004) (English translation)
6. V. Buchstaber, T. Panov, Actions of tori, combinatorial topology and homological algebra. Russian Math. Surveys **55**, 825–921 (2000)
7. L. Cai, On products in a real moment-angle manifold. J. Math. Soc. Japan **69**(2), 503–528 (2017)
8. L. Cai, S. Choi, Integral cohomology groups of real toric manifolds and small covers. https://arxiv.org/pdf/1604.06988.pdf
9. F.R. Cohen, J.P. May, L.R. Taylor, Splitting of certain spaces CX. Math. Proc. Camb. Phil. Soc. **84**, 465–496 (1978)
10. A. Dold, R. Thom, Quasifaserungen und unendliche symmetrische Produkte. Ann. Math. Second Series **67**, 239–281 (1958). https://doi.org/10.2307/1970005
11. M. Franz, On the integral cohomology of smooth toric varieties. Online at: https://arxiv.org/format/math/0308253
12. J. Grbić, S. Theriault, The homotopy type of the complement of a coordinate subspace arrangement. Topology **46**, 357–396 (2007)
13. J. Grbić, S. Theriault, The homotopy type of the polyhedral product for shifted complexes. Adv. Math. **245**, 690–715 (2013). https://doi.org/10.1016/j.aim.2013.05.002
14. K. Iriye, D. Kishimoto, Decompositions of polyhedral products for shifted complexes. Adv. Math. **245**, 716–736 (2013)
15. K. Iriye, D. Kishimoto, Fat wedge filtrations and decomposition of polyhedral products. Kyoto J. Math. (2019). https://doi.org/10.1215/21562261-2017-0038
16. S. López de Medrano, *Topology of the Intersection of Quadrics in* $\mathbb{R}^n$. Algebraic Topology, Arcata California. Lecture Notes in Mathematics, vol. 1370 (Springer, Berlin, 1989), pp. 280–292
17. J. Mukai, Some homotopy groups of the double suspension of the real projective space $\mathbb{R}P^6$, 10th Brazilian Topology Meeting, (São Carlos, 1996). Mat. Contemp. **13**, 235–249 (1997)
18. Q. Zheng, The cohomology algebra of polyhedral product spaces. J. Pure Appl. Algebra **220**, 3752–3776 (2016)
19. G.M. Ziegler, R.T. Živaljević, Homotopy types of subspace arrangements via diagrams of spaces. Math. Ann. **295**, 527–548 (1993). https://doi.org/10.1007/BF01444901

Multiparameter Persistent Homology via Generalized Morse Theory

Peter Bubenik and Michael J. Catanzaro

1 Introduction

Persistent homology is an important tool in topological data analysis, whose goal is to use ideas from topology to understand the "shape of data" [8, 13].

An important example of persistent homology starts with a smooth, compact manifold M and a Morse function $f : M \to \mathbb{R}$. Classical Morse theory concerns itself with the study of M via the critical points of f by analyzing the sublevel sets $F(a) = f^{-1}(\infty, a]$ and how their topology changes as a varies. The subspaces $\{F(a)\}_{a \in \mathbb{R}}$ and their inclusion maps may be used to define a functor $F : \mathbf{R} \to \mathbf{Top}$, where $\mathbf{R}$ is the category given by the linear order on $\mathbb{R}$ and $\mathbf{Top}$ is the category of topological spaces and continuous maps. Composing with singular homology in some degree j and coefficients in a field k, we obtain a functor $H_j F : \mathbf{R} \to \mathbf{Vect_k}$ with codomain the category of k-vector spaces and k-linear maps. Such a functor is called a persistence module.

Let β_j denote the jth Betti number of M, and let M_j denote the number of critical points of index of j of f. Let $M(t) = \sum_j M_j t^j$ and $\beta(t) = \sum_j \beta_j t^j$. Morse observed that $M(t) - \beta(t) = (1 + t)D(t)$ for some polynomial $D(t)$ with non-negative coefficients [28]. That is, the excess of critical points of the Morse function comes in pairs that differ in index by one. A strengthening of this observation is a central result in persistent homology. The persistence module $H_j F$ decomposes into a direct sum of indecomposable summands given by one-dimensional vector spaces supported on an interval. The end points of these intervals are exactly the critical

P. Bubenik (✉)
Department of Mathematics, University of Florida, Gainesville, FL, USA
e-mail: peter.bubenik@ufl.edu

M. J. Catanzaro
Department of Mathematics, Iowa State University, Ames, IA, USA
e-mail: mjcatanz@iastate.edu

A. Bahri et al. (eds.), *Toric Topology and Polyhedral Products*, Fields Institute Communications 89, https://doi.org/10.1007/978-3-031-57204-3_4

values of the paired critical points in Morse's theorem. This pairing of critical values, called the persistence diagram, is central to persistent homology.

While this setting has been very successful, in many applications, the data are best described not by a single function $f : M \to \mathbb{R}$ but by a one-parameter family of functions $f_t : M \to \mathbb{R}$, where $t \in I = [0, 1]$. For example, one may handle noise in the data with a procedure dependent on a parameter t. The resulting homological data may be encoded in a multiparameter persistence module. However, in general, this module does not decompose into one-dimensional summands, and there is no complete invariant analogous to the persistence diagram [9].

We approach multiparameter persistent homology using two distinct generalizations of Morse theory. The first is a parametrized approach to Morse theory, known as Cerf theory. Cerf theory was initiated by J. Cerf in his celebrated proof of the pseudo-isotopy theorem [10]. One outcome of his work was a useful stratification on the space of all smooth functions on a smooth compact manifold, stratified by singularity type. The existence of this stratification implies that generic, one-parameter families of smooth functions are almost always Morse, except for finitely many parameter values at which the function may have cubic, or "birth-death" type, singularities. Cerf developed a convenient framework for understanding how singularities merge, split, and pass one another in families. This understanding is paramount for our analysis.

The second variant is Morse theory adapted to the case of manifolds with boundary. This was developed around the same time as Cerf theory but by several authors independently [3, 15, 20]. Many statements in classical Morse theory can be adapted to manifolds with boundary, so long as the gradient or gradient-like flow used is tangent to the boundary. Critical points occurring on the interior behave as one expects, but on the boundary, they come in two distinct flavors, either stable or unstable, depending on the local flow. Altogether, Morse theory for manifolds with boundary is a powerful extension of its classical analog. We have only touched the surface of using this subject in multiparameter persistence.

1.1 Our Contributions

Given a topological space X and a one-parameter family of (not necessarily continuous) functions $\tilde{f} : I \times X \to \mathbb{R}$, we define a fibered version of $\tilde{f}$ by letting $f : I \times X \to I \times \mathbb{R}$ be given by $f(t, x) = (t, \tilde{f}(t, x))$. The collection of subspaces

$$F(a, b, c) = f^{-1}([a, b] \times (-\infty, c]) \subset I \times X\,, \tag{1}$$

for $0 \leq a \leq b \leq 1$ and $c \in \mathbb{R}$ are our main objects of study. For $[a, b] \subset [a', b']$ and $c \leq c'$, there is an inclusion of $F(a, b, c)$ into $F(a', b', c')$. These topological spaces and continuous maps describe a functor $F : \mathbf{Int_I} \times \mathbf{R} \to \mathbf{Top}$, where $\mathbf{Int_I} \times \mathbf{R}$ is the category given by the product of the partial order on the closed intervals in I and the linear order on $\mathbb{R}$. Composing with singular homology in some degree j

with coefficients in a field k, we obtain a functor $H_j F : \mathbf{Int_I} \times \mathbf{R} \to \mathbf{Vect_k}$. This functor is a multiparameter persistence module.

We prove that this functor is stable with respect to the interleaving distance for perturbations of the one-parameter family of smooth functions (Theorem 3.2).

We consider several examples of such one-parameter families of functions and give complete descriptions of their multiparameter persistence modules (Sects. 4.1, 4.3, 4.4, and 6). In particular, we give decompositions of these modules into their indecomposable summands and provide corresponding visualizations (Figs. 5, 7, and 10). We also show that indecomposable persistent modules arising from one-parameter families of functions may have arbitrarily large dimension (Sect. 4.2).

Now consider the case where X is a smooth compact manifold and $\tilde{f}$ is smooth. Let $F_0(a, b, c) = f^{-1}([a, b] \times \{c\})$. We prove (Theorem 5.2) that for generic a, b, and c, $(F(a, b, c), F_0(a, b, c))$ forms a cobordism between the manifolds with boundary $(F(a, a, c), F_0(a, a, c))$ and $(F(b, b, c), F_0(b, b, c)$. Furthermore, it is naturally equipped with a Morse function

$$\pi_{[a,b]} : F(a, b, c) \to [a, b]\,,$$

given by projection onto the interval $[a, b]$. This Morse function has no critical points on $F(a, b, c) \setminus F_0(a, b, c)$, and the positive and negative critical points on $F_0(a, b, c)$ (Definition 5.1) correspond to boundary stable and boundary unstable critical points, respectively.

We remark that if we restrict our collection of subspaces in (1) to those with $a = 0$, then we obtain a multiparameter persistence module indexed by $I \times \mathbb{R} \subset \mathbb{R}^2$, with the usual product partial order. However, this two-parameter persistence module is a weaker invariant than our three-parameter persistence module. For example, if X is the one-point space, then our persistence module is a complete invariant of one-parameter families of functions, but the weaker invariant is not.

We also remark that in applications computing the full set of critical values (the Cerf diagram—Sect. 2.4) should not be considered to be a prerequisite. In the classical situation of sublevel set persistent homology of a single smooth function (e.g., a sum of a large number of Gaussians), instead of computing the set of critical values, one computes the sublevel set persistent homology of a piecewise linear approximation.

1.2 Related Work

Much of the recent work on multiparameter persistent homology focuses on either its algebraic structure, for example, [4, 6, 16, 26], or its computational challenges, for example, [11, 22, 24, 29]. Other authors have begun to take a more geometric approach similar to our own, such as [14, 25]. There are two recent geometric approaches that are related but still distinct. In [7], the authors use handlebody theory to understand bi-filtrations arising from preimages of 2-Morse functions $f : M \to \mathbb{R}^2$. In a similar vein, the authors of [1] use singularities of maps $M \to \mathbb{R}^2$ to understand preimages in M.

1.3 Motivation

While our framework is theoretical, we are motivated by applications. We highlight two examples: kernel density estimation and kernel regression.

1.3.1 Kernel Density Estimation

Suppose $\{x_1, x_2, \ldots, x_n\} \subset \mathbb{R}^d$ are samples drawn independently from an unknown density function f. An empirical estimator of the density is obtained by the average of bump functions centered at each x_i. The bump functions are translations of a bump function, K, centered at the origin called a *kernel*. That is,

$$\hat{f}_\alpha(x) = \frac{1}{n\alpha} \sum_{i=1}^{n} K\left(\frac{x - x_i}{\alpha}\right),$$

where the parameter α is called the *bandwidth*. A standard choice is the Gaussian kernel, $K(x) = \frac{1}{(2\pi)^{d/2}} \exp(-\|x\|^2/2)$. Other examples include the Epanechnikov and triangular kernels, which appear (up to rescaling) as the functions $g(t)$ and $\tilde{g}(t)$, respectively, of Sect. 4.1.

Properties of the kernel density estimator $\hat{f}$, such as the number of modes (i.e., local maxima), depend on the bandwidth α. In order to obtain a global understanding of these properties for various of α and how they interact, we consider the one-parameter family of functions $\tilde{g} = -\hat{f} : \mathbb{R} \times I \to \mathbb{R}$, where $\tilde{g}(t, x) = -\hat{f}_t(x)$ and I is some bounded interval of parameter values. We obtain a collection of spaces, G, given by (1) and its associated multiparameter persistence modules, $H_j G : \mathbf{Int_I} \times \mathbf{R} \to \mathbf{Vect_K}$. We may use $H_0 G$ for a functorial analysis of the estimation of the modes of f. In particular, the dimension of $H_0 G(\alpha, \alpha, -c)$ equals the number of connected components of the superlevel set $f_\alpha^{-1}([c, \infty))$. Furthermore, the linear maps $H_0 G(a, a, -c) \to H_0 G(a, b, -c) \leftarrow H_0 G(b, b, -c)$ allow one to study the persistence of these connected components.

1.3.2 Kernel Regression

Closely related to kernel density estimation is kernel regression. Suppose we are given data $\{(x_1, y_1), \ldots, (x_n, y_n)\} \subset \mathbb{R}^d \times \mathbb{R}$ sampled from the graph of some unknown function $f : \mathbb{R}^d \to \mathbb{R}$. Consider the Nadaraya-Watson estimator

$$\hat{f}_\alpha(x) = \frac{\sum_{i=1}^n K_\alpha(x - x_i) y_i}{\sum_{i=1}^n K_\alpha(x - x_i)}.$$

In the same way as for kernel density estimation, we obtain a one-parameter family of functions and associated persistence modules.

1.4 Outline

The paper is organized as follows. In Sect. 2, we recall definitions from geometric topology and Cerf theory. We define our primary objects of study including our multiparameter persistence modules in Sect. 3. In Sect. 4, we provide several examples of one-parameter families of functions on manifolds and visualizations of the relevant cobordisms and analyze the multiparameter persistence modules. Finally, in Sect. 5, we prove our main theoretical result that $F(a, b, c)$ is generically equipped with a Morse function and analyze its critical points.

2 Background

We start with providing some background from geometric topology.

2.1 Manifolds with Corners

There are several different, inequivalent notions of manifolds with corners and smooth maps between them in the differential topology literature. The following is a brief summary of [21]. Let $H_k^n = \{(x_1, x_2, \dots, x_n) \mid x_1, x_2, \dots, x_k \geq 0\}$. In particular, $H_0^n = \mathbb{R}^n$ and $H_1^n = [0, \infty) \times \mathbb{R}^{n-1}$.

Definition 2.1 ([21, Definition 2.1]) Let M be a second countable Hausdorff space.

- An *n-dimensional chart on M without boundary* is a pair (U, ψ), where U is an open subset of $\mathbb{R}^n$ and $\psi : U \to M$ is a homeomorphism onto a nonempty open set $\psi(U)$.
- An *n-dimensional chart on M with boundary* for $n \geq 1$ is a pair (U, ψ), where U is an open subset in $\mathbb{R}^n$ or H_1^n and $\psi : U \to M$ is a homeomorphism onto a nonempty open set $\psi(U)$.
- An *n-dimensional chart of M with corners* for $n \geq 1$ is a pair (U, ψ), where U is an open subset of H_k^n for $0 \leq k \leq n$ and $\psi : U \to M$ is a homeomorphism onto a nonempty open subset $\psi(U)$.

Definition 2.2 For $X \subset \mathbb{R}^n$ and $Y \subset \mathbb{R}^m$, a map $f : X \to Y$ is *smooth* if it can be extended to a smooth map between open neighborhoods of X and Y. If $m = n$ and f^{-1} is also smooth, then f is a *diffeomorphism*.

Definition 2.3 An *n-dimensional atlas for M without boundary, with boundary, or with corners* is a collection of n-dimensional charts without boundary, with boundary, or with corners $\{(U_j, \psi_j) \mid j \in J\}$ on M such that $M = \cup_j \psi_j(M)$ and are compatible in the following sense: $\psi_j \circ \psi_k^{-1} : \psi_k^{-1}(\psi_j(U_j) \cap \psi_k(U_k)) \to \psi_j^{-1}(\psi_j(U_j) \cap \psi_k(U_k))$ is a diffeomorphism. An atlas is *maximal* if it is not a proper subset of any other atlas.

Definition 2.4 An *n-dimensional manifold without boundary, with boundary, or with corners* is a second countable Hausdorff space M together with a maximal n-dimensional atlas of charts without boundary, with boundary, or with corners.

Example 2.5 The space Ω of Fig. 2 provides an example of a manifold with corners. There are six corner points (with neighborhoods homeomorphic to H_2^2) at the intersections of V_0, V_1, and Y. The spaces V_0, V_1, and Y are examples of manifolds with boundary. Their interiors, as well as the interior of Ω, are examples of manifolds without boundary.

2.2 Generalized Morse Functions

Morse theory provides powerful methods for understanding manifolds through the lens of smooth functions. Classical Morse theory concerns the study of smooth, compact manifolds without boundary and allows for a transformation from smooth, continuous data (manifolds) to discrete data (critical points and values). An adaptation to Morse theory for manifolds with boundary extends this to the setting of cobordisms. Another generalization we will consider, known as Cerf theory, generalizes this to the study of one-parameter families of functions. The remainder of this subsection is a summary and restatement of ideas from [12, §1] and [27].

Let M and Q be smooth, compact manifolds of dimensions n and q, respectively, and let $f : M \to Q$ be a smooth map. A point $p \in M$ is a *critical point* or *singular point*, if rank $d_p f = 0$ or

$$\operatorname{rank} d_p f < \min(n, q)\,.$$

The set of all critical points of f is denoted $\Sigma(f)$.

Assume $n \geq q$. A point $p \in \Sigma(f)$ is a *fold singularity of index j* (see Fig. 1a) if for some choice of local coordinates near p, the map f is given by

$$\phi : \mathbb{R}^{q-1} \times \mathbb{R}^{n-q+1} \to \mathbb{R}^{q-1} \times \mathbb{R}$$

$$(t, x) \mapsto \left(t, -\sum_{i=1}^{j} x_i^2 + \sum_{i=j+1}^{n-q+1} x_i^2 \right), \tag{2}$$

where $t \in \mathbb{R}^{q-1}$ and $x = (x_1, x_2, \ldots, x_{n-q+1}) \in \mathbb{R}^{n-q+1}$. Let $\Sigma^{10}(f)$ be the set of all fold singularities.

For $q > 1$, a point $p \in \Sigma(f)$ is a *cusp singularity of index* $j + 1/2$ (see Fig. 1b) if for some choice of local coordinates near p, the map f is given by

$$\psi : \mathbb{R}^{q-1} \times \mathbb{R} \times \mathbb{R}^{n-q} \to \mathbb{R}^{q-1} \times \mathbb{R}$$

$$(t, z, x) \mapsto \left(t, x^3 + 3t_1 z - \sum_{i=1}^{j} x_i^2 + \sum_{i=j+1}^{n-q} x_i^2 \right), \tag{3}$$

where $t = (t_1, t_2, \ldots, t_{q-1}) \in \mathbb{R}^{q-1}$, $z \in \mathbb{R}$, and $x = (x_1, x_2, \ldots, x_{n-q}) \in \mathbb{R}^{n-q}$. Set $\Sigma^{11}(f)$ to be the set of all cusp singularities. Finally, let $\Sigma^1(f) = \Sigma^{10}(f) \cup \Sigma^{11}(f)$.

Remark 2.6 Consider the case $q = 1$ and $Q \subset \mathbb{R}$, so that all terms of Eq. (2) involving t vanish. In this case, the fold singularities of $f : M \to Q$ coincide with nondegenerate critical points as in usual Morse theory. If such an f has only fold singularities, then f is known as a Morse function.

Remark 2.7 Both fold and cusp singularities are locally fibered over $\mathbb{R}^{q-1}$ in the sense that the following commute:

$$\begin{array}{ccc} \mathbb{R}^{q-1} \times \mathbb{R}^{n-q+1} & \xrightarrow{\phi} & \mathbb{R}^{q-1} \times \mathbb{R} \\ & \searrow^{\pi} & \downarrow \pi \\ & & \mathbb{R}^{q-1} \end{array} \qquad \begin{array}{ccc} \mathbb{R}^{q-1} \times \mathbb{R} \times \mathbb{R}^{n-q} & \xrightarrow{\psi} & \mathbb{R}^{q-1} \times \mathbb{R} \\ & \searrow^{\pi} & \downarrow \pi \\ & & \mathbb{R}^{q-1}, \end{array}$$

where π is the projection onto $\mathbb{R}^{q-1}$. A single fibered function can be interpreted as a family of functions $\phi_t : \mathbb{R}^{n-q+1} \to \mathbb{R}$ or $\psi_t : \mathbb{R} \times \mathbb{R}^{n-q} \to \mathbb{R}$, indexed over $t \in \mathbb{R}^{q-1}$. In this language, the folds are constant families (see Fig. 1a). The cusps consist of families of functions with two nondegenerate critical points of index j and $j + 1$ for $t_1 < 0$, no critical points for $t_1 > 0$, and a cubic or "birth-death" singularity of index $j + 1/2$ for $t_1 = 0$ (see Fig. 1b).

2.3 Cobordisms

In this section, we recall some basic notions from Morse theory for manifolds with boundary from the well-written summary of [2]. The reader may also consult some

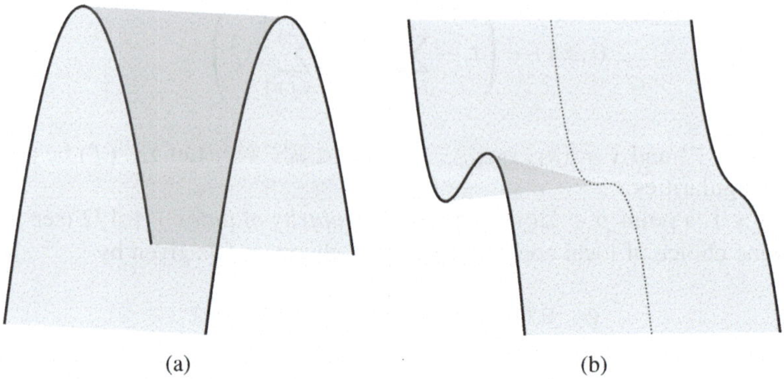

Fig. 1 Local models for (**a**) a fold singularity and (**b**) a cusp singularity. The middle slice of the cusp singularity has a cubic singularity. This is often referred to as a "birth–death" singularity, since the two critical points to the left can be viewed as being "born" (moving right to left) or as "dying" (moving left to right)

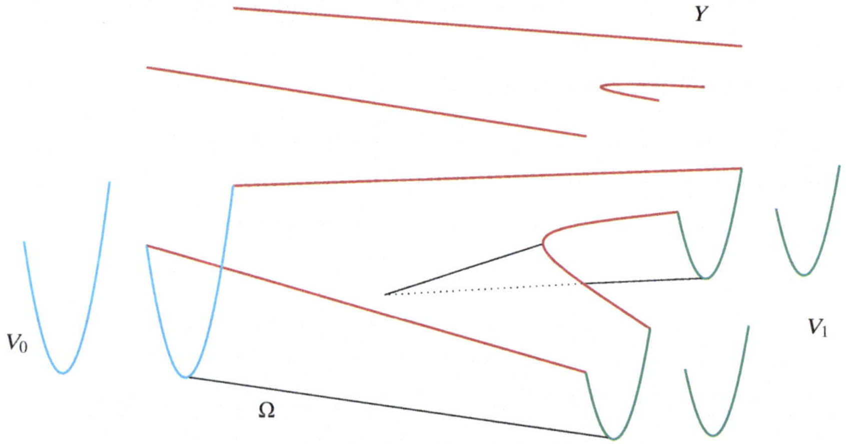

Fig. 2 A manifold with corners Ω provides a cobordism between two manifolds with boundary V_0 and V_1. The boundary ∂V_0 consists of two points and ∂V_1 consists of four points. The manifold with boundary Y can be viewed as a cobordism between ∂V_0 and ∂V_1. Furthermore, Ω is a left-product cobordism

of the original sources, such as [3, 15, 20]. We will assume that all manifolds (with or without boundaries or corners) are smooth.

Let V_0 and V_1 denote two compact n-manifolds with boundaries ∂V_0 and ∂V_1, respectively. Let Ω be a compact $(n+1)$-manifold with corners, $\partial\Omega = Y \cup V_0 \cup V_1$, where $V_0 \cap V_1 = \varnothing$, and $Y \cap V_0 = \partial V_0$, $Y \cap V_1 = \partial V_1$. In this case, we say (Ω, Y) is a *cobordism* between $(V_0, \partial V_0)$ and $(V_1, \partial V_1)$. See Fig. 2. Such a cobordism is a *left-product cobordism* if Ω is diffeomorphic to $V_0 \times [0, 1]$ or a *right-product cobordism* if Ω is diffeomorphic to $V_1 \times [0, 1]$.

Fixing a Riemannian metric on Ω allows us to consider the gradient ∇F of a smooth function $F : \Omega \to [a, b]$. A critical point z of F is *Morse* if the Hessian of F at z is nondegenerate. The function F is a *Morse function* on the cobordism (Ω, Y) if $F^{-1}(a) = V_0$, $F^{-1}(b) = V_1$, there are no critical points on $V_0 \cup V_1$, F only has Morse critical points, and ∇F is everywhere tangent to Y.

The *unstable manifold* W_z^u of a critical point z is the set of all points, which flow out from z under ∇F:

$$W_z^u = \{x \mid \lim_{t \to -\infty} \Phi_t(x) = z\},$$

where Φ_t is the flow generated by ∇F. With the same notation, the *stable manifold* W_z^s of a critical point z is given by

$$W_z^s = \{x \mid \lim_{t \to \infty} \Phi_t(x) = z\}.$$

The stable and unstable manifolds are locally embedded disks [19].

Unlike usual Morse theory, the critical points for a Morse function on a manifold with boundary come in a variety of types. If z is a critical point and $z \in Y$, then z is called a *boundary critical point*. Otherwise, z is called an *interior critical point*. We are primarily interested in boundary critical points, of which there are again two types, determined by the gradient flow. A boundary critical point is *boundary stable* if $T_z W_z^u \subset T_z Y$; otherwise, it is *boundary unstable*.

As usual, the *index* of a boundary critical point z is defined as the dimension of the stable manifold W_z^s. If z is boundary stable, then the index of z is the usual index of $F|_Y$ plus one. On the other hand, if z is boundary unstable, then the index of z coincides with the usual notion of index of the restriction $F|_Y$. See Example 2.11.

Remark 2.8 Note that there are no boundary unstable critical points of index $n+1$ or boundary stable critical points of index 0.

Remark 2.9 We consider the flow generated by ∇F, as is frequently used in most mathematics literature. In other areas such as dynamical systems and physics, the flow generated by $-\nabla F$ is commonly used. The two versions are equivalent, since the stable and unstable manifolds swap after replacing the flow generated by ∇F with that generated by $-\nabla F$.

Proposition 2.10 ([2, Lem 2.10, Thm 2.27, Prop 2.38]) *Let* (Ω, Y) *be a cobordism between* $(V_0, \partial V_0)$ *and* $(V_1, \partial V_1)$.

- *If* (Ω, Y) *admits a Morse function whose critical points are all boundary stable, then* (Ω, Y) *is a left-product cobordism.*
- *If* (Ω, Y) *admits a Morse function whose critical points are all boundary unstable, then* (Ω, Y) *is a right-product cobordism.*
- *If* (Ω, Y) *admits a Morse function with no critical points, then* (Ω, Y) *is both a left- and right-product cobordism.*

Example 2.11 In Fig. 2, the projection of Ω onto the horizontal axis yields a Morse function $F : \Omega \to [0, 1]$, in which $F^{-1}(0) = V_0$, $F^{-1}(1) = V_1$. This function has no interior critical points and a single boundary critical point. The boundary critical point is boundary stable and located at the vertex of the parabola of Y in Ω. This is an index 1 critical point. Proposition 2.10 implies Ω is a left-product cobordism, as is evident from Fig. 2.

If we post-compose F with the involution $t \mapsto 1-t$, then we again have a Morse function with no interior critical points. This composition has the same boundary critical point as before but now it is boundary unstable. The index of this critical point is 1.

2.4 *Cerf Theory*

Let X be a smooth, compact n-manifold and let $I = [0, 1]$ denote the unit interval. A one-parameter family of functions on X is a family of smooth functions $\tilde{f}_t : X \to \mathbb{R}$, where $t \in I$, and the family varies smoothly with respect to t. This is equivalent to specifying a single smooth function $\tilde{f} : I \times X \to \mathbb{R}$. In either case, this data gives rise to a map fibered over the interval

$$f : I \times X \to I \times \mathbb{R}, \quad f(t, z) = (t, \tilde{f}(t, z)),$$

in the sense that the following diagram commutes:

$$\begin{array}{ccc} I \times X & \xrightarrow{f} & I \times \mathbb{R} \\ & \searrow^{\pi_I} & \downarrow{\scriptstyle \pi_I} \\ & & I \end{array} \tag{4}$$

where π_I is the projection onto the I factor.

Our primary tool for understanding such families of functions is the Cerf diagram.

Definition 2.12 The *Cerf diagram* (or *Kirby diagram*) of a family of functions $\tilde{f} : I \times X \to \mathbb{R}$ is given by

$$\bigcup_{t \in I,\, x \in \Sigma(\tilde{f}_t)} (t, \tilde{f}_t(x)) \subset I \times \mathbb{R}.$$

We label each nondegenerate critical value of $\tilde{f}_t$ with its corresponding index.

The Cerf diagram encodes the critical value information of a family of functions as the time parameter t varies [10, 17, 23]. A simple Cerf diagram is shown in Fig. 3.

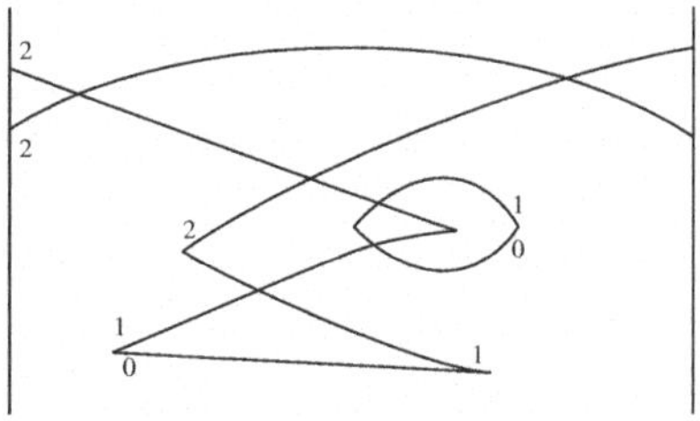

Fig. 3 A Cerf diagram for a certain generic family of smooth functions. Each nondegenerate critical value is labeled by the index of its critical point. See also [18, Section A.3, Figures E, F, G.]

Each (non-end) point on the curves corresponds to a nondegenerate critical value of $\tilde{f}_t$, and the point where two such curves terminate is a cubic singularity of $\tilde{f}_t$.

We will assume that the family $\tilde{f}$ is *generic*, meaning that for all but finitely many t, the fiber $\tilde{f}_t$ has finitely many nondegenerate critical points, each of which has a distinct critical value. Furthermore, we will assume that all remaining fibers have either finitely many nondegenerate critical points exactly two of which have a common critical value or a single cubic singularity and finitely many nondegenerate critical points all of which have distinct critical values.

2.5 Wrinkled Maps

We recall the notion of a wrinkle from [12]. Let

$$w : \mathbb{R}^{q-1} \times \mathbb{R}^{n-1} \times \mathbb{R}^1 \to \mathbb{R}^{q-1} \times \mathbb{R}^1$$

be given by

$$w(t, x, z) = \left(t, z^3 + 3\left(|t|^2 - 1\right) z - \sum_{i=1}^{j} x_i^2 + \sum_{i=j+1}^{n-q} x_i^2 \right),$$

where $|t|^2 = \sum_{i=1}^{q-1} t_i^2$. The set of critical points of w is

$$\Sigma(w) = \{x = 0,\ z^2 + |t|^2 = 1\} \subset \mathbb{R}^{q-1} \times \mathbb{R}^{n-q} \times \mathbb{R},$$

and can be identified with a $(q-1)$ sphere $S^{q-1} \subset \mathbb{R}^{q-1} \times \{0\} \times \mathbb{R}$. This sphere has equator

$$\{x = 0,\ z = 0,\ |t| = 1\} \subset \Sigma^1(w),$$

which we identify with S^{q-2}. This equator consists of cusp singularities of index $j + 1/2$, its upper hemisphere $\Sigma^1(w) \cap \{z > 0\}$ consists of fold singularities of index j, and the lower hemisphere $\Sigma^1(w) \cap \{z < 0\}$ consists of fold singularities of index $j + 1$.

Definition 2.13 ([12]) For an open subset $U \subset M$, a map $f : U \to Q$ is called a *wrinkle of index* $s + 1/2$ if it is equivalent to the restriction of w to an open neighborhood $V \supset D$, where D is the q-dimensional disk $\{z^2 + |y|^2 \leq 1, x = 0\}$ bounded by $\Sigma^1(w)$.

A map $f : M \to Q$ is called *wrinkled* if there exist disjoint open subsets $U_1, U_2, \ldots, U_l \subset M$ such that:

- For each i, $f|_{U_i}$ is a wrinkle.
- If $U = \cup_1^l U_i$, then $f|_{M \setminus U}$ is a submersion.

Definition 2.14 A map $f : M \to Q$ is called *wrinkled with folds* if there exist disjoint open subsets $U_1, U_2, \ldots, U_l \subset M$ such that:

- For each i, $f|_{U_i}$ is a wrinkle.
- If $U = \cup_1^l U_i$, then $f|_{M \setminus U}$ has only fold singularities.

The singular locus of a wrinkled map decomposes into a union of wrinkles $S_i = \Sigma^1(f|_{U_i}) \subset U_i$. As before, each S_i has a $(q-2)$-dimensional equator of cusps, which divides S_i into two hemispheres of folds of adjacent indices. The singular locus of a wrinkled map with folds decomposes into a union of wrinkles and folds.

3 Persistence Modules for One-Parameter Families of Functions

In this section, we define multiparameter persistence modules for *one*-parameter families of functions. The unit interval [0, 1] is denoted by I.

3.1 Indexing Categories

Let $\mathbf{Int_I}$ denote the category whose objects are closed intervals $[a, b] \subset I$ and whose morphisms $[a, b] \to [c, d]$ are inclusions $[a, b] \subset [c, d]$. Let $\Delta^2 = \{(a, b) \,|\, 0 \leq a \leq b \leq 1\}$. The category $\mathbf{Int_I}$ is isomorphic to the category $\mathbf{\Delta}^2$, whose objects are points $(a, b) \in \Delta^2$, and has a unique morphism $(a, b) \to (c, d)$ if and only if $c \leq a \leq b \leq d$. Finally, let $\mathbf{R}$ denote the category corresponding to the poset of real numbers $(\mathbb{R}, \leq)$. Then we have the isomorphic product categories $\mathbf{Int_I} \times \mathbf{R}$ and $\mathbf{\Delta}^2 \times \mathbf{R}$.

Note that there may not exist a map between two objects of $\mathbf{\Delta}^2 \times \mathbf{R}$, in contrast to the (ordinary) sublevel set persistence of Morse functions. There does exist, however, a zigzag of maps between any two objects due to the fact that $\mathbf{Int_I} \cong \mathbf{\Delta}^2$ is a join-semilattice. In particular, $[a, b] \subset [\min(a, a'), \max(b, b')] \supset [a', b']$; for example, see the two arrows in the third triangular slice in Fig. 5.

For $n \geq 1$, let $(\mathbb{R}^n, \leq)$ be the set $\mathbb{R}^n$ together with the product partial order. That is, $(x_1, \ldots, x_n) \leq (y_1, \ldots, y_n)$ if and only if $x_k \leq y_k$ for all $1 \leq k \leq n$. Then the poset Δ^2 includes in the poset $(\mathbb{R}^2, \leq)$ under the mapping $(a, b) \mapsto (-a, b)$. It follows that the product poset $\Delta^2 \times \mathbb{R}$ includes in the poset $(\mathbb{R}^3, \leq)$ under the mapping $(a, b, c) \mapsto (-a, b, c)$. Thus, we have an inclusion of categories $\mathbf{\Delta}^2 \times \mathbf{R} \hookrightarrow \mathbf{R}^3$ where $\mathbf{R}^3$ denotes the category corresponding to the poset $(\mathbb{R}^3, \leq)$. For a poset P and $p \in P$, let $U_p = \{q \in P \mid p \leq q\}$, called the *up-set* of p. Then our persistence modules may also be considered to be $\mathbb{R}^3$-graded modules over the monoid ring $K[U_0]$, where U_0 is the up-set of $0 \in \mathbb{R}^3$ (see [4, 26]).

3.2 Diagrams of Spaces

Let **Top** denote the category of topological spaces and continuous maps. Let X be a topological space, and let $\tilde{f} : I \times X \to \mathbb{R}$ be a (not necessarily continuous in either variable) real-valued function on $I \times X$, which corresponds to a one-parameter family of real-valued functions on X, given by $\tilde{f}_t(x) = \tilde{f}(t, x)$. Let $f : I \times X \to I \times \mathbb{R}$ be the function given by $f(t, z) = (t, \tilde{f}(t, z))$. Then we have a *diagram of spaces* of X given by $F : \mathbf{Int_I} \times \mathbf{R} \to \mathbf{Top}$ or equivalently $F : \mathbf{\Delta}^2 \times \mathbf{R} \to \mathbf{Top}$ given by $F([a, b], c) = f^{-1}([a, b] \times (-\infty, c])$ or $F(a, b, c) = f^{-1}([a, b] \times (-\infty, c])$ and morphisms given by inclusions of the corresponding inverse images. For any subcategory C of $\mathbf{Int_I} \times \mathbf{R}$, we can restrict a diagram of space F to C, forming a *subdiagram of spaces indexed on* C; we omit C if it is clear from context. If the subcategory is finite, we say the diagram of spaces is *finite*.

Remark 3.1 The target category of a diagram of spaces of X can be restricted to $\mathbf{Sub}(I \times X)$, the category whose objects are subspaces of $I \times X$ and whose morphisms are given by inclusion.

3.3 Multiparameter Persistence Modules

Let $\mathbf{Vect_k}$ denote the category of vector spaces over a field k and k-linear maps.

Given a one-parameter family $\tilde{f}$ of real-valued functions on a topological space X as in Sect. 3.2, we have the corresponding diagram of topological spaces F. For $j \geq 0$, let $H_j = H_j(-; k)$ denote the singular homology functor in degree j with coefficients in the field k. The *multiparameter persistence module* corresponding to $\tilde{f}$ is given by the functor $H_j F : \mathbf{Int_I} \times \mathbf{R} \to \mathbf{Vect_k}$ or equivalently $H_j F : \mathbf{\Delta}^2 \times \mathbf{R} \to \mathbf{Vect_k}$.

3.4 Betti and Euler Characteristic Functions

For applied mathematicians, it is sometimes preferable to ignore persistence entirely (i.e., the morphisms in the persistence module) and only compute the pointwise Betti numbers or, cruder still, the pointwise Euler characteristic. While much of the mathematical structure is lost, being able to complete computations on vastly larger datasets may be more important. In this section, we show how these coarser invariants fit within our framework.

Whenever they are well defined, we have the following. For $j \geq 0$, the *jth Betti function* $\beta_j : \Delta^2 \times \mathbb{R} \to \mathbb{Z}$ is given by

$$\beta_j(a, b, c) = \operatorname{rank}(H_j F(a, b, c)).$$

The *Euler characteristic function* $\chi : \Delta^2 \times \mathbb{R} \to \mathbb{Z}$ is given by

$$\chi(a, b, c) = \sum_j (-1)^j \beta_j(a, b, c).$$

In cases where F is given by a cellular complex, the Euler characteristic equals the alternating sum of the number of cells in each dimension.

3.5 Stability

We prove that our multiparameter persistence modules are stable with respect to perturbations of the underlying one-parameter family of functions.

Let X be a topological space and consider two one-parameter families of (not necessarily continuous) functions, $\tilde{f}, \tilde{g} : I \times X \to \mathbb{R}$. Let $F, G : \mathbf{\Delta^2} \times \mathbf{R} \to \mathbf{Top}$ be the corresponding diagrams of spaces defined in Sect. 3.2, and for $j \geq 0$, let $H_j F, H_j G : \mathbf{\Delta^2} \times \mathbf{R} \to \mathbf{Vect_K}$ be the corresponding multiparameter persistence modules defined in Sect. 3.3. Let $d_\infty(\tilde{f}, \tilde{g}) = \sup_{(t,x)\in I\times X} |\tilde{f}_t(x) - \tilde{g}_t(x)|$.

We define a *superlinear family of translations* [5, Section 3.5] on $\mathbf{\Delta^2} \times \mathbf{R}$ given by $\Omega_\varepsilon(a, b, c) = (a, b, c + \varepsilon)$ for $\varepsilon \geq 0$. The corresponding *interleaving distance* [5, Definition 3.20], d_I, is given by the infimum of all ε for which two diagrams or persistence modules indexed by $\mathbf{\Delta^2} \times \mathbf{R}$ are Ω_ε-interleaved [5, Definitions 3.4 and 3.5].

Theorem 3.2 $d_I(H_j F, H_j G) \leq d_\infty(\tilde{f}, \tilde{g})$.

Proof Let $\varepsilon = \sup_{(t,x)\in I\times X} |\tilde{f}_t(x) - \tilde{g}_t(x)|$. It follows from the definitions that F and G are Ω_ε-interleaved. By Bubenik et al. [5, Theorem 3.23], $H_j F$ and $H_j G$ are also Ω_ε-interleaved.

4 Examples I: Basic Examples

In this section, we illustrate our results with several examples. Recall that a one-parameter family of functions on X is a function $\tilde{f} : I \times X \to \mathbb{R}$ consisting of functions $\tilde{f}_t : X \to \mathbb{R}$, indexed by $t \in I$. This gives rise to a map fibered over the interval I

$$f : I \times X \to I \times \mathbb{R}, \quad f(t, z) = (t, \tilde{f}(t, z)). \tag{5}$$

We will replace smooth functions by piecewise linear approximations to make the associated multiparameter persistence module easier to describe—for an example, see Fig. 4. This replacement does not affect the qualitative structure of the module but does change the support of its indecomposable summands.

4.1 Persistence Modules of Graphs of Functions

We begin by considering a one point space $X = \{*\}$. A one-parameter family of functions $\tilde{f}_t : X \to \mathbb{R}$ is equivalent to a function $g : I \to \mathbb{R}$, where $g(t) = \tilde{f}_t(*)$. Hence, the image of the corresponding fibered function $f : I \times X \to I \times \mathbb{R}$ is just the graph of g. Furthermore, since $*$ is a critical point of $\tilde{f}_t$ for all t, the Cerf diagram of f coincides with the graph of g.

For example, let $g(t) = 4t(1-t)$, plotted in Fig. 4. For convenience, we will instead consider the piecewise linear function $\tilde{g}(t) = 2\min(t, 1-t)$. This function is no longer smooth in t, but its simplicity will make it easier to give a complete analysis (see the comment following Eq. (5)).

We have a diagram of topological spaces $F : \mathbf{\Delta}^2 \times \mathbf{R} \to \mathbf{Top}$ given by $F(a, b, c) = f^{-1}([a, b] \times (-\infty, c])$, where $f : I \times X \to I \times \mathbb{R}$ is given by

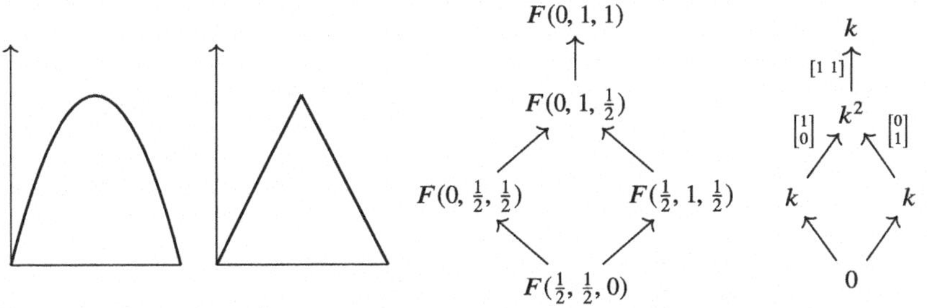

Fig. 4 Left: the graph of $g(t) = 4t(1-t)$ for $t \in [0, 1]$. Middle left: the graph of $\tilde{g}(t) = 2\min(t, 1-t)$ for $t \in [0, 1]$. This is also the image of the map $f : I \times \{*\} \to I \times \mathbb{R}$ given by $f(t, *) = (t, \tilde{g}(t))$. Middle right: a subdiagram of the diagram of spaces $F : \mathbf{\Delta}^2 \times \mathbf{R} \to \mathbf{Top}$ given by $F(a, b, c) = f^{-1}([a, b] \times (-\infty, c])$. Right: the corresponding subdiagram of the persistence module $H_0F : \mathbf{\Delta}^2 \times \mathbf{R} \to \mathbf{Vect_k}$

$(t, *) \mapsto (t, \tilde{g}(t))$. The space $F(a, b, c)$ is empty if $c < 2\min(a, 1 - b)$. That is, $\frac{c}{2} < a \leq b < 1 - \frac{c}{2}$. The space $F(a, b, c)$ is contractible if $c \geq 1$ or if $2\min(a, 1 - b) \leq c < 2\max(a, 1 - b)$. Equivalently, $a \leq \frac{c}{2}$ and $b < 1 - \frac{c}{2}$, or $\frac{c}{2} < a$ and $1 - \frac{c}{2} \leq b$. In the remaining case, $2\max(a, 1 - b) \leq c < 1$, we find $F(a, b, c) \simeq S^0$, two disjoint points. That is, $0 \leq c < 1$, $0 \leq a \leq \frac{c}{2}$, and $1 - \frac{c}{2} \leq b \leq 1$.

The persistence module $H_0 F : \mathbf{\Delta}^2 \times \mathbf{R} \to \mathbf{Vect_k}$ satisfies

$$\dim H_0 F = \begin{cases} 1 & \text{if } c \geq 1 \\ 2 & \text{if } 2\max(a, 1 - b) \leq c < 1 \\ 1 & \text{if } 2\min(a, 1 - b) \leq c < 2\max(a, 1 - b) \\ 0 & \text{if } c < 2\min(a, 1 - b)\,, \end{cases} \tag{6}$$

while the persistence modules $H_j F$ are the trivial K-vector space for all $j > 0$. See Fig. 5 for a visualization of β_0.

The diagram of spaces F has the subdiagram given in Fig. 4, which has a corresponding indecomposable persistence module, also in Fig. 4. This submodule of $H_0 F$ is also visualized in Fig. 5. It follows that $H_0 F$ is an indecomposable persistence module.

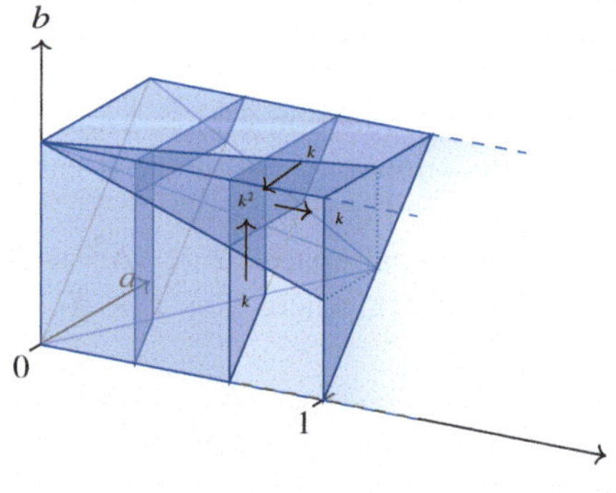

Fig. 5 The multiparameter persistence module $H_* F : \mathbf{\Delta}^2 \times \mathbf{R} \to \mathbf{Vect_k}$ defined by $H_* F(a, b, c) = H_*(f^{-1}([a, b] \times (-\infty, c]))$ where $f : I \times \{*\} \to I \times \mathbb{R}$ is given by $f(t, *) = (t, \tilde{g}(t))$ from Fig. 4. For $j \leq 0$, $H_j(F) = 0$. We have $\beta_0 = 2$ in the open square pyramid given by $0 \leq c < 1$, $0 \leq a \leq \frac{c}{2}$, and $1 - \frac{c}{2} \leq b \leq 1$. Furthermore, $\beta_0 = 1$ in the semi-infinite triangular cylinder given by $0 \leq a \leq b \leq 1$ and $c \geq 1$. For $0 \leq c < 1$, we also have $\beta_0 = 1$ in the region given by $0 \leq a \leq \frac{c}{2}$ and $a \leq b < 1 - \frac{c}{2}$ and the region $\frac{c}{2} < a \leq b$ and $1 - \frac{c}{2} \leq b \leq 1$. Everywhere else, $\beta_0 = 0$. That is, for $c < 0$ and $0 \leq a \leq b \leq 1$ and for $0 \leq c < 1$ and $\frac{c}{2} < a \leq b < 1 - \frac{c}{2}$. The right-hand diagram in Fig. 4 is embedded in $H_0 F$ as indicated. It follows that $H_0 F$ is indecomposable

4.2 Indecomposable Persistence Modules with Arbitrary Maximum Dimension

The example in the previous section can be generalized to produce an indecomposable persistence module arising from a one-parameter family of functions, with arbitrarily large maximum dimension.

For $n > 0$, let $\tilde{g}_n : I \to \mathbb{R}$ be the piecewise linear function obtained by linear interpolation between the values $\tilde{g}_n(\frac{i}{n}) = 0$ for $0 \leq i \leq n$ and $\tilde{g}_n(\frac{2i-1}{2n}) = 1$ for $1 \leq i \leq n$ (the example of Sect. 4.1 is the case $n = 1$). Then we have the corresponding diagram of topological spaces $F : \mathbf{\Delta}^2 \times \mathbf{R} \to \mathbf{Top}$ given by $F(a, b, c) = f^{-1}([a, b] \times (-\infty, c])$, where $f : I \times \{*\} \to I \times \mathbb{R}$ is given by $(t, *) \mapsto (t, \tilde{g}(t))$. Now F has a finite subdiagram $\hat{F}$ given by $F(\frac{i}{n}, \frac{j}{n}, \frac{1}{2})$, where $0 \leq i \leq j \leq n$.

Applying H_0, we obtain the persistence module $H_0 F$, which contains the following persistence module $H_0 \hat{F}$:

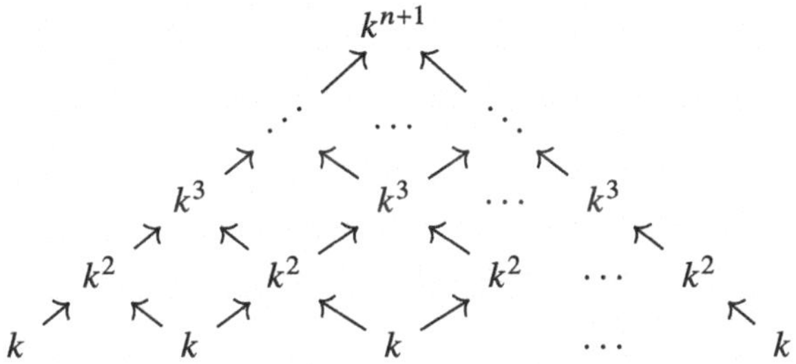

Each linear map $k^m \to k^{m+1}$ pointing up and to the right is given by the inclusion $k^m \to k^m \oplus k \cong k^{m+1}$, and each linear map $k^m \to k^{m+1}$ pointing up and to the left is given by the inclusion $k^m \to k \oplus k^m \cong k^{m+1}$. This persistence module is decomposable into $(n + 1)$ one-dimensional summands, whose support is given by the up-set of one of the $(n + 1)$ minimal elements.

Now append the terminal element $F(0, 1, 1)$ to the diagram $\hat{F}$ to obtain the diagram $\check{F}$, which is also a subdiagram of F. Then we have the persistence module $H_0 \check{F}$

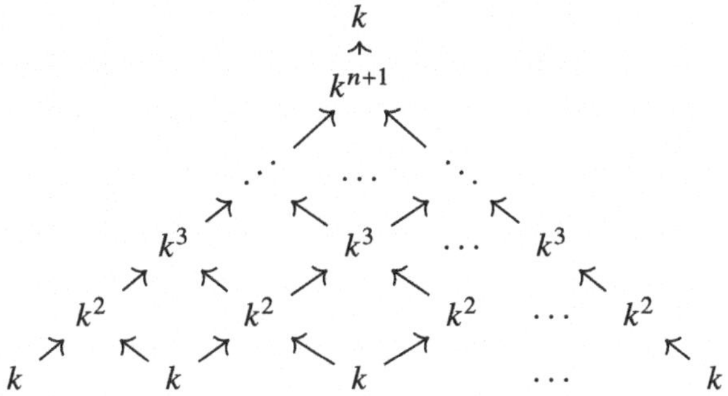

where the linear map $k^{n+1} \to k$ is given by summing the coordinates. This persistence module is indecomposable since the up-set of every minimal element contains the terminal element $H_0F(0, 1, 1) \cong k$.

4.3 A Class of Indecomposable Persistence Modules

Let $g : I \to \mathbb{R}$ be any (not necessarily continuous) bounded real-valued function on the unit interval. Let $f : I \times \{*\} \to I \times \mathbb{R}$ be given by $f(t, *) = (t, g(t))$ and let $F : \mathbf{\Delta}^2 \times \mathbf{R} \to \mathbf{Top}$ be given by $F(a, b, c) = f^{-1}([a, b] \times (-\infty, c])$.

Theorem 4.1 *Let $f_t(*) = g(t)$ be any uniformly bounded one-parameter family of functions on a one point space $\{*\}$. Then the corresponding persistence module H_jF is indecomposable for every $j \geq 0$.*

Proof For all $(a, b, c)) \in \Delta^2 \times \mathbb{R}$, $F(a, b, c)$ deformation retracts to a subset of I, so $H_k(F) = 0$ for $k \geq 1$. Recall (Sect. 3.1) that for $p \in \Delta^2 \times \mathbb{R}$, U_p denotes the up-set of p.

Assume that $H_0F \cong M \oplus N$ is a nontrivial decomposition of H_0F. Then there are nonzero maps $p : H_0F \to M$, $q : H_0F \to N$, $i : M \to H_0F$, and $j : N \to H_0F$ such that $ip + jq = 1_{H_0F}$. Choose $B \in \mathbb{R}$ such that $g(t) \leq B$ for all $t \in I$. Let $T = (0, 1, B)$. Then $(H_0F)_T = k$. It follows that either i_T or j_T is the zero map. Assume without loss of generality that $i_T = 0$.

By definition, we have that for all $t \in I$,

$$F(t, t, c) = \begin{cases} (t, c) & \text{if } c \geq g(t) \\ \emptyset & \text{if } c < g(t). \end{cases}$$

So, in particular, $H_0F(t, t, g(t)) = k$. Furthermore, we have a surjection of persistence modules

$$\bigoplus_{t \in I} k[U_{(t,t,g(t))}] \xrightarrow{\varphi} H_0F.$$

Since φ is surjective and p is nonzero, it follows that $p \circ \varphi$ is nonzero. Therefore, there exists an $a = (t_0, t_0, g(t_0))$ such that $k[U_a] \xrightarrow{p\varphi} M$ is nonzero, which forces $(p\varphi)_a : k[U_a]_a \to M_a$ to also be nonzero. Since $k[U_a]_a \cong k$ and $(H_0F)_a \cong k$, it follows that $p_a : (H_0F)_a \to M_a$ is injective. Therefore, $q_a = 0$.

Since $ip + jq = 1_{H_0F}$, $(H_0F)_{a \leq T} = i_T M_{a \leq T} p_a + j_T N_{a \leq T} q_a = 0$, which is a contradiction. □

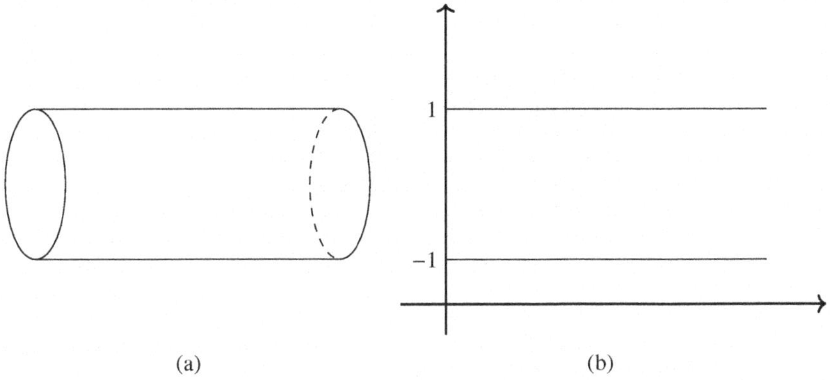

Fig. 6 Left: the cylinder $I \times S^1$. Right: the Cerf diagram of the constant one-parameter family of height functions on the circle

4.4 The Cylinder

Increasing the dimension of the manifold in our examples, consider $X = S^1$. Let $\tilde{f}_t : S^1 \to \mathbb{R}$ be the constant family of height functions on the circle; $\tilde{f}_t(\theta) = \sin\theta$. The corresponding function fibered over the interval $f : I \times S^1 \to I \times \mathbb{R}$ has domain the cylinder and is given by $f(t, \theta) = (t, \sin\theta)$. See Fig. 6a. The corresponding Cerf diagram is shown in Fig. 6b. It consists of two horizontal lines, corresponding to the fold singularities of $\tilde{f}$ given by the global minimum and global maximum of the height function.

By definition, $F(a, b, c) = f^{-1}([a, b] \times (-\infty, c]) = [a, b] \times \{\theta \mid \sin\theta \leq c\}$. Therefore, $F(a, b, c)$ is empty if $c < 0$, $F(a, b, c)$ is contractible if $0 \leq c < 1$, and $F(a, b, c)$ is homotopy equivalent to S^1 if $c \geq 1$. Thus, we find

$$\dim H_0F = \begin{cases} 1 & \text{if } c \geq 0 \\ 0 & \text{if } c < 0, \end{cases} \quad \text{and} \quad \dim H_1F = \begin{cases} 1 & \text{if } c \geq 1 \\ 0 & \text{if } c < 1. \end{cases}$$

This multiparameter persistence module can also be visualized as shown in Fig. 7. The blue region is the support of H_0F, and the red region is the support of H_1F. These two regions are unbounded, analogous to sublevel set persistence of the standard height function on S^1. Since each of H_0F and H_1F are indecomposable and at most one-dimensional, this visualization also shows the structure of the multiparameter persistence module H_*F.

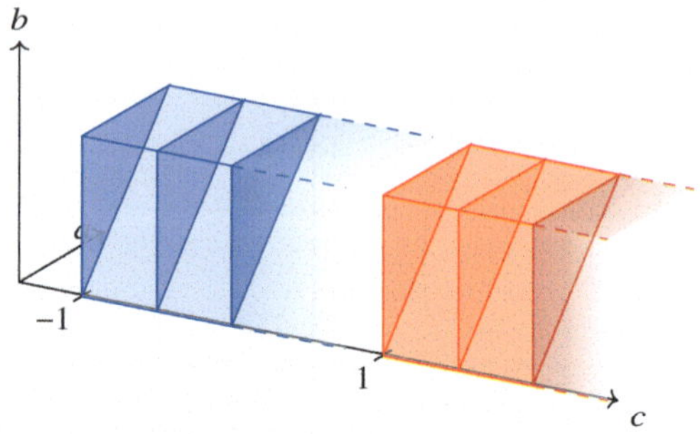

Fig. 7 The multiparameter persistence module of the constant one-parameter family of height functions on the circle. The Betti functions are constant for $0 \leq a \leq b \leq 1$. $\beta_0 = 0$ for $c < -1$ and $\beta_0 = 1$ for $c \geq -1$ (blue). $\beta_1 = 0$ for $c < 1$ and $\beta_1 = 1$ for $c \geq 1$ (red). For H_0F, all linear maps within the blue region are the identity map. Similarly, for H_1F, all linear maps with the red region are the identity map. That is, both H_0F and H_1F are one-dimensional persistence modules supported on semi-infinite triangular prisms in which all nontrivial maps are identity maps

5 Analyzing Diagrams of Spaces

Let X be a smooth, compact manifold. For generic one-parameter families of smooth functions $\tilde{f} : I \times X \to \mathbb{R}$, the nondegenerate critical points of the fibers $\tilde{f}_t$ occur in families themselves, as can be seen in the arcs of the Cerf diagrams of Sect. 4.

Definition 5.1 We say that such a critical point is *positive* if the curve in the Cerf diagram containing its value has positive slope (is locally strictly increasing). Similarly, say that such a critical point is *negative* if the curve in the Cerf diagram containing its value has negative slope (is locally decreasing). There can be points that are neither positive nor negative, e.g., the maximum or minimum of the singular locus of a wrinkle.

Recall that $f : I \times X \to I \times \mathbb{R}$ is given by $f(t, x) = (t, \tilde{f}(t, x)) = (t, \tilde{f}_t(x))$ and $F : \mathbf{\Delta^2} \times \mathbf{R} \to \mathbf{Top}$ is given by $F(a, b, c) = f^{-1}([a, b] \times (-\infty, c])$. For $(a, b, c) \in \Delta^2 \times \mathbb{R}$, let $F_0(a, b, c) = f^{-1}([a, b] \times \{c\})$.

Theorem 5.2 *Suppose $\tilde{f} : I \times X \to \mathbb{R}$ is a generic one-parameter family of functions on a smooth, compact manifold X. Let $0 \leq a < b \leq 1$ and $c \in \mathbb{R}$. Then $(F(a, b, c), F_0(a, b, c))$ is a cobordism between $(F(a, a, c), F_0(a, a, c))$ and $(F(b, b, c), F_0(b, b, c))$. Assume that there are no critical points in $F_0(a, a, c)$ and $F_0(b, b, c)$. Then the projection onto $[a, b]$*

$$\pi_{[a,b]} : F(a, b, c) \to [a, b] \tag{7}$$

is a Morse function on the cobordism $(F(a, b, c), F_0(a, b, c))$. Furthermore, $\pi_{[a,b]}$ has no interior critical points. In addition, positive and negative critical points in $F_0(a, b, c)$ are boundary stable and boundary unstable critical points of $\pi_{[a,b]}$, respectively.

Proof The projection $\pi_{[a,b]} : [a, b] \times X \to [a, b]$ is a submersion and hence has no critical points. Therefore, all critical points of the restriction $\pi_{[a,b]} : F(a, b, c) \to [a, b] \subset I$ must lie on the boundary $Y = F_0(a, b, c)$.

Consider a nondegenerate critical point z of $\tilde{f}_t$ with $z \in Y$. Near this point, there exists a coordinate system for which it is a fold singularity of $\tilde{f}$ given by Eq. (2). In this coordinate system, $\pi_{[a,b]}(t, x) = t$, so $d\pi_{[a,b]} = [\,1\,0 \cdots 0\,]$. For simplicity, assume that $(z, f(z))$ is at the origin. Then the level set $\{0\} \times \{x_1^2 + \cdots + x_j^2 - x_{j+1}^2 - \cdots - x_{n-1}^2 = 0\}$ has tangent space contained in $\{0\} \times \mathbb{R}^{n-1}$ and hence lies $\ker d\pi_{[a,b]}$. Therefore, z is a critical point of $\pi_{[a,b]}$.

Suppose z is a negative critical point of $\tilde{f}_t$ and $z \in Y$. There exists a path $\alpha : \mathbb{R} \to I \times X$ whose image consists of points (t, x) where x is a critical point of $\tilde{f}_t$ so that $\alpha(0) = z$, $\tilde{f}(\alpha(-t)) > c$, and $\tilde{f}(\alpha(t)) < c$ for $t > 0$ (i.e., a parametrization of the preimage of an arc in the Cerf diagram containing the image of z). Thus, α restricted to $[0, \infty)$ provides a path in $F(a, b, c)$ so that $f(\alpha(t)) \notin Y$ for $t > 0$, and therefore, $T_z W_z^u \not\subset T_z Y$. Hence, z is a boundary unstable critical point for $\pi_{[a,b]}$.

On the other hand, suppose z is a positive critical point of $\tilde{f}_t$ and $z \in Y$. Near z, there exists a coordinate system on $I \times X$ of the form prescribed by Eq. (2). In this coordinate system, we find $\pi_{[a,b]}(t, x) = t$ and thus $d\pi_{[a,b]} = [\,1\,0 \cdots 0\,]$. Since z is a positive critical point, if we take a sufficiently small such neighborhood U, then the $\tilde{f}$ function values will increase along the flow lines of $\nabla\pi_{[a,b]}$. Precisely, if $\xi : \mathbb{R} \times I \times X \to I \times X$ denotes the flow generated by $\nabla\pi_{[a,b]}$, then we have $\tilde{f}(\xi(\epsilon, t, x)) \geq \tilde{f}(t, x)$ for $\epsilon \geq 0$ and $(t, x) \in U$. This inequality holds in the restriction to $U \cap F(a, b, c)$. Hence, points on $Y \cap U = F_0(a, b, c) \cap U$ must flow to other points on Y under ξ. In particular, $U \cap W_z^s \subset Y$ and hence $T_z W_z^s \subset T_z Y$. □

Remark 5.3 Theorem 5.2 does not address the case when $F_0(a, b, c)$ contains nondegenerate critical points that are neither positive nor negative or the case that $F_0(a, b, c)$ contains cusp singularities.

The remainder of this section and the next section are dedicated to showing how the theory developed thus far and, in particular, Theorem 5.2, can be applied to examples.

Example 5.4 Consider the function $\tilde{g}(t) = 2\min(t, 1-t)$ in Sect. 4.1. Let $X = \{*\}$ and define $\tilde{f} : I \times X \to \mathbb{R}$ to be given by $\tilde{f}(t, *) = \tilde{g}(t)$ and define $f : I \times X \to I \times \mathbb{R}$ to be given by $f(t, *) = (t, \tilde{g}(t))$. For f, we have the associated diagram of topological spaces $F : \mathbf{\Delta}^2 \times \mathbf{R} \to \mathbf{Top}$.

By definition, $*$ is a critical point of $\tilde{g}_t$ for all $t \in I$. Let $0 \leq a < b \leq 1$ and let $c \in \mathbb{R}$. In the case $a < \frac{1}{2}$ and $2a < c < \tilde{g}(b)$, then $F_0(a, b, c)$ has a positive critical point, and Proposition 5.2 implies this intersection coincides with a boundary stable critical point of $\pi_{[a,b]}$. Proposition 2.10 implies $F(a, b, c)$ is a left-product cobordism. Note that $F(b, b, c)$ is empty. Similarly, if $b > \frac{1}{2}$ and $2(1 - b) < c < \tilde{g}(a)$, then we have a single boundary unstable critical point for $\pi_{[a,b]}$, and $F(a, b, c)$ is a right-product cobordism. In the case that $a < \frac{1}{2} < b$ and $2a, 2(1 - b) < c < 1$, we have that $F(a, b, c)$ is a cobordism between the singletons $F(a, a, c)$ and $F(b, b, c)$. This is not a product cobordism, however, since the projection $\pi_{[a,b]}$ has both a boundary stable and a boundary unstable critical point. Note that $F(\frac{1}{2}, \frac{1}{2}, c)$ is empty.

6 Examples II: The Wrinkled Cylinder

We modify the constant height function on the cylinder of Sect. 4.4 by introducing a wrinkle (Sect. 2.5), as shown in Fig. 8. The wrinkle creates two additional critical points for all times strictly between $t = p$ and $t = q$. As before, the two horizontal lines of the Cerf diagram correspond to fold singularities of $\tilde{f}$, which are the global minimum and global maximum. The functions $\tilde{f}_p$ and $\tilde{f}_q$ have cubic singularities, corresponding to birth and death singularities, respectively, of $\tilde{f}$ at $t = p$ and $t = q$ (see Remark 2.7). The birth singularity at $t = p$ gives rise to a pair of critical points of index 0 and 1, and these two critical points merge together at $t = q$. For all times t distinct from c and d, the function $\tilde{f}_t$ is a Morse function, with either two or four critical points.

The associated diagram of spaces $F : \mathbf{\Delta}^2 \times \mathbf{R} \to \mathbf{Top}$ takes some care to analyze. The input parameters a, b, and c define a semi-infinite strip $[a, b] \times (-\infty, c]$, which can be overlaid on the Cerf diagram (Fig. 9). Call the component of the singular locus containing cubic singularities the *wrinkle envelope*. If the top edge ($[a, b] \times \{c\}$) of the semi-infinite strip ($[a, b] \times (-\infty, c]$) lies in the interior of the wrinkle

Fig. 8 Left and middle: the wrinkled cylinder from two different angles. The left graphic contains the image of $\tilde{f}_t$ for some $t \in (p, q)$, a wrinkled circle, which has four critical points. Right: the Cerf diagram for the wrinkled cylinder

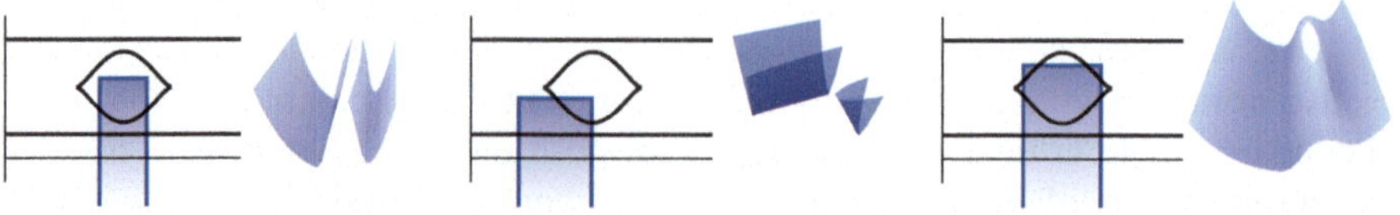

Fig. 9 Left, middle, right: three cases of the semi-infinite strip $[a, b] \times (-\infty, c]$ are shown in blue overlaid on the Cerf diagram from Fig. 8. The corresponding spaces $F(a, b, c)$ are shown to the right of each

envelope, then $F(a, b, c)$ is homotopy equivalent to S^0 (Fig. 9 left). If the top edge of the semi-infinite strip intersects the wrinkle envelope once and $c < m$, then $F(a, b, c)$ is homotopy equivalent to S^0 (Fig. 9 middle). If the top edge of the semi-infinite strip intersects the wrinkle envelope twice and $c \geq m$, then $F(a, b, c)$ is homotopy equivalent to S^1 (Fig. 9 right). In all other cases, $F(a, b, c)$ is homotopy equivalent to the corresponding space for the cylinder (Sect. 4.4).

This topological analysis can be made precise using the language of Theorem 5.2. Consider the three examples shown in Fig. 9. In the leftmost display of Fig. 9, the Cerf diagram does not intersect $[a, b] \times \{c\}$, and according to Theorem 5.2, the projection $\pi_{[a,b]} : F(a, b, c) \to [a, b]$ has no critical points. By Proposition 2.10, $F(a, b, c)$ is a product cobordism diffeomorphic to both $F(a, a, c) \times [a, b]$ and $F(b, b, c) \times [a, b]$. In the middle display of Fig. 9, there is a single negative critical point in $F_0(a, b, c)$. Theorem 5.2 implies $\pi_{[a,b]}$ has a single boundary unstable critical point. By Proposition 2.10, $F(a, b, c)$ is a right-product cobordism, as is evident from the displayed space. Finally, in the rightmost display of Fig. 9 contains both a positive and a negative critical point. Thus, $\pi_{[a,b]}$ has a boundary stable and boundary unstable critical point, so we cannot conclude that $F(a, b, c)$ is either a left- or right-product cobordism.

To aid in visualization of the persistence module, we again linearize the wrinkle (see the comment at the beginning of Sect. 4). In Fig. 10, the blue regions correspond to H_0F, and the red regions correspond to H_1F. Both H_0F and H_1F contain an unbounded region, arising from the global maxima and minima (fold) singularities, and a finite region, due to the wrinkle. Note that both H_0F and H_1F decompose into one-dimensional persistence modules, in contrast to the persistence modules of Theorem 4.1.

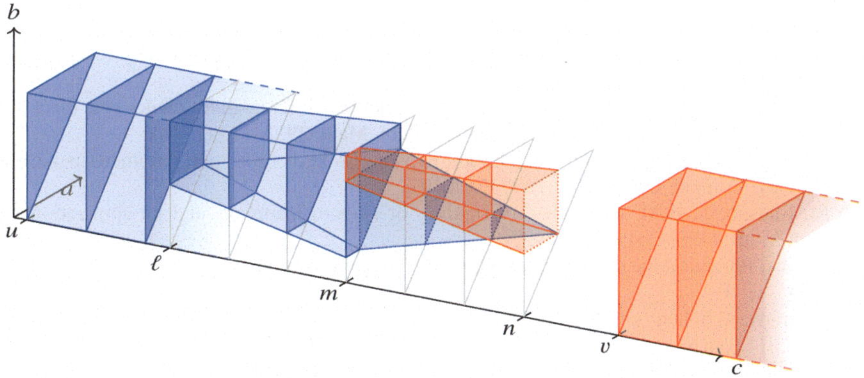

Fig. 10 The persistence module for the wrinkled cylinder. Both H_0F (blue) and H_1F (red) decompose into two one-dimensional persistence modules, one bounded and one unbounded, in which all nontrivial maps are the identity map. The unbounded components, $0 \leq a \leq b \leq 1$ together with $c \geq u$ and $c \geq v$, respectively, are exactly the persistence modules of the cylinder (Fig. 6). The bounded components are due to the wrinkle

For precise formulas, assume $p = \frac{1}{4}$, $q = \frac{3}{4}$. Then the bounded component of $H_1 F$ has support $m \leq c < n$, $0 \leq a \leq \frac{1}{2} - \frac{1}{4}\frac{n-c}{n-m}$, and $\frac{1}{2} + \frac{1}{4}\frac{n-c}{n-m} \leq b \leq 1$. The bounded component of $H_0 F$ has support given by the union of (i) $\ell \leq c < m$, $0 \leq a \leq \frac{1}{2} + \frac{1}{4}\frac{c-\ell}{m-\ell}$, and $\frac{1}{2} - \frac{1}{4}\frac{c-\ell}{m-\ell} \leq b \leq 1$ and (ii) $m \leq c < n$ and $\frac{1}{2} - \frac{1}{4}\frac{n-c}{n-m} < a \leq b < \frac{1}{2} + \frac{1}{4}\frac{n-c}{n-m}$.

Acknowledgments This research was partially supported by the Southeast Center for Mathematics and Biology, an NSF-Simons Research Center for Mathematics of Complex Biological Systems, under National Science Foundation grant no. DMS-1764406 and Simons Foundation grant no. 594594. This material is based upon work supported by, or in part by, the Army Research Laboratory and the Army Research Office under contract/grant number W911NF-18-1-0307.

References

1. M. Assif P K, Y. Baryshnikov, Biparametric persistence for smooth filtrations. arXiv:2110.09602
2. M. Borodzik, A. Némethi, A. Ranicki, Morse theory for manifolds with boundary. Algebraic Geom. Topol. **16**(2), 971–1023 (2016)
3. D. Braess, Morse-Theorie für berandete Mannigfaltigkeiten. Mathematische Annalen **208**, 122–148 (1974)
4. P. Bubenik, N. Milićević, Homological algebra for persistence modules. Found. Comput. Math. **21**(5), 1233–1278 (2021)
5. P. Bubenik, V. de Silva, J. Scott, Metrics for generalized persistence modules. Found. Comput. Math. **15**(6), 1501–1531 (2015)
6. P. Bubenik, J.A. Scott, D. Stanley, Exact weights, path metrics, and algebraic Wasserstein distances. J Appl. and Comput. Topology **7**, 185–219 (2023).
7. R. Budney, T. Kaczynski, Bifiltrations and persistence paths for 2-Morse functions. Alg. & Geom. Topology **23**(6), 2895–2924 (2023).
8. G. Carlsson, Topology and data. Bull. Am. Math. Soc. (N.S.) **46**(2), 255–308 (2009)
9. G. Carlssonn, A. Zomorodian, The theory of multidimensional persistence. Discrete Comput. Geom. **42**(1), 71–93 (2009)
10. J. Cerf, La stratification naturelle des espaces de fonctions différentiables réelles et le théorème de la pseudo-isotopie. Inst. Hautes Études Sci. Publ. Math. **39**, 5–173 (1970)
11. R. Corbet, U. Fugacci, M. Kerber, C. Landi, B. Wang, A kernel for multi-parameter persistent homology. Comput. Graph. X **2**, 100005 (2019)
12. Y.M. Eliashberg, N.M. Mishachev, Wrinkling of smooth mappings and its applications. I. Invent. Math. **130**(2), 345–369 (1997)
13. R. Ghrist, Barcodes: the persistent topology of data. Bull. Am. Math. Soc. (N.S.) **45**(1), 61–75 (2008)
14. R.E. Grady, A. Schenfisch, Natural Stratifications of Reeb Spaces and Higher Morse Functions. arXiv:2011.08404
15. B. Hajduk, Minimal m-functions. Fund. Math. **111**, 179–200 (1981)
16. H.A. Harrington, N. Otter, H. Schenck, U. Tillmann, Stratifying multiparameter persistent homology. SIAM J. Appl. Algebra Geometry, **3**(3), 439–471 (2019).
17. A. Hatcher, J. Wagoner, *Pseudo-Isotopies of Compact Manifolds* (Société Mathématique de France, Paris, 1973). Astérisque, No. 6
18. K. Igusa, The space of framed functions. Trans. Am. Math. Soc. **301**(2), 431–477 (1987)
19. M.C. Irwin, On the stable manifold theorem. Bull. London Math. Soc. **2**(2), 196–198 (1970)

20. A. Jankowski, R. Rubinsztein, Functions with non-degenerate critical points on manifolds with boundary. Comment. Math. Prace Mat. **16**, 99–112 (1972)
21. D. Joyce, *On Manifolds with Corners*. Advances in Geometric Analysis, Advanced Lectures in Mathematics, vol. 21 (International Press, Boston, 2012), pp. 225–228. Available arXiv: 0910.3518
22. M. Kerber, M. Lesnick, S. Oudot, Exact computation of the matching distance on 2-parameter persistence modules. J. Comput. Geom. **11**(2), 4–25 (2020)
23. R. Kirby, A calculus for framed links in S3. Invent. Math. **45**, 35–56 (1978)
24. M. Lesnick, M. Wright, Interactive visualization of 2-D persistence modules. arXiv:1512.00180
25. R. MacPherson, A. Patel, Persistent local systems. Adv. Math. **386**, 107795 (2021)
26. E. Miller, Homological algebra of modules over posets. arXiv:2008.00063
27. J. Milnor, *Morse Theory* (Princeton University Press, Princeton) (1963)
28. M. Morse, Relations between the critical points of a real function of n independent variables. Trans. Am. Math. Soc. **27**(3), 3345–396 (1925)
29. O. Vipond, Multiparameter persistence landscapes. J. Mach. Learn. Res. **21**(61), 1–38 (2020)

Compact Torus Action on the Complex Grassmann Manifolds

Victor M. Buchstaber and Svjetlana Terzić

1 Introduction

Complex Grassmann manifolds $G_{n,k}$ are widely known mathematical objects connected with many fundamental results in algebraic topology and complex, algebraic, and symplectic geometry. In this survey, we present results concerning the canonical torus action on complex Grassmann manifolds, classical as well as recently obtained based on approaches from toric geometry and toric topology. The methods of toric geometry use the action of algebraic torus $(\mathbb{C}^*)^n$ on $G_{n,k}$. The approaches developed in the framework of toric topology are based on the induced action of the standard compact torus $\mathbb{T}^n$ and the moment map $\mu : G_{n,k} \to \Delta_{n,k}$, where $\Delta_{n,k}$ is the standard hypersimplex.

By definition, a toric manifold is a smooth toric variety defined as an algebraic manifold, which can be obtained as the closure of an algebraic torus action orbit. Any Grassmann manifold $G_{n,k}$ is a smooth complex Kähler manifold, but for $1 < k < n-1$, it is not a toric manifold since it is a union of family of toric varieties. Note that this family contains toric varieties, which are not smooth. Therefore, it is natural to introduce the notion of the space of parameters of a family of toric varieties. The objects mainly studied in toric topology are quasitoric manifolds M^{2n} with an effective action of a compact torus T^n and an important assumption that the orbit space M^{2n}/T^n is homeomorphic to a simple n-dimensional polytope P^n. For both toric and quasitoric manifolds, a notable and fundamental property is that for every one of them, there exists a uniquely defined moment map whose image

V. M. Buchstaber
Steklov Mathematical Institute, Russian Academy of Sciences, Moscow, Russia
e-mail: buchstab@mi.ras.ru

S. Terzić (✉)
Faculty of Science and Mathematics, University of Montenegro, Podgorica, Montenegro
e-mail: sterzic@ucg.ac.me

A. Bahri et al. (eds.), *Toric Topology and Polyhedral Products*, Fields Institute Communications 89, https://doi.org/10.1007/978-3-031-57204-3_5

can be identified with the aforementioned simple convex polytope [4, 5, 12]. In the case of Grassmann manifold $G_{n,k}$, the image of the moment map is $\Delta_{n,k}$, which is not a simple polytope for $1 < k < n-1$. The goal of our survey is to show that, nevertheless, the methods and results of toric geometry and toric topology allow to obtain deep results on equivariant topology of the Grassmann manifolds $G_{n,k}$, $1 \le k \le n-1$. It is known that the equivariant structure of Grassmannians $G_{n,2}$ is more regular than in the case of the manifolds $G_{n,k}$ with $n > 5, k > 2$ (for details, see Paragraph 1.11, Adjacency of strata, in [19]). We show that, in the case of $G_{n,2}$, much stronger results can be achieved by exploiting the special case $k = 2$.

In more general situation, one considers an effective T^k-action on a manifold M^{2n} endowed with the moment map $\mu : M^{2n} \to P^k$ for a k-dimensional polytope P^k. An important feature of these actions is their complexity, which is equal to the number $d = n - k$. The torus action in the case of toric and quasitoric manifolds has zero complexity. In this case, the situation is in some sense quite simple, as one can describe the equivariant topology of these manifolds M^{2n} in terms of the combinatorics of the polytope P^n. The equivariant topology of the Grassmannian $G_{4,2}$ with effective T^3-action of complexity one is described in our paper [6]. This result stimulated further exploration of complexity-one torus actions on a wider class of manifolds; see recent papers [2, 25]. It is natural that the complexity of the algebraic topology of the orbit space M^{2n}/T^k grows simultaneously with the complexity of the torus action. As such, the exploration of torus actions of complexity $d \ge 2$ is recognized in the literature as quite a difficult problem. In our relevant paper [7], we describe the equivariant topology of the orbit space $G_{5,2}$ with effective T^4-action of complexity *two*. We believe that the results of the paper [7] can be adapted to study complexity-two torus actions on a wider class of manifolds. Our results on the equivariant topology of $G_{5,2}$ stimulated further investigation of the subject, such as the paper [34], which establishes deep connection between algebraic topology problems of the orbit space $G_{5,2}/\mathbb{T}^5$ and well-known constructions and results of algebraic geometry.

In the paper [10], we have developed the foundations of the theory of $(2n, k)$-manifolds, whose axiomatics appeals to our results on $\mathbb{T}^n$-equivariant structure of $G_{n,k}$, obtained toward the classical open problem of an effective description of the orbit space $G_{n,k}/T^n$. These results allowed us to introduce new notions: universal space of parameters, complex of admissible polytopes and their virtual spaces of parameters, and, as the result, the notion of singular points of the action (for details, see Paragraph 2.8 of the current survey).

The complexity of the effective T^{n-1}-action on $G_{n,k}$ is $d = (n-k-1)(k-1)$, which rapidly grows as a function in n and k. The effective action of T^{n-1} on $G_{n,2}$ gives a family of manifolds with the actions of growing complexity $d = n - 3$. For $G_{n,2}$, we construct an explicit model of the equivariant topology and the orbit space $G_{n,2}/\mathbb{T}^n$. This model resolves the singular points of $\mathbb{T}^n$-action on $G_{n,2}$ (see Sect. 3).

In the final Sect. 4, we establish the relation between the universal space of parameters for $G_{n,2}$ and the Chow quotient $G_{n,2}//(\mathbb{C}^*)^n$ described by Kapranov in [24]. We obtained this result using the technique of wonderful compactification; see [26, 31]. Note that the notion of wonderful compactification first appeared in the paper [16] of De Concini-Procesi in the context of an equivariant compactification of the symmetric spaces G/H; see also [32] and [35] for a comprehensive overview of the subject. This idea has been further developed and applied in many areas, such as Fulton-MacPherson compactification in [18], De Concini-Procesi wonderful models [16, 17], the wonderful compactification of Li [31], and, more recently, the projective wonderful models of toric arrangements by De Concini-Gaiffi and others [13–15]. Furthermore, our result from [9] shows the advantage of wonderful compactification when it comes to the problem of describing the structure of $G_{n,2}/\mathbb{T}^n$. More explicitly, it turns out that the wonderful compactification of arrangements of subvarieties from [31] can be successfully applied for the compactification of the space of parameters $F_n = W_n/(\mathbb{C}^*)^n$ of the main stratum $W_n \subset G_{n,2}$. This compactification uses the points of the Chow variety of $G_{n,2}$ as the points of the build-up or outgrow components.

2 General Results on Grassmannians $G_{n,k}$

The complex Grassmann manifold $G_{n,k} = G_{n,k}(\mathbb{C})$ consists of all k-dimensional complex subspaces in $\mathbb{C}^n$. The manifolds $G_{n,k}$ and $G_{n,n-k}$ are diffeomorphic since for the standard scalar product in $\mathbb{C}^n$, there exists a canonical diffeomorphism $c_{nk} : G_{n,k} \to G_{n,n-k}$ induced by the map, which sends any k-dimensional subspace of $\mathbb{C}^n$ to its orthogonal complement. The coordinate-wise action of the compact torus $\mathbb{T}^n$ on $\mathbb{C}^n$ given by $(t_1, \dots, t_n) \cdot (z_1, \dots, z_n) = (t_1 z_1, \dots, t_n z_n)$ induces the canonical action of $\mathbb{T}^n$ on $G_{n,k}$. The canonical diffeomorphism c_{nk} is equivariant for this action. The canonical action of $\mathbb{T}^n$ on $G_{n,k}$ induces the effective action of $T^{n-1} = \mathbb{T}^n/\text{diag}(\mathbb{T}^n)$ on $G_{n,k}$ with $\binom{n}{k}$ fixed points.

2.1 *Plücker Coordinates and Plücker Embedding*

We recall the definition of Plücker coordinates, which are a classical notion in complex Grassmann manifolds theory. From the canonical basis in $\mathbb{C}^n$, induce a basis in $L \in G_{n,k}$. In this basis, L can be represented by a $(n \times k)$-matrix A_L. Given $I \subset \{1, \dots, n\}$, $|I| = k$, denote by $P^I(A_L)$ the $(k \times k)$-minor of A_L formed by the rows indexed by the elements of I. The complex numbers $(P^I(A_L))$, where I runs through all subsets of $\{1, \dots n\}$ such that $|I| = k$, are known as the Plücker coordinates of the subspace $L \subset \mathbb{C}^n$. The Plücker coordinates $(P^I(L))$ are defined uniquely up to common scalar, and therefore, we can consider the Plücker embedding $G_{n,k} \to \mathbb{C}P^{N-1}$, $N = \binom{n}{k}$, defined by

$$L \to P(L) = (P^I(A_L)), \quad I \subset \{1, \dots n\}, \; |I| = k. \tag{1}$$

The Plücker embedding provides $G_{n,k}$ with a structure of a $k(n-k)$-dimensional complex Kähler manifold. This embedding satisfies the well-known Plücker relations [29], which are quadratic.

The image of the Plücker embedding of $G_{n,2}$, $n \geq 4$, in $\mathbb{C}P^{N-1}$, $N = \binom{n}{2}$, is the intersection of the $\binom{n}{4}$ quadratic hypersurfaces

$$z_{ij}z_{kl} + z_{jk}z_{il} = z_{ik}z_{jl}, \quad 1 \leq i < j < k < l \leq n.$$

The Grassmann manifold $G_{n,2}$ is a smooth algebraic variety embedded in $\mathbb{C}P^{N-1}$. Note that the normal bundle of $G_{n,2}$ in $\mathbb{C}P^{N-1}$ is a complex vector bundle of the dimension $\binom{n-2}{2}$. For $n > 4$, we have $\binom{n-2}{2} < \binom{n}{4}$, and therefore, $G_{n,2}$ is not a complete intersection in $\mathbb{C}P^{N-1}$.

Moreover, since the Plücker relations are quadratic, we can effectively describe the standard local structure of a Grassmann manifold $G_{n,2}$ as well as its normal bundle.

Let us consider the representation $\rho_{n,k} : \mathbb{T}^n \to \mathbb{T}^N$, $N = \binom{n}{k}$, given by the kth exterior power

$$(t_1, \dots, t_n) \to (t_1 \cdots t_k, \dots, t_{n-k+1} \cdots t_n).$$

The Plücker embedding is equivariant for this representation, that is,

$$\mathbb{T}^n \curvearrowright G_{n,k} \to \mathbb{C}P^{N-1} \curvearrowleft \mathbb{T}^N$$

is $\rho_{n,k}$-equivariant.

2.2 *Moment Map and Hypersimplex*

The complex projective space $\mathbb{C}P^{N-1}$ is a toric manifold with the canonical action of the standard compact torus $\mathbb{T}^N$ and the corresponding moment map

$$\mathbb{C}P^{N-1} \to \Delta^{N-1}, \quad z \to \frac{1}{|z|^2} \sum_{i=1}^{N} |z_i|^2 e_i,$$

where $\{e_i\}$ is the standard basis in $\mathbb{R}^N$ and $\Delta^{N-1} \subset \mathbb{R}^N$ is the regular simplex.

We enumerate the coordinates in $\mathbb{R}^N$ by (z_J), $J \subset \{1, \dots, n\}$, $|J| = k$, according to the standard ordering of the set of all k-subsets of the set of n elements. For example, for $n = 4$, $k = 2$, and $N = 6$, we consider the correspondence $1 \leftrightarrow (1,2)$, $2 \leftrightarrow (1,3)$, $3 \leftrightarrow (1,4)$, $4 \leftrightarrow (2,3)$, $5 \leftrightarrow (2,4)$, and $6 \leftrightarrow (3,4)$. Then, one can rewrite the moment map on $\mathbb{C}P^{N-1}$ as

$$z \to \frac{1}{|z|^2} \sum_J |z_J|^2 e_J. \tag{2}$$

Let $\mathbb{R}^n$ be n-dimensional real vector space with the standard basis. The weight vectors of the representation $\rho_{n,k}$ are

$$\Lambda_I \in \mathbb{Z}^n \subset \mathbb{R}^n, \quad (\Lambda_I)_j = 1 \text{ for } j \in I, \quad (\Lambda_I)_j = 0 \text{ for } j \notin I,$$

where $I \subset \{1, \dots, n\}$, $|I| = k$. In other words, Λ_I has k ones and $n - k$ zeros.

The moment map $\mu : G_{n,k} \to \mathbb{R}^n$ [19, 28] is defined by

$$\mu(L) = \frac{1}{|P(L)|^2} \sum |P^I(A_L)|^2 \Lambda_I, \quad |P(L)|^2 = \sum |P^I(A_L)|^2, \tag{3}$$

where the sum goes over all subsets $I \subset \{1, \dots, n\}$, $|I| = k$.

The image of μ is the standard hypersimplex $\Delta_{n,k}$, which is obtained as a convex hull of the vectors Λ_I. More precisely,

$$\Delta_{n,k} = I^n \cap \{(x_1, \dots, x_n) \in \mathbb{R}^n, \sum_{i=1}^{n} x_i = k\}.$$

It follows that $\Delta_{n,k}$ belongs to the hyperplane $x_1 + \cdots + x_n = k$ in $\mathbb{R}^n$; thus, $\Delta_{n,k}$ is a $(n-1)$-dimensional polytope. The hypersimplex $\Delta_{n,k}$ has $\binom{n}{k}$ vertices and $2n$ facets. Every one of its vertices has degree $k(n-k)$. Its boundary consists of two disjoint unions, one comprising n copies of the hypersimplex $\Delta_{n-1,k}$ and the other comprising n copies of the hypersimplex $\Delta_{n-1,k-1}$. The hypersimplex $\Delta_{n,k}$ is simple if and only if $k = 1$ or $k = n - 1$, that is, when it coincides with the simplex Δ^{n-1}.

The edge graph of the hypersimplex $\Delta_{n,.k}$ is the Johnson graph. The Johnson graph is well-known not only in graph theory but also in coding theory; see [23].

Denote the coordinates of $\mathbb{R}^N$ by z_J, $J \subset \{1, \dots, n\}$, $|J| = k$. Define the map

$$\mathbb{R}^N \to \mathbb{R}^n \text{ by } e_J \to \sum_{j \in J} e_j.$$

It gives a bijection between the set of vertices of the simplex Δ^{N-1} and the set of vertices of $\Delta_{n,k}$. Moreover, the composition $G_{n,k} \to \mathbb{C}P^N \to \Delta^{N-1} \to \Delta_{n,k}$ coincides with the moment map for the Grassmannian $G_{n,k}$.

2.3 $\mathbb{T}^n$-Equivariant Automorphisms of $G_{n,k}$

It is known [11] that the group of holomorphic automorphisms $\mathrm{Aut}(G_{n,k})$ is isomorphic to the projective group $PU(n)$ if $n \neq 2k$, while for $G_{2k,k}$, it is isomorphic to $\mathbb{Z}_2 \times PU(2k)$. The group $\mathbb{Z}_2$ is defined by the duality diffeomorphism $c_{n,k} : G_{n,k} \to G_{n,n-k}$. Note also that the symmetric group S_n acts on $\mathbb{C}^n$ by permuting the coordinates, that is, S_n is a subgroup of $\mathrm{Aut}(G_{n,k})$.

Taking into account the canonical $\mathbb{T}^n$-action on $G_{n,k}$, one can prove [8]:

- Let H be a subgroup of $\mathrm{Aut}(G_{n,k})$ such that every element of it commutes with the canonical $\mathbb{T}^n$-action on $G_{n,k}$. Then $H = T^{n-1} \rtimes S_n$ for $n \neq 2k$ and $H = \mathbb{Z}_2 \times (T^{2k-1} \rtimes S_{2k})$ for $n = 2k$.
- Let H be a subgroup of $\mathrm{Aut}(G_{n,k})$ such that the set of fixed points of the $\mathbb{T}^n$-action on $G_{n,k}$ is invariant under the action of every element of H. Then $H = T^{n-1} \rtimes S_n$ for $n \neq 2k$ and $H = \mathbb{Z}_2 \times (T^{n-1} \rtimes S_n)$ for $n = 2k$.

Furthermore, the relationship between the group $\mathrm{Aut}(G_{n,k})$ and the moment map $\mu : G_{n,k} \to \Delta_{n,k}$ is characterized as follows [8]:

Theorem 2.1 *The subspaces $\mu_{n,k}^{-1}(x)/\mathbb{T}^n$ and $\mu_{n,k}^{-1}(\mathfrak{s}(x))/\mathbb{T}^n \subset G_{n,k}/\mathbb{T}^n$ are homeomorphic for any $x \in \Delta_{n,k}$ and any $\mathfrak{s} \in S_n$. In addition, for $n = 2k$, the subspaces $\mu_{2k,k}^{-1}(x)/\mathbb{T}^{2k}$ and $\mu_{2k,k}^{-1}(\bar{f}(x))/\mathbb{T}^{2k}$ are homeomorphic for any $x \in \Delta_{2k,k}$, where $\bar{f}(x) = \mathbf{1} - x$.*

2.4 Charts and Strata

The Plücker coordinates define a smooth atlas on $G_{n,k}$. The charts are given by $M_I = \{L \in G_{n,k} : P^I(L) \neq 0\}$, $I \subset \{1, \dots, n\}$, $|I| = k$, and the homeomorphisms $u_I : M_I \to \mathbb{C}^{k(n-k))}$ are constructed as follows. Any $L \in M_I$ can be uniquely represented by the $(n \times k)$-matrix A_L whose submatrix determined by the rows indexed by I is the identity $k \times k$-matrix. The matrix A_L has $k(n-k)$ nonzero entries $a_{ij}(L)$, and the homeomorphism $u_I : M_I \to \mathbb{C}^{k(n-k)}$ is given by $u_I(L) = (a_{ij}(L))$, $i \notin I$. It can be easily seen that any chart M_I contains exactly one fixed point given by the element L such that $u_i(L) = 0 \in \mathbb{C}^{k(n-k)}$. The number of charts is $N = \binom{n}{k}$, and it coincides with the number of fixed points for the canonical $\mathbb{T}^n$-action on $G_{n,k}$. The charts M_I are open, $\mathbb{T}^n$-invariant dense sets in $G_{n,k}$. It implies that the sets $Y_I = G_{n,k} \setminus M_I$ are closed and $\mathbb{T}^n$-invariant.

We enumerate the charts by $(M_{I_1}, u_{I_1}), \dots, (M_{I_N}, u_{I_N})$. For $\sigma = \{I_1, \dots, I_l\}$, $I_i \subset \{1, \dots, n\}$ such that $|I_i| = k$, $1 \leq i \leq l$ and $1 \leq l \leq N$, define the spaces W_σ by

$$W_\sigma = M_{I_1} \cap \cdots M_{I_l} \cap Y_{I_{l+1}} \cap \cdots Y_{I_N}, \tag{4}$$

where $\{I_{l+1}, \dots, I_N\} = \{I_1, \dots, I_N\} \setminus \{I_1, \dots, I_l\}$. More precisely,

$$W_\sigma = \{L \in G_{n,k} : P^I(L) \neq 0 \text{ for } I \in \sigma \text{ and } P^I(L) = 0 \text{ for } I \notin \sigma\}, \tag{5}$$

Definition 2.2 A non-empty space W_σ is called a *stratum*.

The stratum $W_\sigma = M_{I_1} \cap \dots \cap M_{I_N}$, where $\sigma = \{I_1, \dots, I_N\}$, is called the *main stratum* and denoted by W.

Example 2.3 For $\sigma = \{I\}$, W_σ is a fixed point.

The strata have the following properties [7]:

- They are disjoint, $\mathbb{T}^n$-invariant subspaces and give the equivariant subdivision of $G_{n,k}$, that is, $G_{n,k} = \cup_\sigma W_\sigma$.
- $\mu(W_\sigma) = \overset{\circ}{P}_\sigma$, where P_σ is the convex hull of the vertices $\Lambda_{I_1}, \dots, \Lambda_{I_l}$ and $\sigma = \{I_1, \dots, I_l\}$. In particular, for $\sigma = \{I\}$, $\mu(W_\sigma) = \Lambda_I$.

Definition 2.4 A polytope P_σ is called an *admissible polytope* if $\overset{\circ}{P}_\sigma = \mu(W_\sigma)$ for some stratum W_σ.

2.5 $(\mathbb{C}^*)^n$-Action

Many important properties of the canonical $\mathbb{T}^n$-action on $G_{n,k}$ can be deduced from the fact that the canonical $\mathbb{T}^n$-action on $G_{n,k}$ extends to the canonical action of the algebraic torus $(\mathbb{C}^*)^n$.

When it comes to the relation between $(\mathbb{C}^*)^n$-orbits and the moment map, the classical convexity theorem of [1, 22] states:

Theorem 2.5 *Let $O_{\mathbb{C}}(L)$ be the orbit of an element $L \in G_{n,k}$ of the canonical $(\mathbb{C}^*)^n$-action. Then $\mu(\overline{O_{\mathbb{C}}(L)})$ is a convex polytope in $\mathbb{R}^n$ whose vertex set is given by $\{\Lambda_I | P^I(L) \neq 0\}$. The mapping μ gives a bijection between p-dimensional orbits of the group $(\mathbb{C}^*)^n$ in $\overline{O_{\mathbb{C}}(L)}$ and p-dimensional open faces of the polytope $\mu(\overline{O_{\mathbb{C}}(L)})$.*

It is easy to verify that the strata W_σ are $(\mathbb{C}^*)^n$-invariant as well. Moreover, all $(\mathbb{C}^*)^n$-orbits in a stratum W_σ map on $\overset{\circ}{P}_\sigma$ by the moment map. Therefore, a stratum W_σ can be considered as the collection of $(\mathbb{C}^*)^n$-orbits mapping to the same polytope by the moment map.

Remark 2.6 Note that this is only one of the three equivalent ways the strata were defined in [19, 20]. The other two approaches to the notion of strata are given by the rank functions and by the refinement of the Schubert cell decompositions; see also [6]. We emphasize here that our approach to the notion of strata given by (4) is more general since it does not use the existence of $(\mathbb{C}^*)^n$-action on $G_{n,k}$ such that it extends the $\mathbb{T}^n$-action. This approach turns out to be fundamental for developing our theory of $(2n, k)$-manifolds [10].

Definition 2.7 The orbit space $F_\sigma = W_\sigma/(\mathbb{C}^*)^n$ is referred to as the *space of parameters* of the stratum W_σ.

The properties of the $(\mathbb{C}^*)^n$-action on a stratum W_σ are described by Theorem 2.10 below.

The space of parameters of the main stratum W is denoted by F.

Remark 2.8 The closure $\overline{O_{\mathbb{C}}(L)}$ is a toric variety for any $L \in G_{n,k}$. This toric variety is smooth if and only if its corresponding admissible polytope is a simple polytope. The following problems naturally arise:

1. Describe all admissible polytopes that are simple and describe the corresponding smooth toric varieties.
2. Describe singularities of toric varieties for which the corresponding admissible polytopes are not simple.
3. Describe singular toric varieties for which the corresponding polytopes are simple in m-dimensional faces, $0 < m \le n-2$.

The solutions of the second problem for the manifolds $G_{4,2}$ and $G_{5,2}$ are given in [6] and [7], respectively. The description of all smooth toric varieties for an arbitrary $G_{n,k}$ is given in [33], and smooth algebraic torus orbit closures in $G_{n,k}$ are proved to be products of complex projective spaces.

In connection with the third problem, let us introduce a generalization of the notion of a simple polytope. An n-dimensional polytope is said to be *simple in m-dimensional faces*, $0 < m \le n-2$, if all its m-dimensional faces can be obtained as intersections of $n-m$ facets. For example, a polytope is simple if and only if it is simple in vertices. Octahedron $\Delta_{4,2}$ is not simple, but it is simple in edges.

The $(\mathbb{C}^*)^n$-orbits on $G_{n,k}$ are nicely described in [7] in terms of the Plücker charts. For a subset $J = \{j_1, \dots, j_l\} \subseteq \{1, \dots, k(n-k)\}$, let

$$\mathbb{C}^J = \{(z_1, \dots, z_{k(n-k)}) \in \mathbb{C}^{k(n-k)} | z_j = 0, \;\; j \notin J\}.$$

The coordinate subspaces $\mathbb{C}^J$ are $(\mathbb{C}^*)^n$-invariant and their union is $\mathbb{C}^{k(n-k)}$. Let $\overset{\circ}{\mathbb{C}^J} = \{(z_1, \dots, z_{k(n-k)}) \in \mathbb{C}^J | z_j \neq 0, \;\; j \in J\}$.

Note that any $\overset{\circ}{\mathbb{C}^J}$ is also $(\mathbb{C}^*)^n$-invariant. It follows from the definition: all points of $\overset{\circ}{\mathbb{C}^J}$ have the same stabilizer $(\mathbb{C}^*)_J \subset (\mathbb{C}^*)^n$. Thus, $\overset{\circ}{\mathbb{C}^J}$ is a collection of orbits of the form $(\mathbb{C}^*)^n/(\mathbb{C}^*)_J$. We provide the description of this collection in local coordinates. Depending on J, $\overset{\circ}{\mathbb{C}^J}$ can be one of the following:

- An entire $(\mathbb{C}^*)^n$-orbit

- A collection of $(\mathbb{C}^*)^n$-orbits, which are given by the preimages $F_J^{-1}(c)$, $c = (c_1, \ldots, c_{l-q}) \in (\mathbb{C}^*)^{l-q}$, where $F_J : (\mathbb{C}^*)^J \to (\mathbb{C}^*)^{l-q}$ is a $(\mathbb{C}^*)^n$-invariant algebraic map given by $(z_{j_1}, \ldots, z_{j_l}) \to (z_{j_1}^{\omega_1^1} \cdots z_{j_l}^{\omega_1^l}, \ldots, z_{j_1}^{\omega_{l-q}^1} \cdots z_{j_l}^{\omega_{l-q}^l})$.

  Here $l = |J|$, and q is the rank of the matrix V^J consisting of the weight vectors Λ_j, $j \in J$, of the representation $\rho_{n,k}$. The numbers $\omega_i^j \in \mathbb{Z}$, $1 \le i \le l-q$, $1 \le j \le l$ are determined as follows: the matrix V^J defines a linear map $f_J : \mathbb{R}^n \to \mathbb{R}^l$ such that $f_J(\mathbb{Z}^n)$ is a direct summand in $\mathbb{Z}^l$. Let $\mathcal{L}$ be a lattice in $\mathbb{Z}^l$ spanned by $f_J(\Lambda_j)$, $j \in J$. There exists a unique integral lattice, which is orthogonal to $\mathcal{L}$. By fixing a basis, we obtain the matrix $\hat{V}^J$ of dimension $l \times (l-q)$, the elements of which we denote by ω_i^j.

This leads us to the description of the strata in the Plücker charts. Any stratum W_σ belongs—see (4)—to the intersection of some set of charts. The cardinality of this set is equal to the number of nonzero Plücker coordinates of the points in W_σ.

Let us consider a stratum W_σ, where $|\sigma| = l$ and $1 \le l \le \binom{n}{k} - 1$. It consists of all points from $G_{n,k}$ whose Plücker coordinates P^I are nonzero if and only if $I \in \sigma$. In an arbitrary chart, the Plücker coordinates give rise to polynomials in the local coordinates $z_1, \ldots, z_{k(n-k)}$. Note that the local coordinates $z_1, \ldots z_{k(n-k)}$ represent the corresponding Plücker coordinates. Therefore, $W_\sigma \subset \overset{\circ}{(\mathbb{C})^J}$, for $J = \{j \in \overline{1, k(n-k)} \mid z_j = P^I(W_\sigma) \text{ for some } I \in \binom{n}{k} \setminus \sigma\}$. Let $J = \{j_1, \ldots, j_d\}$, $d = \binom{n}{k} - l$ and let $P_J^I(z_{j_1}, \ldots, z_{j_d})$ be the restriction of P^I on $\overset{\circ}{\mathbb{C}^J}$.

From the given description of $\overset{\circ}{\mathbb{C}^J}$, we obtain that any stratum W_σ in a chart from the set of charts that contain W_σ can only be one of the following:

- The whole $\overset{\circ}{\mathbb{C}^J}$
- The intersection of the entire collection of $(\mathbb{C}^*)^n$-orbits in $\overset{\circ}{\mathbb{C}^J}$ with the family of surfaces defined by the equations imposed by those Plücker coordinates, which are zero or must not be zero for the points from W_σ:

$$u_I(W_\sigma) = \begin{cases} F_J^{-1}(c), \quad c = (c_1, \ldots, c_{l-q}) \in (\mathbb{C}^*)^{l-q} \\ P_J^I(z_{j_1}, \ldots, z_{j_d}) = 0, \quad I \in \binom{n}{k} \setminus \sigma \\ P_J^I(z_{j_1}, \ldots, z_{j_d}) \neq 0, \quad I \in \sigma. \end{cases}$$

Example 2.9 Let us demonstrate these statements in the case of $G_{4,2}$. We consider the chart M_{12} and take into account that its coordinates z_1, z_2, z_3, z_4 are enumerated by $1 \leftrightarrow (1,3)$, $2 \leftrightarrow (1,4)$, $3 \leftrightarrow (2,3)$, $4 \leftrightarrow (2,4)$. It follows from [6] that any $\overset{\circ}{\mathbb{C}^J}$, $J \subset \{(1,3), (1,4), (2,3), (2,4)\}$, such that $|J| < 4$ is an entire $(\mathbb{C}^*)^4$-orbit. For $|J| = 4$, we consider the matrix V^J of the corresponding weight vectors Λ_j, $j \in J$. Its rank is $q = 3$, it defines the map $f_J : \mathbb{R}^4 \to \mathbb{R}^4$, and $f_J(\Lambda_{(1,3)}) = (2,1,1,0)$, $f_J(\Lambda_{(1,4)}) = (1,2,0,1)$, $f_J(\Lambda_{(2,3)}) = (1,0,2,1)$, and $f_J(\Lambda_{(2,4)}) = (0,1,1,2)$.

We obtain the matrix $\hat{V}^J = (1, -1, -1, 1)$. Thus, the map $F : (\mathbb{C}^*)^4 \to \mathbb{C}^*$ is given by $F(z_1, z_2, z_3, z_4) = z_1 z_2^{-1} z_3^{-1} z_4$. It follows that any stratum in the chart M_{12} is the whole $\overset{\circ}{\mathbb{C}^J}$ for $|J| < 4$, while for $|J| = 4$, it is given by the points of the family of surfaces $F^{-1}(c), c \in \mathbb{C}^*$. This family contains two strata determined by the two possible conditions $z_1 z_4 = z_2 z_3$ or $z_1 z_4 \neq z_2 z_3$, depending if the Plücker coordinates P^{34} for the points of a stratum are zero or not. In the first case, it is given by the points in $(\mathbb{C}^*)^4$ of surfaces $z_1 z_4 = z_2 z_3$, while in the second case, it is given by the family, which can be written as, compared to [6]

$$\frac{z_1 z_4}{z_2 z_3} = c, \ c \in \mathbb{C} \setminus \{0, 1\}.$$

Let $\hat{\mu} : W_\sigma / T^n \to \overset{\circ}{P}_\sigma$ be the map induced by the moment map μ. Using the above description of strata in a chart, [7] proves:

Theorem 2.10

1. *All points of a stratum W_σ in $G_{n,k}$ have the same stabilizer $\mathbb{C}^*_\sigma \subset (\mathbb{C}^*)^n$; thus, the torus $(\mathbb{C}^*)^\sigma = (\mathbb{C}^*)^n / \mathbb{C}^*_\sigma$ is the maximal one acting freely on W_σ.*
2. *The map $\hat{\mu} : W_\sigma / T^\sigma \to \overset{\circ}{P}_\sigma$ is a locally trivial fiber bundle whose fiber is an open algebraic manifold F_σ. Thus, there is a canonical trivialization h_σ : $W_\sigma / T^n \cong \overset{\circ}{P}_\sigma \times F_\sigma$.*

It follows that, for the Grassmann manifolds, we have the $\mathbb{T}^n$-invariant, moreover $(\mathbb{C}^*)^n$-invariant, stratification $G_{n,k} = \cup W_\sigma$, which gives the stratification of the orbit space $G_{n,k}/\mathbb{T}^n = \cup W_\sigma/\mathbb{T}^n$. From Theorem 2.10, we obtain the following bijection:

$$G_{n,k}/\mathbb{T}^n = \bigcup_\sigma \overset{\circ}{P}_\sigma \times F_\sigma, \tag{6}$$

where the union is the set-theoretical union. This bijection gives rise to a problem of describing how the components of this union are glued. First note that the polytopes $\overset{\circ}{P}_\sigma$ must be glued together along their common faces. It is clear that the way the spaces of parameters F_σ are glued should be in accordance with the topology of the orbit space $G_{n,k}/\mathbb{T}^n$. One of the approaches to this problem is as follows: since the main stratum W is a dense set in $G_{n,k}$, it follows that the orbit space $G_{n,k}/\mathbb{T}^n$ is a compactification of $\overset{\circ}{\Delta}_{n,k} \times F$. Therefore, we need to find a compactification $\mathcal{F}$ of the space of parameters F such that it is compatible with (6).

This problem can be formulated more precisely as
Problem (★):

- Find a projection

$$\pi_r : \Delta_{n,k} \times \mathcal{F} \to \overline{\Delta_{n,k} \times F} \cong G_{n,k}/\mathbb{T}^n, \tag{7}$$

such that $\mu \circ \pi_r = \pi_1$ for the projection $\pi_1 : \Delta_{n,k} \times \mathcal{F} \to \Delta_{n,k}$.

- For a space $\tilde{F}_\sigma$ determined by

$$\pi_r^{-1}(W_\sigma/\mathbb{T}^n) \cong \overset{\circ}{P}_\sigma \times \tilde{F}_\sigma, \tag{8}$$

find the projection $p_\sigma : \tilde{F}_\sigma \to F_\sigma$ such that the composition

$$\overset{\circ}{P}_\sigma \times \tilde{F}_\sigma \overset{\pi_r}{\to} W_\sigma/T^\sigma \overset{h_\sigma}{\to} \overset{\circ}{P}_\sigma \times F_\sigma$$

coincides with

$$\overset{\circ}{P}_\sigma \times \tilde{F}_\sigma \overset{(I_d, p_\sigma)}{\to} \overset{\circ}{P}_\sigma \times F_\sigma.$$

The solution of this problem gives compactification $\mathcal{F}$ of F with the desired properties. This compactification is called the *universal space of parameters* for the canonical $\mathbb{T}^n$-action on $G_{n,k}$, and the spaces $\tilde{F}_\sigma$ are called the *virtual spaces of parameters* for the strata W_σ.

2.6 Critical and Singular Points

Consider the moment map $\mu : G_{n,k} \to \mathbb{R}^n$. It is a smooth map, and it has critical and regular points and critical and regular values defined in a standard way. It is proved in [7] that the following holds:

- The rank of the differential of the moment map $\mu : G_{n,k} \to \mathbb{R}^n$ at a point L is given by

$$\text{rk}\, d\mu(L) = \dim P_\sigma,$$

where P_σ is an admissible polytope for the stratum W_σ such that $L \in W_\sigma$.
- A point $L \in G_{n,k}$ is a regular point for the moment map $\mu : G_{n,k} \to \mathbb{R}^n$ if and only if its stationary subgroup $T_L \subset \mathbb{T}^n$ is trivial.
- A point $x \in \overset{\circ}{\Delta}_{n,k}$ is a regular value for the moment map $\mu : G_{n,k} \to \mathbb{R}^n$ if and only if the preimage $\mu^{-1}(x)$ consists of regular points only.

The complement to the set of regular points in $G_{n,k}$ is called the set of critical points. This set is $\mathbb{T}^n$-invariant, so one can introduce the notion of a critical point of the orbit space $G_{n,k}/\mathbb{T}^n$.

The regular values of the moment map μ are nicely characterized for general $(2n, k)$-manifold in [10]. In the case of Grassmannians, these properties can be formulated as follows:

- A point x is a regular value of the moment map $\mu : G_{n,k} \to \Delta_{n,k}$ if $\dim P_\sigma = n-1$ for any P_σ such that $x \in \overset{\circ}{P}_\sigma$.
- Consider a point $x \in \overset{\circ}{\Delta}_{n,k}$ such that, for any $P_\sigma \ni x$, $\dim P_\sigma = n-1$. Then the preimage $Q_x = \mu^{-1}(x)$ is a closed smooth submanifold in $G_{n,k}$ of dimension $2k(n-k)-(n-1)$. Consider the torus $T^{n-1} = \mathbb{T}^n/\mathrm{diag}(\mathbb{T}^n)$. Q_x/T^{n-1} is a manifold of dimension $2k(n-k)-2(n-1)$, and, moreover, it is a compactification $\mathcal{F}_x$ of the space of parameters F of the main stratum. Furthermore, $Q_x \cong \mathcal{F}_x \times T^{n-1}$.
- The set of regular values of the moment map $\mu : G_{n,k} \to \Delta_{n,k}$ is a dense set in $\Delta_{n,k}$.
- The manifolds Q_x and Q_y and consequently $\mathcal{F}_x$ and $\mathcal{F}_y$ are diffeomorphic for any x, y belonging to the same connected component of the set of regular values of $\Delta_{n,k}$.

These properties give rise to the following problems:

1. Describe all smooth submanifolds in $G_{n,k}$ with the free T^{n-1}-action that can be realized by Q_x.
2. Describe all smooth submanifolds Q_x/T^{n-1} in $G_{n,k}/\mathbb{T}^n$, and consequently, describe all smooth compactifications $\mathcal{F}_x$ of F in $G_{n,k}/\mathbb{T}^n$.

Remark 2.11 For the Grassmannian $G_{4,2}$, the manifolds $\mathcal{F}_x$ do not depend on a point $x \in \overset{\circ}{\Delta}_{4,2}$, while for the Grassmannian $G_{5,2}$, they do depend on $x \in \overset{\circ}{\Delta}_{5,2}$; see [6, 7].

The tubular neighborhood theorem—see [3]—makes it possible to describe the local structure of the orbit space $G_{n,k}/\mathbb{T}^n$. As explained in [8], any point of $G_{n,k}/\mathbb{T}^n$ that corresponds to a point of $G_{n,k}$ with the trivial stabilizer is a smooth point, while any point of $G_{n,k}/\mathbb{T}^n$ that corresponds to a point of $G_{n,k}$ with a non-trivial stabilizer has a neighborhood with cone-like singularities. In this way, in [6], all smooth and singular points of the orbit space $G_{4,2}/\mathbb{T}^4 \cong S^5$ are described.

Assuming that the problem (★) is solved, to any point $[L]$ of $G_{n,k}/\mathbb{T}^n$ for which $[L] \in W_\sigma/\mathbb{T}^n$, we can assign not only the space of parameters F_σ but also a virtual space of parameters $\tilde{F}_\sigma$ with a projection $p_\sigma : \tilde{F}_\sigma \to F_\sigma$. In [8], the problem (★) is solved for $G_{n,2}/\mathbb{T}^n$, and the projection p_σ is showed *not* to be a homeomorphism in general. As a consequence, in [8], a new notion of a singular point for $G_{n,2}/\mathbb{T}^n$ was introduced.

Definition 2.12 A point $[L] = \mathbb{T}^n \cdot L \in G_{n,2}/\mathbb{T}^n$ is said to be a singular point for the standard $\mathbb{T}^n$-action on $G_{n,2}$ if the space of parameters F_σ of the stratum $W_\sigma \ni L$ is not homeomorphic to the virtual space of parameters $\tilde{F}_\sigma$.

In [8], the following results were proved:

- In the case $n = 4$, all singular points of $G_{4,2}/\mathbb{T}^4$ are critical points, but there are critical points that are not singular. More precisely, all points in $G_{4,2}$ that have exactly two nonzero Plücker coordinates are critical points, which are not singular.

- For $G_{n,2}/\mathbb{T}^n$ with $n \geq 5$, all critical points are singular.
- The set $G_{n,2}/\mathbb{T}^n \setminus \mathrm{Sing}(G_{n,2}/\mathbb{T}^n)$ is an open dense set in $G_{n,2}/\mathbb{T}^n$ and, moreover, a manifold.

2.7 *Orbit Spaces $G_{4,2}/T^4$ and $G_{5,2}/T^5$*

In this subsection, we explain how the notions introduced above can be used to describe the structure of the orbit spaces $G_{4,2}/\mathbb{T}^4$ and $G_{5,2}/\mathbb{T}^5$.

2.7.1 $G_{4,2}/T^4$

The standard action of $\mathbb{T}^4$ on $G_{4,2}$ gives a seminal example of complexity-one torus action. The topology of the orbit space $G_{4,2}/\mathbb{T}^4$ is explicitly described in [6], where the following result is proved:

$$G_{4,2}/T^4 \cong S^5.$$

More precisely, the image of the moment map is the octahedron $\Delta_{4,2}$, and the admissible polytopes are faces Q of octahedron (see the picture below) and six four-sided pyramids P_{ij}, $1 \leq i < j \leq 4$, based in three diagonal squares of the octahedron $K_{ij,pq}$, $\{i, j\} \cap \{p, q\} = \emptyset$, $i, j, p, q \in \{1, \dots 6\}$.

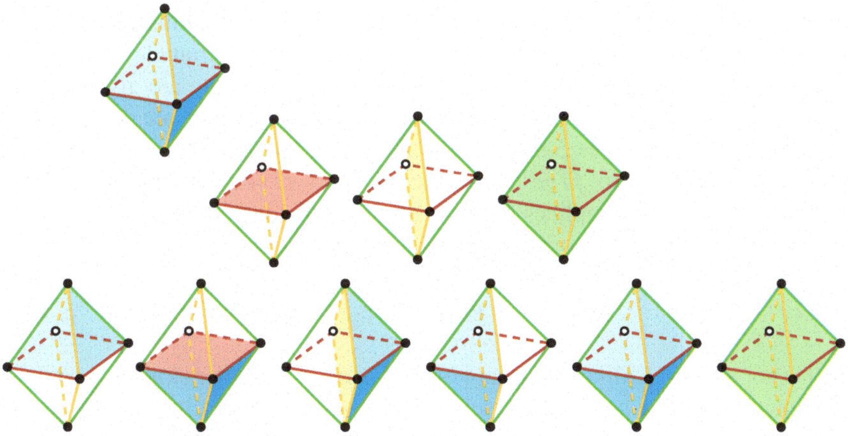

The space of parameters of the main stratum W is $F = \mathbb{C}P^1 \setminus \{(1:0), (0:1), (1:1)\}$, and the space of parameter of any other stratum is a point, so $W/\mathbb{T}^4 \cong \overset{\circ}{\Delta}_{4,2} \times (\mathbb{C}P^1 \setminus \{(1:0), (0:1), (1:1)\})$, and $W_\sigma/\mathbb{T}^4 \cong \overset{\circ}{P}_\sigma$ for any such stratum W_σ. Thus,

$$G_{4,2}/\mathbb{T}^4 = (\overset{\circ}{\Delta}_{4,2} \times(\mathbb{C} \setminus \{0, 1\})) \cup \bigcup_{P_\sigma \neq \Delta_{4,2}} \overset{\circ}{P}_\sigma . \tag{9}$$

We have proved that the compactification of $\overset{\circ}{\Delta}_{4,2} \times(\mathbb{C} \setminus \{0, 1\})$ that is compatible with the bijection (9) gives the universal space of parameters $\mathcal{F} \cong \mathbb{C}P^1$. Furthermore, it gives

$$G_{4,2}/\mathbb{T}^4 \cong (\Delta_{4,2} \times \mathbb{C}P^1)/_{((x,c_1)\sim(y,c_2)\Leftrightarrow x=y\in\partial\Delta_{4,2})} = S^2 * S^2 \cong S^5.$$

Now, it is easy to see that any point of $\overset{\circ}{\Delta}_{4,2} \setminus\{\overset{\circ}{K}_1, \overset{\circ}{K}_2, \overset{\circ}{K}_3\}$, where K_i are the diagonal squares of the octahedron $\Delta_{4,2}$, is a regular value of the moment map. In addition, the singular points of $G_{4,2}/\mathbb{T}^4$ are given by $\hat{\mu}^{-1}(\partial\Delta_{4,2})$.

2.7.2 $G_{5,2}/T^5$

The canonical action of the torus $\mathbb{T}^5$ on $G_{5,2}$ is a seminal example of complexity-two torus action. The topology of the orbit space $G_{5,2}/\mathbb{T}^5$ is described up to homotopy in [7], where this space is proved to be homotopy equivalent to the space X obtained by attaching the disk D^8 to the space $\Sigma^4\mathbb{R}P^2$ along a generator of the group $\pi_7(\Sigma^4\mathbb{R}P^2)$ that is known to be isomorphic to $\mathbb{Z}_4$ [27, 30].

The image of the moment map is $\Delta_{5,2}$, and the family of strata and their admissible polytopes is much wider in this case. The admissible polytopes of dimension 4 are $\Delta_{5,2}$ and 10 polytopes with 9 vertices, 15 polytopes with 8 vertices, and 10 polytopes with 7 vertices. In addition, there are *ten* admissible polytopes inside $\Delta_{5,2}$ of dimension 3, each with *six* vertices, that is, triangle pyramids P_i. The remaining admissible polytopes belong to $\partial\Delta_{5,2}$, and they are given by *five* octahedra, *five* simplices, and their admissible polytopes, which are determined by the fact that $\mu^{-1}(\partial\Delta_{5,2}) = 5\#G_{4,2} \cup 5\#\mathbb{C}P^3$, i.e., a union of *five* disjoint copies of $G_{4,2}$ corresponding to *five* octahedra $\Delta_{4,2}$ and *five* disjoint copies of $\mathbb{C}P^3$ corresponding to *five* simplices Δ_3 that form the boundary of $\Delta_{5,2}$.

The space of parameters of the main stratum is

$$F=\{((c_1 : c_1^{'}), (c_2 : c_2^{'}), (c_3 : c_3^{'})) \in (\mathbb{C}P^1)^3 | c_1c_2^{'}c_3=c_1^{'}c_2c_3^{'},\ c_i, c_i^{'} \neq 0, 1,\ c_i \neq c_i^{'}\},$$

which is not a point. Unlike in the case of $G_{4,2}$, there exist other strata whose spaces of parameters are not points. Precisely, for the strata whose admissible polytopes have *nine* vertices, the spaces of parameters are homeomorphic to $\mathbb{C}P^1 \setminus \{(1 : 0), (0 : 1), (1 : 1)\}$. The strata whose admissible polytopes are the octahedra on the boundary of $\Delta_{5,2}$ have the same space of parameters. For any other strata, its space of parameters is a point.

We proved in [7] that the compactification of $\overset{\circ}{\Delta}_{5,2} \times F$ that is compatible with the bijection (6) is the universal space of parameters $\mathcal{F}$. It is, moreover, a smooth surface obtained by the blow-up of the surface $\bar{F} = \{((c_1 : c_1'), (c_2 : c_2'), (c_3 : c_3')) \in (\mathbb{C}P^1)^3 | c_1 c_2' c_3 = c_1' c_2 c_3'\}$ at the point $((1 : 1), (1 : 1), (1 : 1))$. In birational geometry, this surface is known as the Del Pezzo surface of degree 5; see [34] for details.

All points of $\overset{\circ}{\Delta}_{5,2} \setminus \cup_{i=1}^{10} \overset{\circ}{P}_i$, where P_i are triangle pyramids—see [7]—are regular values of the moment map $\mu : G_{5,2} \to \Delta_{5,2}$. The singular points of $G_{5,2}/\mathbb{T}^5$ in the sense of Definition 2.12 are given by $\hat{\mu}^{-1}(\partial\Delta_{5,2})$ and by those points of the strata whose admissible polytopes are the prisms P_i, $1 \leq i \leq 10$.

2.8 Relation to the Theory of $(2n, k)$-Manifolds

The properties of $\mathbb{T}^n$-action on $G_{n,k}$ that are fundamental for the description of the equivariant topology of the orbit space $G_{n,k}/\mathbb{T}^n$ mostly come from the fact that the $\mathbb{T}^n$-action has an extension to $(\mathbb{C}^*)^n$-action. These properties form the basis of the theory of $(2n, k)$-manifolds [10]. A $(2n, k)$-manifold is a smooth closed oriented simply connected manifold M^{2n} with:

- Smooth effective action θ of the torus T^k on M^{2n}, where $1 \leq k \leq n$ such that the stabilizer of any point is connected
- Open smooth θ-equivariant map $\mu : M^{2n} \to \mathbb{R}^k$ whose image is a k-dimensional convex polytope (here we consider the action of T^k on $\mathbb{R}^k$ to be trivial)

such that it satisfies six axioms formulated in [10]. We emphasize that these axioms actually arise from properties of the $\mathbb{T}^n$-action on $G_{n,k}$ that are crucial for the description of the orbit space $G_{n,k}/\mathbb{T}^n$:

- Axiom 1 is a generalization of the following property of Grassmann manifolds: all charts in the Plücker atlas are $\mathbb{T}^n$-invariant open dense sets each containing exactly one fixed point.
- Axiom 2 is a generalization of the following property of Grassmann manifolds: the moment map gives a bijection between the set of fixed points and the set of vertices.
- Axiom 3 is a generalization of the following property of Grassmann manifolds: all points of a stratum W_σ have the same stabilizer $T_\sigma \subset T^n$ (see the first statement of Theorem 2.10).
- Axiom 4 is a generalization of the second statement of Theorem 2.10.
- Axiom 5 is a generalization of the properties of $(\mathbb{C}^*)^n$-orbits of a stratum given by Theorem 2.5.
- Axiom 6 generalizes notions of the universal space of parameters $\mathcal{F}$ and virtual spaces of parameters $\tilde{F}_\sigma$ of the strata, together with projections $p_\sigma : \tilde{F}_\sigma \to F_\sigma$.

The class of $(2n, k)$-manifolds comprises, among others, quasitoric manifolds, which are $(2n, n)$-manifolds; complex Grassmann manifolds $G_{n,k}$, which are

$(2k(n-k), n-1)$-manifolds; the complete complex flag manifolds F_n, which are $(n(n-1), n-1)$-manifolds; etc.

The notions of strata and admissible polytopes for a $(2n, k)$-manifold are defined analogous to the case of Grassmann manifolds. The notion of the space of parameters of a stratum generalizes the notion of the space of parameters of a stratum in the case of $G_{n,k}$ and is defined by Axiom 3 since the T^k-action on a $(2n, k)$-manifold in general *does not* have an extension to $(\mathbb{C}^*)^k$-action. Examples of $(2n, k)$-manifolds for which T^k-action is not in general induced by an $(\mathbb{C}^*)^k$-action are quasitoric manifolds M^{2n} and standard spheres S^{2n}, which are $(2n, 1)$-manifolds.

Using these axioms from [10], one can construct a model for the orbit space M^{2n}/T^k for an arbitrary $(2n, k)$-manifold M^{2n}. The complex of admissible polytopes $C(M^{2n}, P^k)$ is defined as the formal union of all admissible polytopes. The canonical map $\hat{\pi} : C(M^{2n}, P^k) \to P^k$ defines the canonical map $f : M^{2n} \to C(M^{2n}, P^k)$ by $\hat{\pi} \circ f = \mu$, so, with the help of the map f, we endow $C(M^{2n}, P^k)$ with the quotient topology. We denote $P_\sigma^{'}$ the relative interiors of the polytopes from the complex $C(M^{2n}, P^k)$ to distinguish them from their copies in P^k. Now, consider the space

$$\mathcal{E} = \cup P_\sigma^{'} \times \tilde{F}_\sigma \subset C(M^{2n}, P^k) \times \mathcal{F}.$$

There is a natural map $H : \mathcal{E} \to M^{2n}/T^k = \cup W_\sigma/T^\sigma = \cup \overset{\circ}{P}_\sigma \times F_\sigma$ defined by $H(x_\sigma, \tilde{c}_\sigma) = (x_\sigma, p_\sigma(\tilde{c}_\sigma))$, and Axiom 6 requires it to be a continuous map.

In [10], $(\mathcal{E}, H)$ is proved to be a model for M^{2n}/T^k, that is,

Theorem 2.13 *The orbit space M^{2n}/T^k is homeomorphic to the quotient space of $\mathcal{E}$, and the homeomorphism is given by the map H.*

Example 2.14 Let us consider the space F_3 of complete flags in $\mathbb{C}^3$. In [10], we have applied methods of the theory of $(2n, k)$-manifolds to prove that $F_3/\mathbb{T}^3 \cong S^4$. The $\mathbb{T}^3$-action on F_3 induces an effective action of T^2 on F_3, which is an example of complexity-one torus action. The ideas of the theory of $(2n, k)$-manifolds were also exploited in [6] to describe the orbit space $\mathbb{C}P^5/T^4 \cong S^2 * \mathbb{C}P^2$, which is an example of complexity-two torus action. Here the action of the torus T^4 on $\mathbb{C}P^5$ is given by the second exterior power embedding of T^4 into T^6 and the canonical action of T^6 on $\mathbb{C}P^5$.

3 $\mathbb{T}^n$-Action on Grassmann Manifolds $G_{n,2}$

Among all Grassmann manifolds $G_{n,k}$, the manifolds $G_{n,2}$ stand out, first because of their application in mathematics and mathematical physics, but also because of regularity of their stratification, which we previously defined in Sect. 2.4. The word "regularity" here means the following: for a Grassmann manifold $G_{n,k}$, $1 \leq k \leq$

$n-1$, the boundary of any stratum W_σ is contained in the union of the strata $W_{\bar{\sigma}}$ such that $\bar{\sigma} \subset \sigma$, that is,

$$\partial W_\sigma \subseteq \bigcup_{\bar{\sigma}} W_{\bar{\sigma}}. \tag{10}$$

A critical property of this inclusion is that it is generally strict. Gel'fand and Serganova [19] (see also [10]) showed that, in the case of the Grassmann manifold $G_{7,3}$, one can find and describe explicitly a stratum W_σ such that:

- Its space of parameters is a point, meaning that it consists of exactly one $(\mathbb{C}^*)^7$-orbit.
- There exists a stratum $W_{\bar{\sigma}}$ for some $\bar{\sigma} \subset \sigma$ whose space of parameters is *not* a point.
- The boundary ∂W_σ has non-empty intersection with $W_{\bar{\sigma}}$.

For such a stratum, inclusion (10) is strict.

As shown in [19], this does not happen to the strata in Grassmannians $G_{n,2}$, $n \geq 2$, that is, in this case, inclusion (10) is an equality. It is in this sense that the equivariant structure of any Grassmannian $G_{n,2}$ is regular.

The $\mathbb{T}^n$-action on Grassmannians $G_{n,2}$ is studied in detail in [8], where we have constructed a nice, almost smooth model for its orbit space.

The image of the moment map $\mu : G_{n,2} \to \mathbb{R}^n$ is the hypersimplex $\Delta_{n,2}$, and its boundary is the union of two families of polytopes, the first of which consists of n copies of the hypersimplex $\Delta_{n-1,2}$ and the second one of n copies of the simplex Δ^{n-2}. In other words,

$$\mu^{-1}(\partial \Delta_{n,2}) = n\#G_{n-1,2} \cup n\#\mathbb{C}P^{n-2}.$$

Therefore, the equivariant topology of the Grassmann manifold $G_{n,2}$ can be inductively described in the class of $G_{m,2}$-manifolds, $3 \leq m \leq n$, by studying the equivariant topology of $\mu^{-1}(\overset{\circ}{\Delta}_{m,2}) \subset G_{n,2}$.

The induction step is based on describing all admissible polytopes that intersect $\overset{\circ}{\Delta}_{n,2}$. In [8], we have proved that these admissible polytopes can be described using nice arrangement of hyperplanes in $R^{n-1} = \{(x_1, \ldots, x_n) \in \mathbb{R}^n | x_1 + \ldots + x_n = 2\}$. In [8], we have proven:

- All admissible polytopes with dimension $\leq n-3$ belong to the boundary $\partial \Delta_{n,2}$.
- The admissible polytopes of dimension $n-2$ that have non-empty intersection with $\overset{\circ}{\Delta}_{n,2}$ are given by the intersection of $\Delta_{n,2}$ with the planes of the form

$$\sum_{i \in S, \|S\|=p} x_i = 1, \quad \text{where } S \subset \{1, \ldots, n\},\ 2 \leq p \leq [\frac{n}{2}]. \tag{11}$$

- An admissible polytope P_σ such that $\dim P_\sigma = n-1$ is either $\Delta_{n,2}$ or the intersection of $\Delta_{n,2}$ with the collection $\mathcal{H} = \{H_{S_1}, \ldots, H_{S_l}\}$ consisting of the half-spaces of the form

$$H_S : \sum_{i \in S} x_i \leq 1, \quad S \subset \{1, \ldots, n\}, \ \|S\| = k, \ 2 \leq k \leq n-2,$$

  such that $S_i \cap S_j = \emptyset$ whenever $H_{S_1}, H_{S_2} \in \mathcal{H}$.

This shows that it is important here to consider the hyperplane arrangement Π that consists of the planes given by the Eq. (11). Now, consider a hyperplane arrangement in $R^{n-1} = \{(x_1, \ldots, x_n) \in \mathbb{R}^n | x_1 + \ldots + x_n = 2\}$ given by

$$\mathcal{A}_n = \Pi \cup \{x_i = 0, \ 1 \leq i \leq n\} \cup \{x_i = 1, \ 1 \leq i \leq n\}$$

and the face lattice $L(\mathcal{A}_n)$ of this arrangement. The intersection $L(\mathcal{A}_{n,2}) = L(\mathcal{A}_n) \cap \Delta_{n,2}$ produces a decomposition of $\Delta_{n,2}$ into closed polytopes, whose faces belong to the lattice $L(\mathcal{A}_{n,2})$. The decomposition of $\Delta_{n,2}$ into interiors of these polytopes and the interiors of their faces we call a *chamber decomposition* and denote by $\overset{\circ}{L}(\mathcal{A}_{n,2})$. We call an element $C \in \overset{\circ}{L}(\mathcal{A}_{n,2})$ a *chamber*. In [8], we have proven:

- The chamber decomposition $\overset{\circ}{L}(\mathcal{A}_{n,2})$ induces on $\overset{\circ}{\Delta}_{n,2}$ the decomposition which coincides with the decomposition given by the intersections of relative interiors of all admissible polytopes in $\overset{\circ}{\Delta}_{n,2}$.

We denote the chambers from $\overset{\circ}{L}(\mathcal{A}_{n,2}) \cap \overset{\circ}{\Delta}_{n,2}$ by C_ω, where ω consists of all σ such that $C_\omega \subset \overset{\circ}{P}_\sigma$. For the preimages of chambers $\hat{C}_\omega = \mu^{-1}(C_\omega)$, we have proven that:

- There exists a canonical homeomorphism

$$h_\omega : \hat{C}_\omega \to C_\omega \times F_\omega,$$

  where F_ω is a compactification of the space of parameters F of the main stratum, which is given by the spaces F_σ such that $C_\omega \subset \overset{\circ}{P}_\sigma$.

When $\dim C_\omega = n-1$, this follows from [7], where the spaces $\hat{\mu}^{-1}(x) = \cup_{\sigma \in \omega} F_\sigma \subset G_{n,2}/\mathbb{T}^n$ are showed to be smooth manifolds, diffeomorphic for all $x \in C_\omega$, that is, diffeomorphic to some manifold F_ω. For a chamber of an arbitrary dimension, it follows from an observation from [21].

The main stratum is given by the following system of equations in the local coordinates $(z_3, \ldots, z_n, w_3, \ldots, w_n)$ of an arbitrary chart:

$$c_{ij}^{'} z_i w_j = c_{ij} z_j w_i, \quad 3 \leq i < j \leq n,$$

where the parameters are $(c_{ij} : c'_{ij}) \in \mathbb{C}P^1$, $c_{ij}, c'_{ij} \neq 0$, and $c_{ij} \neq c'_{ij}$. The number of parameters is $N = \binom{n-2}{2}$. The parameters satisfy the relations

$$c_{ij}c'_{il}c_{jl} = c'_{ij}c_{il}c'_{jl}, \quad 3 \leq i < j < l \leq n. \tag{12}$$

We have obtained an embedding of the orbit space $F = W/(\mathbb{C}^*)^n$ into $(\mathbb{C}P^1)^N$. Its image is an open algebraic manifold in $(\mathbb{C}P^1)^N$ given by the intersection of cubic surfaces (12) and the conditions that $(c_{ij}; c'_{ij}) \in \mathbb{C}P^1_A = \mathbb{C}P^1 \setminus A$, where $A = \{(1:0), (0:1), (1:1)\}$.

For the space of parameters F_σ of an arbitrary stratum W_σ such that $\overset{\circ}{P}_\sigma \subset \overset{\circ}{\Delta}_{n,2}$:

- If $\dim P_\sigma = n - 2$, then F_σ is a point.
- If $\dim P_\sigma = n - 1$, and F_σ is not a point, then the space F_σ is obtained by restricting the intersection of the cubic surfaces (12) to some q factors $\mathbb{C}P^1_B = \mathbb{C}P^1 \setminus B$ in $(\mathbb{C}P^1)^N$, where $B = \{(1:0), (0:1)\}$, and $0 \leq q \leq l$.

As for the universal space of parameters, we have proven [8] that the space $\mathcal{F}_n$, which is explicitly described in [9] by the techniques of the wonderful compactification, is the universal space of parameters for the canonical $\mathbb{T}^n$-action on $G_{n,2}$. To elaborate, we start from the closure $\bar{F}$ of the space F in $(\mathbb{C}P^1)^N$, that is,

$$\bar{F} = \{((c_{ij} : c'_{ij})) \in (\mathbb{C}P^1)^N | c_{ij}c'_{il}c_{jl} = c'_{ij}c_{il}c'_{jl}, \ 3 \leq i < j < l \leq n\},$$

which is a smooth variety [9]. Consider the subvarieties given by

$$\hat{F}_I = \bar{F}_n \cap \{(c_{ik} : c'_{ik}) = (c_{il} : c'_{il}) = (c_{kl} : c'_{kl}) = (1:1)\}, \tag{13}$$

for $I = \{i, k, l\} \in \{I \subset \{1, \dots, n\}, \ |I| = 3\}$ and $n \geq 5$.

In terminology of wonderful compactification [9, 31], one takes the building set $\mathcal{G}_n$ to be:

- $\mathcal{G}_n = \emptyset$ for $n = 4$.
- $\mathcal{G}_n = \{G = \bigcap_I \hat{F}_I \subset \bar{F}_n\}$, that is, all possible non-empty intersections of $\hat{F}_I$'s.

Following [9], denote by $\mathcal{F}_n$ a smooth, compact manifold, which is obtained as the wonderful compactification of $\bar{F}$ with the building set $\mathcal{G}_n$. In [8], we have proven that $\mathcal{F}_n$ is a universal space of parameters for the canonical $\mathbb{T}^n$-action on $G_{n,2}$.

Having a universal space of parameters $\mathcal{F}_n$, to any stratum W_σ, one can assign a virtual space of parameters $\tilde{F}_\sigma \subset \mathcal{F}_n$. In [8], these virtual spaces of parameters are proved to have the following very nice properties:

- For $x \in \overset{\circ}{\Delta}_{n,2}$, denote by

$$\tilde{F}_x = \bigcup_{x \in \overset{\circ}{P}_\sigma} \tilde{F}_\sigma .$$

Then

$$\tilde{F}_x = \mathcal{F}_n \text{ for any } x \in \overset{\circ}{\Delta}_{n,2} .$$

- For any chamber C_ω, one has $\tilde{F}_\sigma \cap \tilde{F}_{\sigma'} = \emptyset$ whenever $\sigma, \sigma' \in \omega$.
- The union

$$\mathcal{F}_n = \bigcup_{C_\omega \subset \overset{\circ}{P}_\sigma} \tilde{F}_\sigma , \tag{14}$$

is a disjoint union for any chamber C_ω.

Since, by definition, for any virtual space of parameters $\tilde{F}_\sigma$, there exists a projection $p_\sigma : \tilde{F}_\sigma \to F_\sigma$, it follows from (14) that for any chamber C_ω, there exists a projection

$$p_\omega : \mathcal{F}_n \to F_\omega \text{ defined by } p_\omega(y) = p_\sigma(y) \text{ for } y \in \tilde{F}_\sigma . \tag{15}$$

Denote by $\overset{\circ}{G}_{n,2} = \mu^{-1}(\overset{\circ}{\Delta}_{n,2})$. The following disjoint decomposition holds:

$$\overset{\circ}{G}_{n,2} /\mathbb{T}^n \cong \bigcup_\omega \hat{C}_\omega \cong \bigcup_\omega (C_\omega \times F_\omega). \tag{16}$$

where the topology on the right-hand side is given by the induced moment map $\hat{\mu} : G_{n,2}/\mathbb{T}^n \to \Delta_{n,2}$ and the natural projection $G_{n,2}/\mathbb{T}^n \to G_{n,2}/(\mathbb{C}^*)^n$.

Altogether, we have

$$G_{n,2}/\mathbb{T}^n \cong \overset{\circ}{G}_{n,2} /\mathbb{T}^n \cup (\bigcup_{q=1}^{n} G_{n-1,2}(q)/T^{n-1}) \cup (\bigcup_{q=1}^{n} \Delta^{n-2}(q)). \tag{17}$$

The topology on the right-hand side of (17) is defined by the canonical embeddings $G_{n-1,2}(q) \to G_{n,2}$ and $\mathbb{C}P^{n-2}(q) \to G_{n,2}$, $1 \le q \le n$.

The universal space of parameters for $\mathbb{C}P^{n-2}(q)$, $1 \le q \le n$, is a point. The canonical embeddings $\hat{i}_q : G_{n-1,2}(q) \to G_{n,2}$ are defined by the inclusions $i_q : \mathbb{C}^{n-1} \to \mathbb{C}^n$, $i_q(z_1, \ldots, z_{n-1}) = (z_1, \ldots z_{q-1}, 0, z_q, \ldots z_{n-1})$, $1 \le q \le n$. Therefore, the relation between the universal spaces of parameters $\mathcal{F}_n$ for $G_{n,2}$ and $\mathcal{F}_{n-1,q}$ for $G_{n-1,2}(q)$, $1 \le q \le n$, is given by

$$\mathcal{F}_{n-1,q} = \mathcal{F}|_{\{(c_{ij}:c'_{ij}), i,j \neq q\}}, \tag{18}$$

which defines the restriction $r_q : \mathcal{F}_n \to \mathcal{F}_{n-1,q}$.

Thus, all previous constructions apply to $\mathcal{F}_{n-1,q}$ and $\Delta_{n-1,2}(q) \subset \partial\Delta_{n,2}$, which is defined by $\Delta_{n-1,2}(q) = \Delta_{n,2} \cap \{x_q = 0\}$, $1 \leq q \leq n$. Denote by $p_\omega^q : \mathcal{F}_{n-1,q} \to F_\omega$ the map given by (15) for the Grassmannian $G_{n-1,2}(q)$, $1 \leq q \leq n$. Altogether, we obtain:

- For any chamber $C_\omega \subset \partial\Delta_{n,2}$, one can define the projection $p_\omega : \mathcal{F}_n \to F_\omega$. If $C_\omega \subset \Delta^{n-1}(q)$, this projection maps $\mathcal{F}_n$ to a point, while for $C_\omega \subset \Delta_{n-1,2}(q)$, this projection is defined by $p_\omega(y) = (p_\omega^q \circ r_q)(y)$.

  Consider the space

$$U_n = \Delta_{n,2} \times \mathcal{F}_n. \tag{19}$$

One can inductively define a projection

$$U_n \to (\overset{\circ}{\Delta}_{n,2} \times \mathcal{F}_n) \cup (\bigcup_{q=1}^{n} U_{n-1,q}) \cup (\bigcup_{q=1}^{n} \Delta^{n-2}(q)), \tag{20}$$

for $U_{n-1,q} = \Delta_{n-1,2}(q) \times \mathcal{F}_{n-1,q}$, which is given by $(x, f) \to (x, r_q(f))$ if $x \in \Delta_{n-1,2}(q)$ and by $(x, f) \to x$ if $x \in \Delta^{n-2}(q)$, where $1 \leq q \leq n$.

Altogether, we obtain:

Theorem 3.1 *The map*

$$G : U_n \to G_{n,2}/\mathbb{T}^n, \quad G(x, y) = h_\omega^{-1}(x, p_\omega(y)) \ \textit{if and only if} \ x \in C_\omega, \tag{21}$$

is correctly defined. Moreover, the map G is a continuous surjection, and the orbit space $G_{n,2}/\mathbb{T}^n$ is homeomorphic to the quotient of the space U_n by the map G.

4 Universal Space of Parameters for $G_{n,2}$ and the Chow Quotient

In [9], using the techniques of wonderful compactification, an explicit construction of the space $\mathcal{F}_n$, coming purely from the equivariant topology of the Grassmann manifolds $G_{n,2}$, is provided. The authors obtain a smooth, compact manifold diffeomorphic to the moduli space $\overline{\mathcal{M}}(0, n)$ of genus zero stable curves with n-marked distinct points, that is, to the Chow quotient $G_{n,2}/\!/(\mathbb{C}^*)^n$.

The idea for this construction comes from a problem we have already discussed: knowing that $\overline{W/\mathbb{T}^n} = G_{n,2}/\mathbb{T}^n$ for the main stratum W and $W/\mathbb{T}^n \cong \overset{\circ}{\Delta}_{n,2} \times F$, find a compactification of the space of parameters F of the main stratum that is compatible with the decomposition of the orbit space $G_{n,2}/\mathbb{T}^n = \cup \overset{\circ}{P}_\sigma \times F_\sigma$. To do that, we first note the following property. Consider a stratum W_σ in a chart M_{ij}

and assign to it the space $\tilde{F}_{\sigma,ij} \subset \mathcal{F}_n$ using the fact that $W_\sigma/\mathbb{T}^n$ is in the boundary of $W_n/\mathbb{T}^n \cong \overset{\circ}{\Delta}_{n,2} \times F$. Then the space $\tilde{F}_{\sigma,ij}$ must not depend on a chart $M_{ij} \supset W_\sigma$. Thus, in order to describe $\mathcal{F}_n$, we must start by considering $\bar{F}_n \subset (\mathbb{C}P^1)^N$ and all possible $\tilde{F}_{\sigma,ij} \subset \bar{F}_n$ and make corrections along those $\tilde{F}_{\sigma,ij}$ that are not independent of a chart M_{ij}.

While realizing this approach in [9], the authors note that the main stratum W belongs to any chart of the Plücker atlas for $G_{n,2}$, so the transition maps between the charts induce the family $\mathcal{A}$ of automorphisms of the space of parameters F. This family of automorphisms is explicitly described in [9]. The idea of how to overcome the problem of dependence of the virtual spaces of parameters on the charts is the following: find a compactification $\mathcal{F}_n$ for F such that any automorphism from $\mathcal{A}$ extends to an automorphism of $\mathcal{F}_n$. In that sense, subvarieties in the smooth variety $\bar{F} \subset (\mathbb{C}P^1)^N$ given by

$$\hat{F}_I = \bar{F}_n \cap \{(c_{ik} : c_{ik}^{'}) = (c_{il} : c_{il}^{'}) = (c_{kl} : c_{kl}^{'}) = (1 : 1)\}, \tag{22}$$

for $I = \{i, k, l\} \in \{I \subset \{1, \dots, n\},\ |I| = 3\}$ and $n \geq 5$, are showed [9] to represent the singularities of the extension of the automorphisms from the family $\mathcal{A}$.

In order to resolve these singularities, in [9], the authors apply the method of wonderful compactification with the building set $\mathcal{G}_n$, described in Sect. 3 of the present survey, and obtain a compact smooth manifold $\mathcal{F}_n$, for which all automorphisms of $\mathcal{A}$ are proved to extend to the automorphisms of $\mathcal{F}_n$.

Further on, using the description of the space $\overline{\mathcal{M}}(0, n)$ given by Keel [26], the spaces $\mathcal{F}_n$ and $\overline{\mathcal{M}}(0, n)$ are proved to be diffeomorphic.

On the other hand, there is, due to Kapranov [24], a well-known construction from algebraic geometry known as the Chow quotient. The idea behind the definition of the Chow quotient is the construction from algebraic geometry known as the Chow variety, which is a compact variety whose points parametrize algebraic cycles in a given variety of the same dimension and degree. The corresponding Chow variety $C_{2(n-1)}(G_{n,k}, \delta)$ of a Grassmann manifold $G_{n,k}$ consists of all algebraic cycles in $G_{n,k}$ of dimension $2(n-1)$ whose homology class is δ, where $\delta \in H_{2(n-1)}(G_{n,k}, \mathbb{Z})$ is the homology class of the closure of a generic $(\mathbb{C}^*)^n$-orbit in $G_{n,k}$.

By definition, the Chow quotient $G_{n,k}//(\mathbb{C}^*)^n$ is the closure of the image of the natural map

$$W/(\mathbb{C}^*)^n \to C_{2(n-1)}(G_{n,k}, \delta), \quad x \to \overline{(\mathbb{C}^*)^n \cdot x},$$

where W is the main stratum in $G_{n,k}$.

The build-up or outgrow components to $W/(\mathbb{C}^*)^n$ in $G_{n,k}//(\mathbb{C}^*)^n$ are described in [24] using the fact that:

- The algebraic cycles forming Chow quotient $G_{n,k}//(\mathbb{C}^*)^n$ are of the form $Z = \sum_i Z_i$, where Z_i are the closures of $(\mathbb{C}^*)^n$-orbits in $G_{n,k}$ such that the matroid polytopes $\mu(Z_i)$ give a polyhedral decomposition of $\Delta_{n,k}$.

Here the matroid polytopes defined in [24] in the case of $G_{n,2}$ coincide with our admissible polytopes. Recall that the admissible polytopes for $G_{n,2}$ can be described as well in terms of hyperplane arrangement Π given by (11). In our terminology, the Chow quotient actually gives a compactification of the space of parameters F_n of the main stratum.

In [9], the authors establish a relation between a chamber C_ω in $\Delta_{n,2}$ such that $\dim C_\omega = n-1$ and the Chow quotient $G_{n,2}//(\mathbb{C}^*)^n$. Let $\mathcal{P}_\sigma = \{\mathcal{P}_{\sigma,1}, \dots, \mathcal{P}_{\sigma,s}\} \subset \mathcal{P}$ be the set of all decompositions $\mathcal{P}_{\sigma,i}$ of $\Delta_{n,2}$ that contain P_σ. Let $\tilde{Z}_{\sigma,i} \subset G_{n,2}//(\mathbb{C}^*)^n$ be the family of algebraic cycles determined by the decompositions $\mathcal{P}_{\sigma,i}$. These cycles are of the form

$$Z_{\sigma_{i_1}}(c_{\sigma_{i_1}}) + \dots + Z_{\sigma,i_q}(c_{\sigma_{i_q}})$$

for $(c_{\sigma_{i_1}}, \dots, c_{\sigma_{i_q}}) \in F_{\sigma_{i_1}} \times \dots \times F_{\sigma_{i_q}}$, where $Z_{\sigma_{i_j}}(c_{\sigma_{i_j}})$ is the closure of an algebraic torus orbit in $W_{\sigma_{i_j}}$, and this orbit is determined by $c_{\sigma_{i_j}} \in F_{\sigma_{i_j}} = W_{\sigma_{i,j}}/(\mathbb{C}^*)^n$. Let further

$$\tilde{Z}_\sigma = \bigcup_{i=1}^{s} \tilde{Z}_{\sigma,i}.$$

In [9], we have proven that:

- Any C_ω, $\dim C_\omega = n-1$, defines a decomposition of $G_{n,2}//(\mathbb{C}^*)^n$ into a disjoint union, that is,

$$\bigcup_{\sigma\in\omega} \tilde{Z}_\sigma = G_{n,2}//(\mathbb{C}^*)^n. \tag{23}$$

Remark 4.1 Since we have proven that the universal space of parameters $\mathcal{F}_n$ and $\overline{\mathcal{M}}(0,n)$ are diffeomorphic, and the result of [24] gives that the manifolds $G_{n,2}//(\mathbb{C}^*)^n$ and $\overline{\mathcal{M}}(0,n)$ are isomorphic, it follows that our universal space of parameters $\mathcal{F}_n$ describes the topology of the result of gluing of the outgrow components to F_n in $G_{n,2}//(\mathbb{C}^*)^n$. Using the description of $\mathcal{F}_n$ by the techniques of wonderful compactification, in [9], these outgrow components are explicitly described for $n=4$ and $n=5$.

Example 4.2 In the easiest case of $n=4$, the outgrow components of $G_{4,2}//(\mathbb{C}^*)^4$ consist of three points. These three points correspond to the three decompositions of $\Delta_{4,2}$ into two *four*-sided pyramids; see 2.7.1. These points glue together to $F_4 \cong \mathbb{C}P^1_A$ in $G_{4,2}//(\mathbb{C}^*)^4$ to give the universal space of parameters $\mathbb{C}P^1$. This case is also discussed in [24], but this observation independently follows from [6] and the identification between $\mathcal{F}_4$ and $G_{4,2}//(\mathbb{C}^*)^4$.

References

1. M.F. Atiyah, Convexity and commuting Hamiltonians. Bull. London Math. Soc. **14**(1), 1–15 (1982)
2. A. Ayzenberg, Torus actions of complexity 1 and their local properties. Proc. Steklov Inst. Math. **302**(1), 16–32 (2018). Tr. Mat. Inst. Steklova **302** (2018)
3. G. Bredon, *Introduction to Compact Transformation Groups* (Academic Press, New York, 1972)
4. V.M. Buchstaber, T.E. Panov, *Torus Actions and Their Applications in Topology and Combinatorics*. University Lecture Series 24 (American Mathematical Society, Providence, 2002)
5. V.M. Buchstaber, T.E. Panov, *Toric Topology*. Mathematical Surveys and Monographs, vol. 204 (American Mathematical Society, Providence, 2015)
6. V.M. Buchstaber, S. Terzić, Topology and geometry of the canonical action of T^4 on the complex Grassmannian $G_{4,2}$ and the complex projective space $\mathbb{C}P^5$. Mosc. Math. J. **16**(2), 237–273 (2016)
7. V.M. Buchstaber, S. Terzić, Toric topology of the complex Grassmann manifolds. Mosc. Math. J. **19**(3), 397–463 (2019)
8. V.M. Buchstaber, S. Terzić, A resolution of singularities for the orbit spaces $G_{n,2}/T^n$. arXiv:2009.01580 (2020)
9. V.M. Buchstaber, S. Terzić, The orbit spaces $G_{n,2}/T^n$ and the Chow quotients $G_{n,2}//(\mathbb{C}^*)^n$ of the Grassmann manifolds $G_{n,2}$. arXiv:2104.08858 (2021)
10. V.M. Bukhshtaber, S. Terzich, The foundations of $(2n, k)$-manifolds. Mat. Sb. **210**(4), 41–86 (2019); Sb. Math. **210** (Russian)
11. W.-L. Chow, On the geometry of algebraic homogeneous spaces. Ann. Math. (2) **50**, 32–67 (1949)
12. M.W. Davis, T. Januszkiewicz, Convex polytopes, Coxeter orbifolds and torus actions. Duke Math. J. **62**, 417–451 (1991)
13. C. De Concini, G. Gaiffi, Projective wonderful models for toric arrangements. Adv. Math. **327**, 390–409 (2018)
14. C. De Concini, G. Gaiffi, Cohomology rings of compactifications of toric arrangements. Algebraic Geom. Topol. **19**, 503–532 (2019)
15. C. De Concini, G. Gaiffi, O. Papini, On projective wonderful models for toric arrangements and their cohomology. Eur. J. Math. **6**, 790–816 (2020)
16. C. De Concini, C. Procesi, Wonderful models of subspace arrangements. Sel. Math. New Ser. **1**(3), 459–494 (1995)
17. C. De Concini, C. Procesi, Hyperplane arrangements and holonomy equations. Sel. Math. New Ser. **1**, 495–535 (1995)
18. R. Fulton, R. MacPherson, A compactification of configuration space. Ann. Math. **139**, 183–225 (1994)
19. I.M. Gelfand, V.V. Serganova, Combinatoric geometry and torus strata on compact homogeneous spaces. Russ. Math. Survey **42**(2)(254), 108–134 (1987)
20. I.M. Gelfand, R.M. Goresky, R.D. MacPherson, V.V. Serganova, Combinatorial geometries, convex polyhedra, and schubert cells. Adv. Math. **63**, 301–316 (1987)
21. M. Goresky, M. MacPherson, *On the Topology of Algebraic Torus Actions*. Algebraic Groups. Porc. Symp., Utrecht/Netherlands 1986. Lecture Notes in Mathematics, vol. 1271 (1987), pp. 73–90
22. V. Guillemin, V. Sternberg, Convexity properties of the moment mapping. Invent. Math. **67**, 491–513 (1982)
23. G. Kalai, *Polytope Skeletons and Paths*. Handbook of Discrete and Computational Geometry, 2nd edn. Chapter 20 (CRC Press, Boca Raton, 2004)
24. M.M. Kapranov, *Chow quotients of Grassmannians I*, ed. by I.M. Gel'fand Seminar. Advances in Soviet Mathematics, vol. 16, Part 2 (American Mathematical Society, Providence, 1993), pp. 29–110

25. Y. Karshon, S. Tolman, Topology of complexity one quotients. Pacific J. Math. **308**(2), 333–346 (2020)
26. S. Keel, Intersection theory of moduli space of stable n-pointed curves of genus zero. Trans. Am. Math. Soc. **330**(2), 545–574 (1992)
27. M.A. Kervaire, J. Milnor, Groups of homotopy spheres, I. Ann. Math. **77**, 504–537 (1963)
28. F.C. Kirwan, *Cohomology of Quotients in Symplectic and Algebraic Geometry* (Princeton University Press, Princeton, 1984)
29. S.L. Kleiman, D. Laksov, Schubert calculus. Am. Math. Monthly **79**(10), 1061–1082 (1972)
30. J.P. Levine, Lectures on groups of homotopy spheres, in *Algebraic and Geometric Topology*. Lecture Notes in Mathematics, vol. 1126 (Springer, Berlin, 1985), pp. 62–95
31. L. Li, Wonderful compactification of an arrangement of subvarieties. Michigan Math. J. **58**(2), 535–563 (2009)
32. D. Luna, Th. Vust, Plongements d'espaces homog'enes. Comment. Math. Helv. **58**, 186–245 (1983)
33. M. Noji, K. Ogiwara, The smooth toric orbit closures in the Grassmannians. Proc. Steklov Inst. Math. **305**, 251–261 (2019)
34. H. Suess, Toric topology of the Grassmannian of planes on $\mathbb{C}^5$ and the del Pezzo surface of degree 5. Mosc. Math. J. **21**(3), 639–652 (2021)
35. D.A. Timashev, *Homogeneous Spaces and Equivariant Embeddings*. Encyclopedia of Mathematical Sciences (Springer, Berlin, 2011), 245 pp.

On the Enumeration of Fano Bott Manifolds

Yunhyung Cho, Eunjeong Lee, Mikiya Masuda, and Seonjeong Park

1 Introduction

A Bott manifold of complex dimension n is a smooth projective toric variety whose fan is the normal fan of a polytope combinatorially equivalent to the n-dimensional unit cube $[0, 1]^n$. A family of Bott manifolds was first considered by Grossberg and Karshon [9] in the context of toric degenerations of Bott–Samelson varieties. Since then, topological or geometric properties of Bott manifolds have been intensively studied in [2, 4, 6, 7]. Recently, motivated by Suyama's work [16], we showed the c_1-cohomological rigidity for Fano Bott manifolds, which means that two Fano Bott manifolds are isomorphic if and only if there is a ring isomorphism between their cohomology rings preserving their first Chern classes [5].

It is known that there are only finitely many smooth Fano toric varieties up to isomorphism in each dimension (cf. [14]), and therefore, there are also only finitely many Fano Bott manifolds up to isomorphism in each dimension. Higashitani and Kurimoto [12] associate *signed rooted forests* with Fano Bott manifolds to

Y. Cho
Department of Mathematics Education, Sungkyunkwan University, Seoul, Republic of Korea
e-mail: yunhyung@skku.edu

E. Lee (✉)
Center for Geometry and Physics, Institute for Basic Science, Pohang, Republic of Korea

Department of Mathematics, Chungbuk National University, Cheongju, Republic of Korea
e-mail: eunjeong.lee@chungbuk.ac.kr

M. Masuda
Osaka Central Advanced Mathematical Institute, Osaka, Japan

S. Park
Department of Mathematics Education, Jeonju University, Jeonju, Republic of Korea
e-mail: seonjeongpark@jj.ac.kr

A. Bahri et al. (eds.), *Toric Topology and Polyhedral Products*, Fields Institute Communications 89, https://doi.org/10.1007/978-3-031-57204-3_6

classify Fano Bott manifolds up to diffeomorphism. In this paper, we enumerate the isomorphism classes of Fano Bott manifolds and the diffeomorphism classes of indecomposable Fano Bott manifolds using this correspondence.

To introduce our main result, we prepare some terminologies. Recall that a fan associated with a Bott manifold of complex dimension n is the normal fan of a polytope combinatorially equivalent to $[0, 1]^n$ and so it has $2n$ rays. We denote the primitive ray generators by $\mathbf{v}_1, \ldots, \mathbf{v}_n, \mathbf{w}_1, \ldots, \mathbf{w}_n$, where $\mathbf{v}_i$ and $\mathbf{w}_i$ are pairwise normal vectors of *opposite facets*.

If a Bott manifold is *Fano*, then it is known that the sum $\mathbf{v}_i + \mathbf{w}_i$ is either the zero vector or another ray generator, say $\mathbf{v}_{\varphi(i)}$ or $\mathbf{w}_{\varphi(i)}$, where φ is a permutation on $[n]$. Accordingly, one may associate a signed rooted forest with a Fano Bott manifold as follows:

- The set of vertices can be identified with $[n] = \{1, \ldots, n\}$.
- The vertex i is a root if and only if the sum $\mathbf{v}_i + \mathbf{w}_i$ is the zero vector.
- For each vertex i and its parent, denoted by $\varphi(i)$, the edge $\{i, \varphi(i)\}$ is signed by $+$ if $\mathbf{v}_i + \mathbf{w}_i = \mathbf{v}_{\varphi(i)}$ and by $-$ if $\mathbf{v}_i + \mathbf{w}_i = \mathbf{w}_{\varphi(i)}$.

See Sect. 2 for more precise definition.

For each vertex i of a signed rooted forest, we obtain another signed rooted forest by changing the signs of all the edges connecting i and its children simultaneously. By considering this operation for all vertices, we obtain an equivalence relation $\sim$ on the set SF_n of isomorphism classes of signed rooted forests with vertices $[n]$. It is observed in [12, Remark 5.8] that the isomorphism classes of Fano Bott manifolds of complex dimension n bijectively correspond to the equivalence classes $SF_n/\sim$ of signed rooted forests with n vertices. Now we state our main theorem.

Theorem 1.1 (Corollary 4.4) *The generating function* $F(x) = \sum_{n=0}^{\infty} |SF_n/\sim| x^n$ *satisfies*

$$F(x) = \exp\left(\sum_{k=1}^{\infty} \frac{x^k}{2k}\left(F(x^{2k}) + F(x^k)^2\right)\right).$$

This functional equation determines $F(x)$. Indeed, a straightforward computation shows

$$F(x) = 1 + x + 2x^2 + 5x^3 + 13x^4 + 37x^5 + 111x^6 + 345x^7 + 1105x^8 + 3624x^9 + \cdots$$

The generating function $\Delta(x) = 1 + \sum_{n=1}^{\infty} \Delta_n x^n$ for the number Δ_n of rooted triangular cacti with $2n + 1$ vertices and n triangles satisfies the same functional equation (see [10, 11]). It turns out that there is a bijective correspondence between $SF_n/\sim$ and rooted triangular cacti with $2n + 1$ vertices and n triangles.

The result [12] by Higashitani and Kurimoto implies that the diffeomorphism classes of indecomposable Fano Bott manifolds of complex dimension n bijectively correspond to $SF_{n-1}/\sim$. Here, we say a Fano Bott manifold is *indecomposable* if it is not isomorphic to a product of lower-dimensional Fano Bott manifolds. This provides an enumeration of the diffeomorphism classes of indecomposable Fano Bott manifolds.

This paper is organized as follows. In Sect. 2, we provide the definition of Bott manifolds and their Fano conditions. Moreover, we recall the association of signed rooted forests with Fano Bott manifolds. In Sect. 3, we show that the association induces a bijection between the isomorphism classes in Fano Bott manifolds and the equivalence classes $SF_n/\sim$ of signed rooted forests with n vertices. In Sect. 4, we enumerate the equivalence classes $SF_n/\sim$ of signed rooted forests. In Sect. 5, we give a bijective correspondence between $SF_n/\sim$ and rooted triangular cacti with $2n+1$ vertices and n triangles.

2 Fano Bott Manifolds and Signed Rooted Forests

In this section, we review the definition of Bott manifolds and their fans. We also recall the relation between Fano Bott manifolds and signed rooted forests from [12].

Definition 2.1 ([9, §2.1]) A Bott tower $B_\bullet$ is an iterated $\mathbb{C}P^1$-bundle starting with a point:

$$\begin{array}{ccccccc} B_n & \longrightarrow & B_{n-1} & \longrightarrow \cdots \longrightarrow & B_1 & \longrightarrow & B_0, \\ \| & & & & \| & & \| \\ P(\underline{\mathbb{C}} \oplus \xi_n) & & & & \mathbb{C}P^1 & & \{\text{a point}\} \end{array}$$

where each B_i is the complex projectivization of the Whitney sum of a holomorphic line bundle ξ_i and the trivial line bundle $\underline{\mathbb{C}}$ over B_{i-1}. The total space B_n is called a *Bott manifold*.

A Bott manifold B_n is a smooth projective toric variety by the construction. Its fan Σ has $2n$ rays. We denote by $\{\mathbf{v}_1, \dots, \mathbf{v}_n, \mathbf{w}_1, \dots, \mathbf{w}_n\}$ the ray generators, where a pair of $\mathbf{v}_i$ and $\mathbf{w}_i$ does not span a cone for each i. A subset S of ray generators having n elements form a maximal cone of Σ if and only if

$$\{\mathbf{v}_i, \mathbf{w}_i\} \not\subset S \qquad \text{for any } i \in [n].$$

Because of this description, one may see that the fan Σ is the normal fan of a polytope combinatorially equivalent to the cube $[0, 1]^n$.

For a fan Σ, let $\Sigma(1)$ be the set of all one-dimensional cones in Σ, i.e., the set of rays. Then for each $\rho \in \Sigma(1)$, we denote by $\mathbf{u}_\rho$ the generator of the ray ρ. We call a subset $P \subset \{\mathbf{u}_\rho \mid \rho \in \Sigma(1)\}$ a *primitive collection* if

$$\mathrm{Cone}(P) \notin \Sigma \quad \text{but } \mathrm{Cone}(P \setminus \{x\}) \in \Sigma \quad \text{for every } x \in P.$$

We denote by $\mathrm{PC}(\Sigma)$ the set of primitive collections of Σ. We briefly review Batyrev's criterion [1, Proposition 2.3.6] determining whether a given toric variety is Fano or not. Let Σ be a smooth complete fan. For each primitive collection $P = \{\mathbf{u}_1, \dots, \mathbf{u}_r\}$, there exists a unique cone σ such that $\mathbf{u}_1 + \cdots + \mathbf{u}_r$ is in the relative interior of σ. Let $\mathbf{v}_1, \dots, \mathbf{v}_\ell$ be the primitive generators of σ. Then

$$\mathbf{u}_1 + \cdots + \mathbf{u}_r = a_1\mathbf{v}_1 + \cdots + a_\ell\mathbf{v}_\ell$$

for some positive integers $a_1, \dots, a_\ell$. If the sum of primitive generators is the zero vector, then the cone σ is of zero-dimensional, and the set $\{\mathbf{v}_1, \dots, \mathbf{v}_\ell\}$ is assumed to be empty. We call this relation the *primitive relation* for P, and we define the *degree* of P by

$$\deg(P) := r - (a_1 + \cdots + a_\ell).$$

Here, we note that if the sum of primitive generators of P is the zero vector, then $\deg(P) = r$.

Proposition 2.2 ([1, Proposition 2.3.6]) *Let X_Σ be a nonsingular projective toric variety and $PC(\Sigma)$ be the primitive collection of the fan Σ. Then the toric variety X_Σ is Fano if and only if* $\deg(P) > 0$ *for every $P \in PC(\Sigma)$.*

Now we apply Batyrev's criterion to Bott manifolds. Let Σ be the fan of a Bott manifold B_n. Then the set of primitive collections is

$$\mathrm{PC}(\Sigma) = \{\{\mathbf{v}_i, \mathbf{w}_i\} \mid i \in [n]\}. \tag{1}$$

Using Proposition 2.2, we can see that B_n is Fano if and only if each primitive collection $P = \{\mathbf{v}_i, \mathbf{w}_i\}$ satisfies one of the following:

1. $\mathbf{v}_i + \mathbf{w}_i = \mathbf{0}$ (that is, $\deg(P) = 2 > 0$).
2. $\mathbf{v}_i + \mathbf{w}_i = \mathbf{v}_{\varphi(i)}$ (that is, $\deg(P) = 2 - 1 = 1 > 0$).
3. $\mathbf{v}_i + \mathbf{w}_i = \mathbf{w}_{\varphi(i)}$ (that is, $\deg(P) = 2 - 1 = 1 > 0$).

Here, $\varphi\colon [n] \setminus Z \to [n]$, where $Z := \{i \mid \mathbf{v}_i + \mathbf{w}_i = \mathbf{0}\}$. We also define a *sign map* $\sigma\colon [n] \setminus Z \to \{+, -\}$ by

$$\sigma(i) = \begin{cases} + & \text{if } \mathbf{v}_i + \mathbf{w}_i = \mathbf{v}_{\varphi(i)}, \\ - & \text{if } \mathbf{v}_i + \mathbf{w}_i = \mathbf{w}_{\varphi(i)}. \end{cases}$$

This leads us to the following definition.

Definition 2.3 ([12, Definition 4.1]) Let Σ be the fan of a Fano Bott manifold having (ordered) ray generators $\mathcal{S} = (\mathbf{v}_1, \dots, \mathbf{v}_n, \mathbf{w}_1, \dots, \mathbf{w}_n)$ with the primitive collections as in (1). Let φ and σ be as above. We define the *associated signed rooted forest* $(\mathcal{T}, s) = (\mathcal{T}(\Sigma, \mathcal{S}), s(\Sigma, \mathcal{S}))$ (i.e., rooted forest $\mathcal{T}$ with the sign map $s\colon E(\mathcal{T}) \to \{+, -\}$) to be:

- $V(\mathcal{T}) = [n]$.

- $E(\mathcal{T}) = \{\{i, \varphi(i)\} \mid i \in [n] \setminus Z\}$ and $s(\{i, \varphi(i)\}) = \sigma(i)$.

From the definition, one can see that for a signed rooted forest $(\mathcal{T}, s)$, the set of roots is Z and the parent of each vertex $i \in [n] \setminus Z$ is $\varphi(i)$. We denote the assignment provided in Definition 2.3 by Φ, that is, $\Phi(\Sigma, \mathcal{S}) = (\mathcal{T}(\Sigma, \mathcal{S}), s(\Sigma, \mathcal{S}))$ is the associated signed rooted forest.

Remark 2.4 The association Φ is surjective, that is, for each signed rooted forest $(\mathcal{T}, s)$ with n vertices, there exists a Fano Bott manifold of dimension n whose fan defines $(\mathcal{T}, s)$.

Example 2.5 In this example, we present ray generators of the fan of a Bott manifold using a matrix, i.e., the columns of an $n \times 2n$ matrix are ray generators. Consider the following two matrices:

$$A = \left[\begin{array}{ccc|ccc} 1&0&0 & -1&0&0 \\ 1&1&0 & 0&-1&-1 \\ 0&0&1 & 0&0&-1 \end{array}\right], \quad A' = \left[\begin{array}{ccc|ccc} 1&0&0 & -1&0&0 \\ 1&1&0 & 0&-1&0 \\ 1&0&1 & 0&1&-1 \end{array}\right].$$

Let B be the Bott manifold such that the ray generators $(\mathbf{v}_1, \mathbf{v}_2, \mathbf{v}_3, \mathbf{w}_1, \mathbf{w}_2, \mathbf{w}_3)$ of the fan are the column vectors of A. Then, we have

$$\begin{aligned} \mathbf{v}_1 + \mathbf{w}_1 &= \mathbf{v}_2, \\ \mathbf{v}_2 + \mathbf{w}_2 &= \mathbf{0}, \\ \mathbf{v}_3 + \mathbf{w}_3 &= \mathbf{w}_2. \end{aligned}$$

Therefore, the Bott manifold B is Fano, and moreover, $\varphi(1) = 2, \varphi(3) = 2$, and $\sigma(1) = +$, $\sigma(3) = -$. The associated signed rooted tree is given in Fig. 1(6) (without vertex labeling).

Let B' be the Bott manifold such that ray generators $(\mathbf{v}_1, \mathbf{v}_2, \mathbf{v}_3, \mathbf{w}_1, \mathbf{w}_2, \mathbf{w}_3)$ of the fan are the column vectors of A'. Consider a primitive collection $P = \{\mathbf{v}_1, \mathbf{w}_1\}$. The sum of ray generators is

$$\mathbf{v}_1 + \mathbf{w}_1 = \mathbf{v}_2 + \mathbf{v}_3,$$

so $\deg(P) = 2 - 2 = 0 \not> 0$ and this primitive collection does not satisfy the Fano condition. Therefore, the Bott manifold B' is not Fano.

We provide all signed rooted forests having three vertices in Fig. 1.

Remark 2.6 We say that a signed rooted forest is *binary* if each vertex has at most two children, and when the vertex has two children, the edges connecting the vertex and its children have different signs. In Fig. 1, all but (5) and (7) are binary. The set of binary rooted forests provides a certain family of Fano toric Richardson varieties (called *of Catalan type*) in the full flag variety [13].

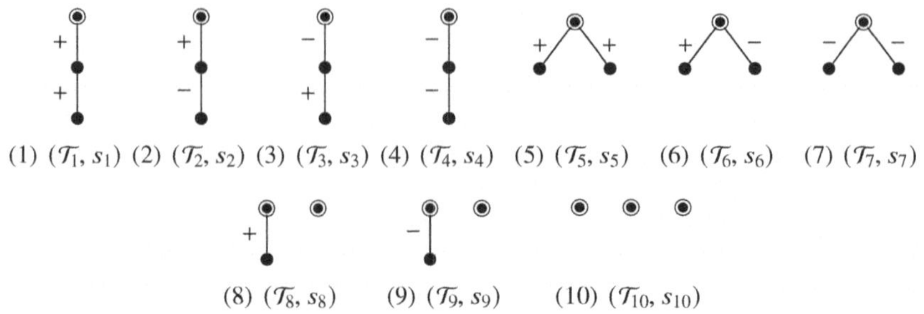

Fig. 1 Signed rooted forests with three vertices

3 Classification of Fano Bott Manifolds

We say that signed rooted forests $(\mathcal{T}, s)$ and $(\mathcal{T}', s')$ with vertices $[n]$ are *isomorphic* if there is a permutation $\pi \in \mathfrak{S}_n$, which sends the roots of $\mathcal{T}$ to the roots of $\mathcal{T}'$ and induces a bijection between the edges preserving the signs. Let SF_n be the isomorphism classes of signed rooted forests with vertices $[n]$. For each vertex $i \in [n]$, we define an operation

$$r_i : SF_n \to SF_n$$

which changes the signs of all edges connecting the vertex i and its children simultaneously. Denote by $\sim$ the equivalence relation on SF_n generated by the operations r_i for all $i \in [n]$. The following is mentioned in [12, Remark 5.8], but we include its proof for readers' convenience.

Theorem 3.1 (cf. [12, Remark 5.8]) *The isomorphism classes in Fano Bott manifolds of complex dimension n bijectively correspond to $SF_n/\sim$.*

Proof Let B be a Fano Bott manifold of complex dimension n, $\Sigma = \Sigma_B$ a fan defining B. We fix an ordering on ray generators $\mathcal{S} = (\mathbf{v}_1, \dots, \mathbf{v}_n, \mathbf{w}_1, \dots, \mathbf{w}_n)$ of Σ. For $A \in \mathrm{GL}(n, \mathbb{Z})$, we denote by $A \cdot \Sigma$ the fan consisting of cones $A \cdot \sigma$'s for $\sigma \in \Sigma$ and denote by $A \cdot \mathcal{S}$ the ordered ray generators of $A \cdot \Sigma$ given by

$$A \cdot \mathcal{S} := (A\mathbf{v}_1, \dots, A\mathbf{v}_n, A\mathbf{w}_1, \dots, A\mathbf{w}_n).$$

By Cho et al. [5, Proposition 3.4], another pair $(\Sigma', \mathcal{S}')$ defines a Fano Bott manifold isomorphic to B if and only if $\Sigma' = A \cdot \Sigma_B$ for some $A \in \mathrm{GL}(n, \mathbb{Z})$, and the set $\mathcal{S}'$ is obtained from $A \cdot \mathcal{S}$ by performing the following two operations on $A \cdot \mathcal{S}$:

(Op.1) Swapping $A\mathbf{v}_i$ with $A\mathbf{w}_i$, that is,

$$\mathcal{S}_i' := (A\mathbf{v}_1, \dots, A\mathbf{v}_{i-1}, A\mathbf{w}_i, A\mathbf{v}_{i+1}, \dots, A\mathbf{v}_n,$$

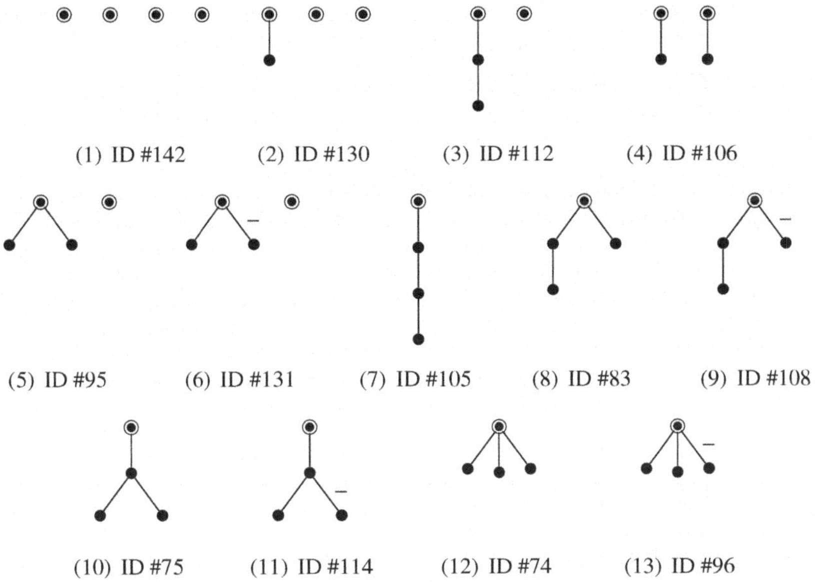

Fig. 2 Representatives of $SF_4/\sim$. Numbers are IDs by Øbro

$$A\mathbf{w}_1, \dots, A\mathbf{w}_{i-1}, A\mathbf{v}_i, A\mathbf{w}_{i+1}, \dots, A\mathbf{w}_n).$$

(Op.2) Reordering $A\mathbf{v}_i$ (as well as $A\mathbf{w}_i$'s), that is, for a permutation $\pi \in \mathfrak{S}_n$,

$$\mathcal{S}'_\pi := (A\mathbf{v}_{\pi(1)}, \dots, A\mathbf{v}_{\pi(n)}, A\mathbf{w}_{\pi(1)}, \dots, A\mathbf{w}_{\pi(n)}).$$

For the ordered ray generators $\mathcal{S}'_i$ obtained by applying **(Op.1)**, we have $\Phi(\Sigma', \mathcal{S}'_i) = r_i(\Phi(\Sigma_B, \mathcal{S}))$. For the ordered ray generators $\mathcal{S}'_\pi$ obtained by applying **(Op.2)**, $\Phi(\Sigma', \mathcal{S}'_\pi)$ is obtained from $\Phi(\Sigma_B, \mathcal{S})$ by changing the labeling on the vertices by the permutation π, so they are isomorphic as signed rooted forests. This finishes the proof. □

Example 3.2 Consider SF_3 described in Fig. 1. We obtain five equivalence classes

$$(\mathcal{T}_1, s_1) \sim (\mathcal{T}_2, s_2) \sim (\mathcal{T}_3, s_3) \sim (\mathcal{T}_4, s_4), \quad (\mathcal{T}_5, s_5) \sim (\mathcal{T}_7, s_7),$$
$$(\mathcal{T}_6, s_6), \quad (\mathcal{T}_8, s_8) \sim (\mathcal{T}_9, s_9), \quad (\mathcal{T}_{10}, s_{10}).$$

All signed rooted forests in SF_4 are illustrated in Fig. 2. Roots of the forests are the top vertices. We omit plus signs on edges and put a minus sign on an edge. We also write ID numbers of the corresponding Fano Bott manifolds according to the list of "smooth toric Fano varieties" [14] in the Graded Ring Database [3].

Higashitani and Kurimoto [12] provide another equivalence relation $\approx$ on the set of signed rooted forests, which is used to consider the diffeomorphism classes in Fano Bott manifolds. The equivalence relation $\approx$ is induced from the relation $\sim$ by neglecting signs on the edges incident on the roots. Using this relation, we recall the following:

Theorem 3.3 ([12, Theorem 1.8 and Remark 6.4]) *The diffeomorphism classes in Fano Bott manifolds of complex dimension n bijectively correspond to* $SF_n/\approx$.

We say that a Bott manifold B is *indecomposable* if it is not isomorphic to a product of lower-dimensional Bott manifolds (as toric varieties). Otherwise, we say that B is *decomposable*.[1]

Corollary 3.4 *The diffeomorphism classes of indecomposable Fano Bott manifolds of complex dimension n bijectively correspond to* $SF_{n-1}/\sim$.

Proof We first notice that by Theorem 3.1, a Fano Bott manifold is indecomposable if and only if the corresponding signed rooted forest is a signed rooted tree, that is, it has only one root vertex. Since the equivalence relation $\approx$ is induced from the relation $\sim$ by neglecting the signs on the edges incident on the root, we obtain the desired bijection by erasing the root vertex. □

Example 3.5 In Fig. 2, three pairs $\{(5), (6)\}$, $\{(8), (9)\}$, $\{(12), 13)\}$ are diffeomorphic to each other, but (10) and (11) are not diffeomorphic to each other. Considering signed rooted trees in Fig. 2, we obtain the five equivalence classes

$$\{(7), (8), (9), (10), (11), (12), (13)\}/\approx = \{[(7)], [(8)], [(10)], [(11)], [(12)]\}.$$

By erasing the root vertex, each of them is associated with an element in $SF_3/\sim$.

$$[(7)] \leftrightarrow [(\mathcal{T}_1, s_1)], \quad [(8)] \leftrightarrow [(\mathcal{T}_8, s_8)], \quad [(10)] \leftrightarrow [(\mathcal{T}_5, s_5)],$$
$$[(11)] \leftrightarrow [(\mathcal{T}_6, s_6)], \quad [(12)] \leftrightarrow [(\mathcal{T}_{10}, s_{10})].$$

4 Counting Signed Rooted Forests in Terms of Signed Rooted Trees

We denote by $ST_n/\sim$ the set of signed rooted trees in $SF_n/\sim$. We set $t_n = |ST_n/\sim|$ and $f_n = |SF_n/\sim|$. Now we let $T(x)$ and $F(x)$ be the generating functions of the sequences $\{t_n\}$ and $\{f_n\}$, respectively, that is,

[1] The notion of indecomposability can have different meanings in different contexts. Especially, in [8], they consider another family of smooth manifolds, called *real Bott manifolds*, and say that a real Bott manifold is indecomposable if it is not *diffeomorphic* to a product of lower-dimensional real Bott manifolds.

$$T(x) = \sum_{n=1}^{\infty} t_n x^n \qquad \text{and} \qquad F(x) = 1 + \sum_{n=1}^{\infty} f_n x^n.$$

In this section, we compute the generating functions $T(x)$ and $F(x)$ and study their relations.

Proposition 4.1 *The generating function $F(x)$ satisfies*

$$F(x) = \prod_{k=1}^{\infty} (1 - x^k)^{-t_k}.$$

Proof Note that from the generalized binomial theorem, for any positive integer m, we have

$$\begin{aligned}(1-x)^{-m} &= \sum_{p=0}^{\infty} \binom{-m}{p} (-x)^p \\ &= \sum_{p=0}^{\infty} \frac{(-m)(-m-1)\cdots(-m-(p-1))}{p!} (-x)^p \\ &= \sum_{p=0}^{\infty} \binom{m-1+p}{p} x^p.\end{aligned}$$

Since

$$(1 - x^k)^{-t_k} = \sum_{p_k=0}^{\infty} \binom{t_k - 1 + p_k}{p_k} x^{kp_k}, \tag{2}$$

the coefficient of x^n in the product $\prod_{k=1}^{\infty} (1 - x^k)^{-t_k}$ is given by

$$\sum_{(p_1,\ldots,p_n)} \prod_{k=1}^{n} \binom{t_k - 1 + p_k}{p_k}, \tag{3}$$

where $(p_1, \ldots, p_n)$ runs over all n-tuples of nonnegative integers with $\sum_{k=1}^{n} kp_k = n$. Here $\binom{t_k-1+p_k}{p_k}$ is the number of signed rooted forests with p_k components such that each component is a signed rooted tree with k vertices by (2), and the sum in (3) counts all decompositions of elements in $SF_n/\sim$ into connected components, so the proposition follows. □

The following is a consequence of the above proposition.

Corollary 4.2 *The generating functions $F(x)$ and $T(x)$ satisfy*

$$F(x) = \exp\left(\sum_{n=1}^{\infty} \frac{T(x^n)}{n}\right).$$

Proof Taking logarithm on both sides of

$$F(x) = \prod_{k=1}^{\infty}(1 - x^k)^{-t_k},$$

we obtain

$$\log F(x) = -\sum_{k=1}^{\infty} t_k \log(1 - x^k) = \sum_{k=1}^{n}\sum_{n=1}^{\infty} t_k \frac{x^{kn}}{n} = \sum_{n=1}^{\infty} \frac{T(x^n)}{n},$$

which implies the corollary. □

Lemma 4.3 (cf. (1) in [11]) *The generating functions $F(x)$ and $T(x)$ satisfy*

$$T(x) = \frac{x}{2}\left(F(x^2) + F(x)^2\right).$$

Proof We set

$$ST/\sim := \bigsqcup_{n=1}^{\infty} ST_n/\sim \qquad \text{and} \qquad SF/\sim := \bigsqcup_{n=0}^{\infty} SF_n/\sim,$$

where $SF_0/\sim$ is understood to be the empty set. Given an unordered pair $\{A, B\}$ of $SF/\sim$, we obtain an element AB of $ST/\sim$ by joining the roots of A and B to a new root v and assign all the new edges joining the roots of A to v, say $+$ sign, and all the new edges joining the roots of B to v, say $-$ sign. We may assign $-$ sign to the former and $+$ sign to the latter. In any case, AB is well-defined in $ST/\sim$. Conversely, given an element T of $ST/\sim$, there is a unique unordered pair $\{A, B\}$ of $SF/\sim$ such that $AB = T$.

This implies the lemma. Indeed, $F(x)^2$ counts unordered pairs $\{A, B\}$ twice when A and B are different but once when $A = B$. This is why we add $F(x^2)$ in the formula. Multiplication by x corresponds to the new vertex v. □

Combining Corollary 4.2 and Lemma 4.3, we obtain the following functional equation mentioned in the introduction.

Corollary 4.4 (cf. (3) in [11]) *The generating function $F(x)$ satisfies*

Table 1 The numbers of equivalence classes of signed rooted trees and signed rooted forests

n	1	2	3	4	5	6	7	8	9	10
t_n	1	1	3	7	21	60	189	595	1948	6455
f_n	1	2	5	13	37	111	345	1105	3624	12,099

$$F(x) = \exp\left(\sum_{n=1}^{\infty} \frac{x^n}{2n}\left(F(x^{2n}) + F(x^n)^2\right)\right).$$

This functional equation determines $F(x)$. Using Lemma 4.3 and Corollary 4.4, we obtain Table 1.

The numbers t_n and f_n in Table 1 satisfy $f_n < 2t_n < 4f_{n-1}$ for $n \leq 10$, and the sequences $\{t_n/t_{n-1}\}_{n=2}^{10}$ and $\{f_n/f_{n-1}\}_{n=2}^{10}$ are both increasing. We leave the following question.

Question 4.5 *Are the sequences t_n/t_{n-1} and f_n/f_{n-1} increasing and bounded above by* 4*?*

5 Rooted Triangular Cacti

The formula in Corollary 4.4 also holds for the generating function of the number of rooted triangular cacti with $2n + 1$ vertices and n triangles. In this section, we give a bijective correspondence between the equivalence classes $SF_n/\sim$ of signed rooted forests and rooted triangular cacti with $2n + 1$ vertices and n triangles.

Definition 5.1 A *cactus* (or a *cactus tree*) is a connected graph in which any two simple cycles have at most one vertex in common; equivalently, no line lies on more than one cycle. A *triangular cactus* (or a 3-*cactus*) is a cactus such that every cycle has length three. A *rooted triangular cactus* is a triangular cactus having a root vertex.

We sometimes call a 3-cycle in a 3-cactus a *triangle*. In Fig. 3, we present rooted 3-cacti having nine vertices and four triangles. The sequence of numbers of rooted 3-cacti with $2n + 1$ vertices and n triangles is Sequence A003080 in [15]. Note that a cactus is also called a *Husimi tree* (see [10, 11]).

Proposition 5.2 *There is a bijective correspondence between $SF_n/\sim$ and the set of rooted* 3*-cacti with $2n + 1$ vertices and n triangles.*

Proof Let $(\mathcal{T}, s)$ be a signed rooted forest. For each root vertex of $(\mathcal{T}, s)$, we draw a triangle and decorate the top vertex of the triangle with a double circle to indicate the root of the triangular cacti. For each child of the root of $\mathcal{T}$, we draw a triangle as follows. If the sign of the edge incident on the root is positive, we attach the new

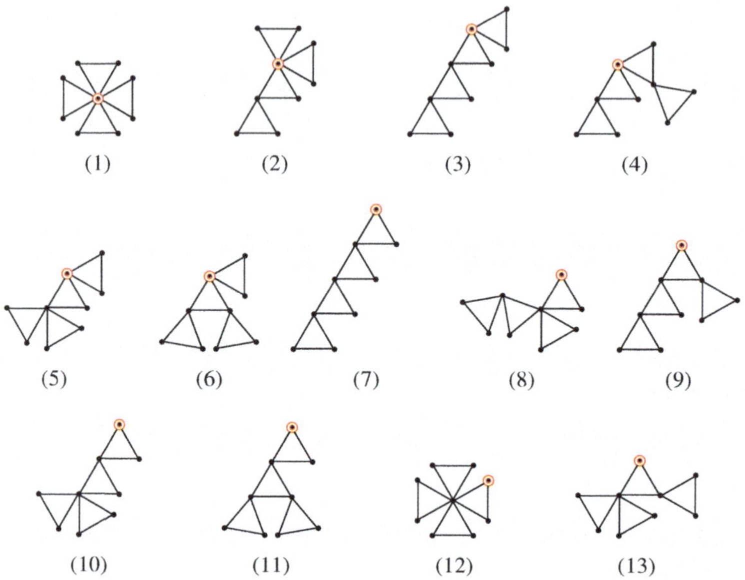

Fig. 3 Rooted triangular cacti

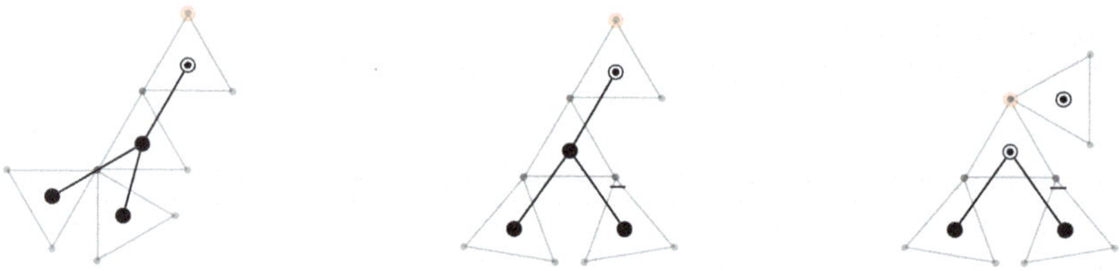

Fig. 4 Construction of triangular cacti from signed rooted forests

triangle to the left bottom vertex; if the sign is negative, we attach the new triangle to the right bottom vertex. Continuing this process to each child vertex, we get a bunch of rooted triangular cacti. Finally, we merge all the root vertices of rooted triangular cacti to one root vertex, so we obtain one rooted triangular cacti. See Fig. 4. Obviously, the rooted triangular cacti corresponding to $(\mathcal{T}, s)$ and $r_i(\mathcal{T}, s)$ are isomorphic to each other. This proves the proposition. □

Acknowledgments Cho was supported by the National Research Foundation of Korea (NRF) grant funded by the Government of Korea (MSIP, Ministry of Science, ICT and Future Planning) (no. 2020R1C1C1A01010972 and no. 2020R1A5A1016126). Lee was supported by the Institute for Basic Science (IBS-R003-D1). Masuda was supported in part by JSPS Grant-in-Aid for Scientific Research 19K03472 and a HSE University Basic Research Program. Park was supported by the Basic Science Research Program through the National Research Foundation of Korea (NRF) funded by the Government of Korea (NRF-2018R1A6A3A11047606). This work was partly supported by Osaka City University Advanced Mathematical Institute (MEXT Joint Usage/Research Center on Mathematics and Theoretical Physics JPMXP0619217849).

References

1. V.V. Batyrev, On the classification of toric Fano 4-folds. Algebraic Geom. **94**(9), 1021–1050 (1999)
2. C.P. Boyer, D.M.J. Calderbank, C.W.T. Friedman, The Kähler geometry of Bott manifolds. Adv. Math. **350**, 1–62 (2019)
3. G. Brown, A. Kasprzyk, Graded ring database, http://www.grdb.co.uk/forms/toricsmooth (accessed May 30, 2024)
4. B.N. Chary, On Mori cone of Bott towers, J. Algebra **507**, 467–501 (2018)
5. Y. Cho, E. Lee, M. Masuda, S. Park, Unique toric structure on a Fano Bott manifold. J. Symplec. Geometry **21**(3), 439–462 (2023). https://www.intlpress.com/site/pub/pages/journals/items/jsg/content/vols/0021/0003/a001/index.php
6. S. Choi, D.Y. Suh, Properties of Bott manifolds and cohomological rigidity. Algebr. Geom. Topol. **11**(2), 1053–1076 (2011)
7. S. Choi, M. Masuda, S. Murai, Invariance of Pontrjagin classes for Bott manifolds. Algebr. Geom. Topol. **15**(2), 965–986 (2015)
8. S. Choi, M. Masuda, S.-i. Oum, Classification of real Bott manifolds and acyclic digraphs. Trans. Am. Math. Soc. **369**(4), 2987–3011 (2017)
9. M. Grossberg, Y. Karshon, Bott towers, complete integrability, and the extended character of representations. Duke Math. J. **76**(1), 23–58 (1994)
10. F. Harary, R.Z. Norman, The dissimilarity characteristic of Husimi trees. Ann. Math. (2) **58**, 134–141 (1953)
11. F. Harary, G.E. Uhlenbeck, On the number of Husimi trees. I. Proc. Nat. Acad. Sci. U.S.A. **39**, 315–322 (1953)
12. A. Higashitani, K. Kurimoto, Cohomological rigidity for Fano Bott manifolds. Math. Z. (2022). https://doi.org/10.1007/s00209-022-02994-w
13. E. Lee, M. Masuda, S. Park, Toric Richardson varieties of Catalan type and Wedderburn–Etherington numbers. Eur. J. Combinat. **108**, 103617 (2023). https://www.sciencedirect.com/science/article/abs/pii/S0195669822001135
14. M. Øbro, An algorithm for the classification of smooth Fano polytopes. Preprint. arXiv:0704.0049v1 (2007)
15. N.J.A. Sloane (ed.), The On-Line Encyclopedia of Integer Sequences (OEIS), published electronically at https://oeis.org (accessed May 30, 2024)
16. Y. Suyama, Fano generalized Bott manifolds. Manuscripta Math. **163**(3–4), 427–435 (2020)

Dga Models for Moment-Angle Complexes

Matthias Franz

1 Introduction

Let Σ be a simplicial complex on the set $[m] = \{1, \ldots, m\}$, containing the empty simplex $\varnothing$ and possibly having ghost vertices, and let

$$\mathcal{Z}(\Sigma) = \mathcal{Z}_\Sigma(D^2, S^1) = \bigcup_{\sigma \in \Sigma} (D^2, S^1)^\sigma \subset (D^2)^m \tag{1}$$

be the associated moment-angle complex, where

$$(D^2, S^1)^\sigma = \left\{ (z_1, \ldots, z_m) \in (D^2)^m \;\middle|\; z_i \in S^1 \text{ if } i \notin \sigma \right\}. \tag{2}$$

Moment-angle complexes play a central role in toric topology; see [8]. Replacing (D^2, S^1) by (D^n, S^{n-1}) for any $n \geq 1$ gives generalized moment-angle complexes. Taking arbitrary CW pairs leads to polyhedral products, which have gained a lot of attention in homotopy theory recently; see [1] for a survey.

The moment-angle complex $\mathcal{Z}(\Sigma)$ is homotopy-equivalent to the complement of a complex coordinate subspace arrangement, which is a smooth toric variety. The integral cohomology ring of $\mathcal{Z}(\Sigma)$ was computed by the author [11, Sect. 4] (using the language of toric varieties) and shortly afterwards by Baskakov–Buchstaber–Panov [4].[1] The result is an isomorphism of graded rings

[1] The argument appearing in [7, Thm. 7.7] and earlier publications by the same authors is incorrect; compare [12, Sect. 1].

M. Franz (✉)
Department of Mathematics, University of Western Ontario, London, ON, Canada
e-mail: mfranz@uwo.ca

A. Bahri et al. (eds.), *Toric Topology and Polyhedral Products*, Fields Institute Communications 89, https://doi.org/10.1007/978-3-031-57204-3_7

$$H^*(\mathcal{Z}(\Sigma)) = \mathrm{Tor}_{\mathbf{R}}(\mathbb{Z}, \mathbb{Z}[\Sigma]), \tag{3}$$

where $\mathbf{R} = \mathbb{Z}[t_1, \ldots, t_m]$ and $\mathbb{Z}[\Sigma]$ is the Stanley–Reisner ring of Σ with generators $t_1, \ldots, t_m$ of degree 2. Taking the Koszul resolution of $\mathbb{Z}$ over $\mathbf{R}$, one can describe the ring (3) as the cohomology of the commutative differential graded algebra (cdga)

$$\mathbf{A}(\Sigma) = \mathbb{Z}[\Sigma] \otimes \bigwedge(s_1, \ldots, s_m), \qquad d\,s_i = t_i, \quad d\,t_i = 0 \tag{4}$$

for $i \in [m]$, where each s_i has degree 1. Dividing out all squares t_i^2 as well as all terms $s_i\,t_i$, one obtains a quasi-isomorphic dga $\mathbf{B}(\Sigma)$. As a cdga, $\mathbf{B}(\Sigma)$ is generated by the s_i and $t_i = d\,s_i$ and has the relations $s_i\,t_i = t_i\,t_i = 0$ for $i \in [m]$ as well as $t_{i_1} \cdots t_{i_k} = 0$ whenever $\{i_1, \ldots, i_k\} \notin \Sigma$.

Theorem 1.1 *The singular cochain algebra $C^*(\mathcal{Z}(\Sigma))$ is quasi-isomorphic to the dgas $\mathbf{A}(\Sigma)$ and $\mathbf{B}(\Sigma)$, naturally with respect to inclusions of subcomplexes.*

Recall that a dga A is called an (integral) *dga model* for a space X if A can be connected to $C^*(X)$ via a zigzag of dga quasi-isomorphisms. In this language, Theorem 1.1 asserts that both $\mathbf{A}(\Sigma)$ and $\mathbf{B}(\Sigma)$ are dga models for $\mathcal{Z}(\Sigma)$.

That $C^*(\mathcal{Z}(\Sigma))$ and $\mathbf{A}(\Sigma)$ are quasi-isomorphic is already implicit in the author's computation of $H^*(\mathcal{Z}(\Sigma))$; see [11, Sect. 4]. A different proof has recently been obtained by the author as a byproduct of his work on the cohomology rings of partial quotients of moment-angle complexes [12, Prop. 6.1]. As remarked there, this result answers a question posed by Berglund [5, Question 5], which was exactly whether $\mathbf{A}(\Sigma)$ is a dga model for $\mathcal{Z}(\Sigma)$. The aim of the present note is to give a much shorter proof for this model. Like Baskakov–Buchstaber–Panov's calculation, it is based on the dga $\mathbf{B}(\Sigma)$. The rational versions of $\mathbf{A}(\Sigma)$ and $\mathbf{B}(\Sigma)$ are (analogously defined) cdga models for the polynomial differential forms on $\mathcal{Z}(\Sigma)$ by a result of Panov–Ray [21, Thm. 6.2].

The proof of Theorem 1.1 appears in the following section and an adaptation to real moment-angle complexes in Sect. 3. In the final section, we relate the resulting cup product formulas for real and complex moment-angle complexes with others appearing in the literature. We in particular provide a proof that has been missing so far for a product formula stated by Gitler and López de Medrano [14].

2 Proof of Theorem 1.1

We will obtain Theorem 1.1 by dualizing the analogous homological result. To state the latter, we need to introduce some terminology. As already done in Theorem 1.1, we write $C(-)$ and $C^*(-)$ for *normalized* singular (co)chains with integral coefficients.

Recall that the normalized singular chain complex of a space X is obtained from the usual non-normalized one by dividing out the subcomplex of degenerate simplices. A singular n-simplex is called degenerated if it factors through an $(n-1)$-dimensional one via a map $\Delta^n \to \Delta^{n-1}$ between standard simplices that in barycentric coordinates is given by $(t_0, \ldots, t_n) \mapsto (t_0, \ldots, t_i + t_{i+1}, \ldots, t_n)$ for some $0 \le i < n$. Projecting non-normalized to normalized singular chains is a homotopy equivalence; compare [17, Sect. VIII.6]. The normalized singular cochain complex $C^*(X)$ is the dual of $C(X)$ with differential

$$(d\,\gamma)(x) = -(-1)^{|\gamma|}\,\gamma(d\,x) \tag{5}$$

for $\gamma \in C^n(X)$ and a singular $(n+1)$-simplex x.

The chain complex $C(X)$ is a differential graded coalgebra (dgc) with diagonal and augmentation given by

$$\Delta x = \sum_{k=0}^{n} x(0\ldots k) \otimes x(k\ldots n) \qquad \text{and} \qquad \varepsilon(x) = \begin{cases} 1 & \text{if } n = 0, \\ 0 & \text{otherwise} \end{cases} \tag{6}$$

for an n-simplex x. Here $x(k_1 \ldots k_2)$ denotes the face of x with vertices $k_1, \ldots, k_2$.

We also recall part of the Eilenberg–Zilber theorem; compare [10, Sect. 17]. Given two spaces X and Y, the shuffle map

$$\nabla\colon C(X) \otimes C(Y) \to C(X \times Y) \tag{7}$$

is a homotopy equivalence of complexes, natural in X and Y. It moreover is associative and a morphism of dgcs. Hence, for any spaces $X_1, \ldots, X_m$, we have a natural quasi-isomorphism of dgcs

$$C(X_1) \otimes \cdots \otimes C(X_m) \to C\bigl(X_1 \times \cdots \times X_m\bigr), \tag{8}$$

again denoted by ∇.

Let $\mathbb{Z}\langle\Sigma\rangle$ be the Stanley–Reisner coalgebra of Σ dual to $\mathbb{Z}[\Sigma]$, *cf.* [8, Sect. 8.2]. The canonical basis for $\mathbb{Z}\langle\Sigma\rangle$, considered as a $\mathbb{Z}$-module, are the monomials u_α indexed by allowed multi-indices $\alpha \in \mathbb{N}^m$. A multi-index α is *allowed* if it is supported on some simplex in Σ, that is, if

$$\operatorname{supp}\alpha := \{\, i \in [m] \mid \alpha_i > 0 \,\} \in \Sigma. \tag{9}$$

The degree of u_α is $2(\alpha_1 + \cdots + \alpha_m)$. The structure maps are given by

$$\Delta u_\alpha = \sum_{\beta+\gamma=\alpha} u_\beta \otimes u_\gamma, \qquad \varepsilon(u_\alpha) = \begin{cases} 1 & \text{if } \alpha = 0, \\ 0 & \text{otherwise.} \end{cases} \tag{10}$$

We consider the tensor product of graded coalgebras

$$\mathbf{K}(\Sigma) = \mathbb{Z}\langle\Sigma\rangle \otimes \bigwedge(v_1, \ldots, v_m), \tag{11}$$

where each v_i is primitive of degree 1. We turn $\mathbf{K}(\Sigma)$ into a dgc by defining

$$d(u_\alpha \otimes v_\tau) = \sum_{\alpha_i>0} u_{\alpha-i} \otimes v_i \wedge v_\tau \tag{12}$$

for allowed multi-indices $\alpha \in \mathbb{N}^m$ and $\tau \subset [m]$. Here we have written $\alpha - i$ for the multi-index that is obtained from α by decreasing the ith component by 1 as well as $v_\tau = v_{i_1} \wedge \cdots \wedge v_{i_k}$ if $\tau = \{i_1 < \cdots < i_k\}$. For $\sigma \in \Sigma$, we also write $u_\sigma = u_\alpha$ where α is the indicator function of $\sigma \subset [m]$,

$$\alpha_i = \begin{cases} 1 & \text{if } i \in \sigma, \\ 0 & \text{if } i \notin \sigma, \end{cases} \tag{13}$$

and we use the abbreviation $u_\varnothing = v_\varnothing = u_\varnothing \otimes v_\varnothing = 1$.

Let $\mathbf{L}(\Sigma)$ be the sub-dgc of $\mathbf{K}(\Sigma)$ spanned by all elements $u_\sigma \otimes v_\tau$ with disjoint subsets $\sigma \in \Sigma$ and $\tau \subset [m]$. The dual of $\mathbf{K}(\Sigma)$ is the dga $\mathbf{A}(\Sigma)$, and that of $\mathbf{L}(\Sigma)$ is $\mathbf{B}(\Sigma)$.

Theorem 2.1 *The dgcs $C(\mathcal{Z}(\Sigma))$, $\mathbf{K}(\Sigma)$, and $\mathbf{L}(\Sigma)$ are quasi-isomorphic, naturally with respect to inclusions of subcomplexes.*

The proof is given in the remainder of this section. Applying the universal coefficient theorem for cohomology then establishes Theorem 1.1.

The following two observations are immediate. We write $\Sigma_{|i}$ for the restriction of Σ to the single vertex $i \in [m]$. It contains either the empty simplex only or additionally the 0-simplex $\{i\}$.

Lemma 2.2 *For any $\sigma \in \Sigma$, there are canonical isomorphisms of dgcs*

$$\mathbf{K}(\sigma) = \bigotimes_{i=1}^{m} \mathbf{K}(\sigma_{|i}), \qquad \mathbf{L}(\sigma) = \bigotimes_{i=1}^{m} \mathbf{L}(\sigma_{|i}).$$

Lemma 2.3 *Let Σ_1, Σ_2 be subcomplexes of Σ. There are short exact sequences*

$$0 \longrightarrow \mathbf{K}(\Sigma_1 \cap \Sigma_2) \longrightarrow \mathbf{K}(\Sigma_1) \oplus \mathbf{K}(\Sigma_2) \longrightarrow \mathbf{K}(\Sigma_1 \cup \Sigma_2) \longrightarrow 0,$$
$$0 \longrightarrow \mathbf{L}(\Sigma_1 \cap \Sigma_2) \longrightarrow \mathbf{L}(\Sigma_1) \oplus \mathbf{L}(\Sigma_2) \longrightarrow \mathbf{L}(\Sigma_1 \cup \Sigma_2) \longrightarrow 0.$$

Let y be the usual parametrization of S^1, considered as a singular 1-simplex. Choose a singular 2-simplex x in D^2 that restricts to y on the edge (12) and maps the other two edges (01) and (02) to the identity element $e \in S^1$. Then

$$
\begin{aligned}
d\,y &= 0, & \Delta y &= y \otimes e + e \otimes y, & (14)\\
d\,x &= x(12) - x(02) + x(01) & \Delta x &= x \otimes e + x(01) \otimes x(12) + e \otimes x & (15)\\
&= y, & &= x \otimes e + e \otimes x.
\end{aligned}
$$

Here $x(01)$ and $x(02)$ drop out because they are degenerate. For this to hold, it is crucial that we work with normalized chains.

We use the singular simplices x and y to define a dgc map

$$\Psi(\Sigma)\colon \mathbf{L}(\Sigma) \to C(\mathcal{Z}(\Sigma)). \tag{16}$$

For $m = 1$, we map $u_1 \mapsto x$, $v_1 \mapsto y$ and $1 \mapsto e \in C(S^1)$; this is well-defined by (14) and (15). For $m > 1$ and $\sigma \subset [m]$, we set

$$
\begin{aligned}
\Psi(\sigma)\colon \mathbf{L}(\sigma) = \bigotimes_{i=1}^{m} \mathbf{L}(\sigma_{|i}) &\xrightarrow{\otimes \Psi(\sigma_{|i})} \bigotimes_{i=1}^{m} C(\mathcal{Z}(\sigma_{|i}))\\
&\xrightarrow{\nabla} C\big(\mathcal{Z}(\sigma_{|1}) \times \cdots \times \mathcal{Z}(\sigma_{|m})\big) = C(\mathcal{Z}(\sigma)),
\end{aligned} \tag{17}
$$

using Lemma 2.2 and the fact that the shuffle map is a morphism of dgcs. In the general case, $\Psi(\Sigma)$ is determined by imposing naturality with respect to inclusions of subcomplexes. In other words, $\Psi(\Sigma)$ agrees on $\mathbf{L}(\sigma) \subset \mathbf{L}(\Sigma)$ with $\Psi(\sigma)$, followed by the inclusion $C(\mathcal{Z}(\sigma)) \hookrightarrow C(\mathcal{Z}(\Sigma))$.

Lemma 2.4 *The map $\Psi(\Sigma)$ is a quasi-isomorphism of dgcs.*

Proof The case $m = 1$ is settled by a direct verification; we therefore assume $m > 1$ and proceed by induction on the size of Σ. If Σ has a single maximal simplex σ, then $\Psi(\Sigma) = \Psi(\sigma)$ is a quasi-isomorphism because so are the shuffle map and, by the Künneth theorem, the tensor product of the maps $\Psi(\sigma_{|i})$.

Otherwise we can split Σ up into two smaller complexes Σ_1 and Σ_2 with intersection $\Sigma_{12} = \Sigma_1 \cap \Sigma_2$. The naturality of Ψ gives us a map between the long exact sequence corresponding to the short exact sequence for $\mathbf{L}$ from Lemma 2.3 and the Mayer–Vietoris sequence for the CW complex $\mathcal{Z}(\Sigma) = \mathcal{Z}(\Sigma_1) \cup \mathcal{Z}(\Sigma_2)$

$$
\begin{array}{ccccccccc}
\to H_{*+1}(\mathbf{L}(\Sigma)) & \to & H_*(\mathbf{L}(\Sigma_{12})) & \to & H_*(\mathbf{L}(\Sigma_1)) \oplus H_*(\mathbf{L}(\Sigma_2)) & \to & H_*(\mathbf{L}(\Sigma)) & \to & \\
\downarrow{\scriptstyle \Psi_{*+1}(\Sigma)} & & \downarrow{\scriptstyle \Psi_*(\Sigma_{12})} & & \downarrow{\scriptstyle (\Psi_*(\Sigma_1), \Psi_*(\Sigma_2))} & & \downarrow{\scriptstyle \Psi_*(\Sigma)} & & \\
\to H_{*+1}(\mathcal{Z}(\Sigma)) & \to & H_*(\mathcal{Z}(\Sigma_{12})) & \to & H_*(\mathcal{Z}(\Sigma_1)) \oplus H_*(\mathcal{Z}(\Sigma_2)) & \to & H_*(\mathcal{Z}(\Sigma)) & \to & ,
\end{array} \tag{18}
$$

where we have written $\Psi_*(\Sigma)$ instead of $H_*(\Psi(\Sigma))$, etc. The five lemma together with induction implies that $\Psi_*(\Sigma)$ is an isomorphism. □

An analogous argument shows that the inclusion map

$$\mathbf{L}(\Sigma) \hookrightarrow \mathbf{K}(\Sigma) \tag{19}$$

is a quasi-isomorphism of dgcs.[2] If Σ has a single maximal simplex, we combine Lemma 2.2 with the Künneth theorem. Otherwise we again write $\Sigma = \Sigma_1 \cup \Sigma_2$ and compare the long exact sequences associated with both short exact sequences in Lemma 2.3.

This completes the proof of Theorem 2.1.

Remark 2.5 Theorems 1.1 and 2.1 remain valid for all generalized moment-angle complexes $\mathcal{Z}_\Sigma(D^n, S^{n-1})$ with even $n \geq 2$, up to the obvious degree shifts. In particular, the generators s_i and t_i in (4) are now of degrees $|s_i| = n-1$ and $|t_i| = n$. The singular n-simplex x is obtained by collapsing all but the last facet of the standard n-simplex to a point, and y is this last facet.

If $n \geq 3$ is odd, then $|y|$ is even and $|x|$ is odd. Proceeding as before, we get a quasi-isomorphism between $C^*(\mathcal{Z}_\Sigma(D^n, S^{n-1}))$ and the cdga $\tilde{\mathbf{B}}(\Sigma)$ with generators s_i of degree $n-1$ and $t_i = d\,s_i$ of degree n as well as relations

$$s_i\, s_i = s_i\, t_i = 0, \qquad \text{and} \qquad t_{i_1} \cdots t_{i_k} = 0 \quad \text{if } \{i_1, \ldots, i_k\} \notin \Sigma. \tag{20}$$

Note that the Stanley–Reisner relations are monomial and therefore independent of the order of the anticommuting variables t_i.

In general, such a quasi-isomorphism does not hold for the case $n = 1$, which we treat in the following section.

3 Real Moment-Angle Complexes

It is not difficult to adapt our approach to real moment-angle complexes

$$\mathcal{Z}_\mathbb{R}(\Sigma) = \mathcal{Z}_\Sigma(D^1, S^0) \subset (D^1)^m. \tag{21}$$

We start with the homological setting. As a chain complex, we define the analogue $L(\Sigma)$ of $\mathbf{L}(\Sigma)$ as before, except that now the degrees are $|u_i| = 1$ and $|v_i| = 0$ for all $i \in [m]$.

Let us consider the case $m = 1$ first. Writing $u = u_1$ and $v = v_1$, we turn $L(\Sigma)$ into a dgc via the augmentation $\varepsilon(v) = \varepsilon(u) = 0$ and the diagonal

$$\Delta v = v \otimes 1 + 1 \otimes v + v \otimes v, \tag{22}$$

$$\Delta u = u \otimes 1 + 1 \otimes u + u \otimes v. \tag{23}$$

Let x be the canonical path from $e = 1$ to $g = -1 \in S^0$, considered as a singular 1-simplex in $D^1 = [-1, 1]$, and let $y = g - e$. Then

[2] It is also a homotopy equivalence of complexes. See [7, Lemma 7.10] or [8, Lemma 3.2.6] for an explicit homotopy inverse.

$$d\,x = y, \qquad d\,y = 0, \tag{24}$$

$$\Delta y = g \otimes g - e \otimes e = y \otimes e + e \otimes y + y \otimes y, \tag{25}$$

$$\Delta x = x \otimes g + e \otimes x = x \otimes e + e \otimes x + x \otimes y. \tag{26}$$

Given that for $m = 1$ we either have $\Sigma = \{\varnothing\}$ or $\Sigma = \{\varnothing, \{1\}\}$, one verifies directly that the map

$$\Psi_{\mathbb{R}}(\Sigma)\colon L(\Sigma) \to C(\mathcal{Z}_{\mathbb{R}}(\Sigma)), \qquad 1 \mapsto e, \quad v \mapsto y, \quad u \mapsto x \tag{27}$$

is a quasi-isomorphism of dgcs. (Since this map is injective, one can also use it to justify that $L(\Sigma)$ with the diagonal given above is indeed a dgc.)

For $m > 1$, we again proceed exactly as before. We use the isomorphism of complexes

$$L(\sigma) = \bigotimes_{i=1}^{m} L(\sigma_{|i}) \tag{28}$$

analogous to Lemma 2.2 to define a dgc structure on $L(\sigma)$ for each simplex $\sigma \in \Sigma$ as well as a dgc map $L(\sigma) \to C(\mathcal{Z}_{\mathbb{R}}(\sigma))$. Then we extend both the dgc structure and the map to the colimit $L(\Sigma)$ of the $L(\sigma)$ over all $\sigma \in \Sigma$. The same proof as for Lemma 2.4 shows that the resulting map

$$\Psi_{\mathbb{R}}(\Sigma)\colon L(\Sigma) \to C(\mathcal{Z}_{\mathbb{R}}(\Sigma)) \tag{29}$$

is a quasi-isomorphism. In this step, one uses the obvious analogue of Lemma 2.3 for L instead of $\mathbf{L}$.

We thus have established the following.

Lemma 3.1 *The map $\Psi_{\mathbb{R}}(\Sigma)$ is a quasi-isomorphism of dgcs.*

We now turn to cohomology. The dual of the dgc $L(\Sigma)$ is the dga $B(\Sigma)$ with generators s_i of degree 0 and t_i of degree 1 satisfying the relations

$$d\,s_i = -t_i, \qquad d\,t_i = 0, \tag{30}$$

$$s_i\,s_i = s_i, \qquad t_i\,s_i = t_i, \qquad s_i\,t_i = 0, \qquad t_i\,t_i = 0, \qquad \prod_{j \in \sigma} t_j = 0 \tag{31}$$

for any $i \in [m]$ and $\sigma \notin \Sigma$ plus the rule that variables corresponding to *distinct* subscripts commute in the graded sense.

The minus sign in $d\,s_i = t_i$ comes from the general definition of the differential on the dual of a complex, *cf.* (5) and [17, eq. (II.3.1)]. For $m = 1$, we have

$$(d\,s)(x) = -(-1)^{|s|}\,s(d\,x) = -s(y) = -1 = -t(x). \tag{32}$$

The minus sign could be removed by replacing t_i with $-t_i$, that is, by mapping u to $-x$. The minus sign does not appear in [9, p. 512] because of a different sign convention for the dual complex.

We can sum up our discussion as follows.

Theorem 3.2 *There is a quasi-isomorphism of dgas*

$$C^*(\mathcal{Z}_{\mathbb{R}}(\Sigma)) \to B(\Sigma),$$

natural with respect to inclusions of subcomplexes.

We in particular recover Cai's isomorphism of graded rings [9, Sects. 3 & 4]

$$H^*(\mathcal{Z}_{\mathbb{R}}(\Sigma)) = H^*(B(\Sigma)). \tag{33}$$

In fact, our proof shares some similarities with Cai's. This would be even more so if we worked with cubical singular chains; compare [19]. We also remark that in the case of real moment-angle complexes, it is not necessary to pass to normalized (singular) chains. (The shuffle map is a morphism of dgcs for non-normalized chains already, and the formulas (24)–(26) do not need normalization, either.)

We discuss the dga $A(\Sigma)$ analogous to $\mathbf{A}(\Sigma)$ only for coefficients in $\mathbb{Z}_2$. General coefficients are considered in [13] where Theorem 3.2 is furthermore extended to real toric spaces, that is, to quotients of $\mathcal{Z}_{\mathbb{R}}(\Sigma)$ by freely acting subgroups of $(\mathbb{Z}_2)^m$.

In characteristic 2, the dga $A(\Sigma)$ has the same generators s_i and t_i as $B(\Sigma)$ and the relations

$$d\,s_i = t_i, \qquad d\,t_i = 0, \tag{34}$$

$$s_i\,s_i = s_i, \qquad t_i\,s_i = s_i\,t_i + t_i, \qquad \prod_{j\in\sigma} t_j = 0 \tag{35}$$

for $i \in [m]$ and $\sigma \notin \Sigma$, again with the additional rule that variables corresponding to different subscripts commute. Observe that the ideal generated by the relations (35) is closed under the differential, so that $A(\Sigma)$ is a well-defined dga. The projection map $A(\Sigma) \to B(\Sigma) \otimes \mathbb{Z}_2$ is again obtained by dividing out the ideal generated by the products $s_i\,t_i$ and t_i^2 for all $i \in [m]$, and it can be seen to be a quasi-isomorphism by an argument analogous to the one given before or to [7, Lemma 7.10].

The Stanley–Reisner ring $\mathbb{Z}_2[\Sigma]$, now with generators of degree 1, is contained in $A(\Sigma)$ as a sub-dga (with trivial differential). Moreover, if $\Sigma = [m]$ is the full simplex, then $A(\Sigma)$ is the Koszul resolution of $\mathbb{Z}_2$ over $R = \mathbb{Z}_2[t_1, \ldots, t_m]$. In general, $A(\Sigma)$ is the tensor product of this resolution and $\mathbb{Z}_2[\Sigma]$ over R, which gives the additive isomorphism

$$H^*(\mathcal{Z}_{\mathbb{R}}(\Sigma); \mathbb{Z}_2) = \operatorname{Tor}_R(\mathbb{Z}_2, \mathbb{Z}_2[\Sigma]). \tag{36}$$

It is not multiplicative for the canonical product on the torsion product, as can be seen for $\Sigma = \{\varnothing\}$ already, *cf.* [12, Sect. 10.3]. In this case, we have $\mathbb{Z}_2[\Sigma] = \mathbb{Z}_2$, so that the torsion product is a strictly exterior algebra. On the other hand, $\mathcal{Z}_{\mathbb{R}}(\Sigma) = (S^0)^m$ is a finite set of points. Hence, any cohomology class on $\mathcal{Z}_{\mathbb{R}}(\Sigma)$ is a $\mathbb{Z}_2$-valued function on these points and therefore squares to itself.

4 Comparison of Several Product Formulas

The aim of this section is to relate the product formula in the cohomology of a (complex) moment-angle complex with Baskakov's formula [3] and also the formula for real moment-angle complexes with the one claimed by Gitler and López de Medrano [14] as well as the one given by Bahri–Bendersky–Cohen–Gitler [2] for arbitrary polyhedral products.[3] We note that another description for a class of polyhedral products including all $\mathcal{Z}_\Sigma(D^n, S^{n-1})$ has been given by Zheng [22, Example 7.12].

We start with a variant of the generalized smash moment-angle complexes introduced in [2, Def. 2.2]. For a closed subset A of a compact Hausdorff space X and a basepoint $* \in A$, we define the space

$$\mathcal{S}(X, A) = \bigl\{\, x \in \mathcal{Z}(X, A) \bigm| x_i = * \text{ for some } i \in [m] \,\bigr\} \tag{37}$$

and based on it the pair

$$\widehat{\mathcal{Z}}_\Sigma(X, A) = \bigl(\mathcal{Z}_\Sigma(X, A), \mathcal{S}_\Sigma(X, A)\bigr). \tag{38}$$

We then have an isomorphism

$$H^*(\widehat{\mathcal{Z}}_\Sigma(X, A)) = H^*_c\bigl(\mathcal{Z}_\Sigma(X, A) \setminus \mathcal{S}_\Sigma(X, A)\bigr) = H^*_c\bigl(\mathcal{Z}_\Sigma(X \setminus *, A \setminus *)\bigr), \tag{39}$$

where $H^*_c(-)$ denotes cohomology with compact supports, *cf.* [18, Part I].

We now specialize to

$$\widehat{\mathcal{Z}}_{\mathbb{R}}(\Sigma) = \widehat{\mathcal{Z}}_\Sigma(D^1, S^0) \tag{40}$$

(where the basepoint is $e = 1 \in S^0$) and observe that

$$\begin{aligned} \mathcal{Z}_\Sigma(D^1, S^0) \setminus \mathcal{S}_\Sigma(D^1, S^0) &= \mathcal{Z}_\Sigma\bigl(D^1 \setminus \{e\}, S^0 \setminus \{e\}\bigr) \\ &= \mathcal{Z}_\Sigma\bigl([-1, 1), \{-1\}\bigr) \approx \mathcal{Z}_\Sigma\bigl([0, \infty), \{0\}\bigr) = C\,\Sigma, \end{aligned} \tag{41}$$

[3] Another cup product formula has been given by de Longueville [15, Thm. 1.1] for complements of real coordinate subspace arrangements. However, his formula is incorrect in general; see [14, Sect. 3] for a counterexample.

where $C\,\Sigma$ is the unbounded cone over the simplicial complex Σ. We think of Σ as embedded into the hyperplane of $\mathbb{R}^m$ with coordinate sum equal to 1.

The analysis of $\mathcal{Z}_{\mathbb{R}}(\Sigma)$ in the preceding section carries over to the present case. One simply ignores the element $e \in S^0$ and the counit 1 in the cochain algebra. (Recall that the cohomology with compact supports is a ring without unit in general.) The result is a quasi-isomorphism between the relative cochain algebra $C^*(\widehat{\mathcal{Z}}_{\mathbb{R}}(\Sigma))$ and the multiplicatively closed subcomplex $\hat{B}(\Sigma) \subset B(\Sigma)$ spanned by all m-fold products

$$a_1 \cdots a_m \qquad \text{where each } a_i = s_i \text{ or } t_i. \tag{42}$$

In particular, there is a multiplicative isomorphism

$$H_c^*(C\,\Sigma) = H^*(C\,\Sigma, \Sigma) \cong H^*(\hat{B}(\Sigma)) \tag{43}$$

where $C\,\Sigma$ denotes the bounded cone over Σ with base Σ. Not surprisingly, $\hat{B}(\Sigma)$ does not have a unit unless $\Sigma = \{\varnothing\}$.

We now compare $\hat{B}(\Sigma)$ to the dgas $\mathbf{B}(\Sigma)$ and $B(\Sigma)$ for complex and real moment-angle complexes, respectively. In the complex case, we have a direct sum decomposition of complexes

$$\mathbf{B}^*(\Sigma) = \bigoplus_{\alpha \subset [m]} \hat{B}^{*-|\alpha|}(\Sigma_\alpha) \tag{44}$$

where Σ_α is the full subcomplex of Σ on the vertex set α. This gives Hochster's formula

$$H^*(\mathcal{Z}(\Sigma)) = \bigoplus_{\alpha \subset [m]} H_c^{*-|\alpha|}(C\,\Sigma_\alpha) = \bigoplus_{\alpha \subset [m]} \tilde{H}^{*-|\alpha|-1}(\Sigma_\alpha), \tag{45}$$

cf. [7, Thm. 3.2.7], where we have used the additive isomorphism

$$H_c^*(C\,\Sigma) = \tilde{H}^{*-1}(\Sigma) \tag{46}$$

between the reduced cohomology of the simplicial complex Σ and the cohomology with compact supports of the unbounded cone over it. (Recall that $\tilde{H}^{-1}(\varnothing) = \mathbb{Z}$.)

The additive isomorphism (45) can be made multiplicative in the following way: For $\alpha \cap \beta \neq \varnothing$, the product

$$H_c^*(C\,\Sigma_\alpha) \otimes H_c^*(C\,\Sigma_\beta) \to H_c^*(C\,\Sigma_{\alpha\cup\beta}), \tag{47}$$

vanishes. For disjoint α, β, we use the cross product via the composition

$$H_c^*(C\,\Sigma_\alpha) \otimes H_c^*(C\,\Sigma_\beta) \xrightarrow{\times} H_c^*(C\,\Sigma_\alpha \times C\,\Sigma_\beta) \xrightarrow{\iota^*} H_c^*(C\,\Sigma_{\alpha\cup\beta}), \tag{48}$$

where $\iota\colon C\,\Sigma_{\alpha\cup\beta} \hookrightarrow C\,\Sigma_\alpha \times C\,\Sigma_\beta$ is the canonical inclusion. This is Baskakov's formula [3], expressed in terms of Cartesian products of cones and cohomology with compact supports instead of joins of simplices and reduced cohomology.

For a real moment-angle complex, we again have a direct sum decomposition

$$B(\Sigma) = \bigoplus_{\alpha\subset[m]} \hat{B}^*(\Sigma_\alpha), \tag{49}$$

hence also a Hochster formula

$$H^*(\mathcal{Z}_{\mathbb{R}}(\Sigma)) = \bigoplus_{\alpha\subset[m]} H_c^*(C\,\Sigma_\alpha) = \bigoplus_{\alpha\subset[m]} \tilde{H}^{*-1}(\Sigma_\alpha). \tag{50}$$

Note that there are no degree shifts by $|\alpha|$ this time. The isomorphism becomes multiplicative if one uses the following generalization of the product (48).

Recall that for any open subsets $U = X \setminus A$ and $V = X \setminus B$ of a compact Hausdorff space X, there is a cup product

$$\begin{aligned} H_c^*(U) \otimes H_c^*(V) &= H^*(X, A) \otimes H^*(X, B) \\ &\xrightarrow{\cup} H^*\big(X, A \cup B\big) = H_c^*(U \cap V), \end{aligned} \tag{51}$$

cf. also [18, Sect. 7.4].

Now let $\pi_\alpha\colon \mathcal{Z}_{\mathbb{R}}(\Sigma_{\alpha\cup\beta}) \to \mathcal{Z}_{\mathbb{R}}(\Sigma_\alpha)$ be the (proper) projection that sends the coordinates z_i with $i \notin \alpha$ to the basepoint e and analogously for $\pi_\beta\colon \mathcal{Z}_{\mathbb{R}}(\Sigma_{\alpha\cup\beta}) \to \mathcal{Z}_{\mathbb{R}}(\Sigma_\beta)$. We define the $*$-product as the composition

$$\begin{aligned} H_c^*(C\,\Sigma_\alpha) \otimes H_c^*(C\,\Sigma_\beta) &\xrightarrow{\pi_\alpha^*\otimes\pi_\beta^*} H_c^*(\pi_\alpha^{-1}(C\,\Sigma_\alpha)) \otimes H_c^*(\pi_\beta^{-1}(C\,\Sigma_\beta)) \\ &\xrightarrow{\cup} H_c^*\big(\pi_\alpha^{-1}(C\,\Sigma_\alpha) \cap \pi_\beta^{-1}(C\,\Sigma_\beta)\big) = H_c^*(C\,\Sigma_{\alpha\cup\beta}). \end{aligned} \tag{52}$$

In term of the isomorphism (43), this exactly means to multiply representatives lying in $\hat{B}(\Sigma_\alpha)$ and $\hat{B}(\Sigma_\beta)$ inside $B(\Sigma_{\alpha\cup\beta})$, which gives elements in $\hat{B}(\Sigma_{\alpha\cup\beta})$. Note that this construction reduces to the product (48) if α and β are disjoint.

The $*$-product is visibly graded commutative, something that was not obvious from the multiplication rules (31). Looking back, we can see that these asymmetric formulas arose from the non-commutativity of the Alexander–Whitney map and the fact that only one of the two vertices of the singular 1-simplex x in $X = D^1$ can be the basepoint e.

The product (52) coincides with the $*$-product given by Bahri–Bendersky–Cohen–Gitler [2, Thm. 1.4] because the former map can be thought of as induced by the partial diagonal

$$\hat{\Delta}_I^{J,L}\colon \hat{Z}(K_I) \to \hat{Z}(K_J) \wedge \hat{Z}(K_L) \tag{53}$$

defined in [2, eq. (1.5)] to construct the $*$-product. In our notation, we have $K = \Sigma$, $J = \alpha$, $L = \beta$, and $I = J \cup L = \alpha \cup \beta$. For the comparison, one uses that the compactly supported cohomology of

$$C\,\Sigma = \mathcal{Z}_\Sigma(X, A) \smallsetminus \mathcal{S}_\Sigma(X, A) \tag{54}$$

with $(X, A) = (D^1, S^0)$ is equal to the reduced cohomology of the quotient

$$\hat{Z}(K) = \hat{Z}(K(X, A)) = \mathcal{Z}_\Sigma(X, A) \,\big/\, \mathcal{S}_\Sigma(X, A). \tag{55}$$

considered in [2].

As remarked earlier, the product formula in [2] is valid for general polyhedral products. We can recover the version for complex moment-angle complexes if we replace the pair $\widehat{\mathcal{Z}}_{\mathbb{R}}(\Sigma) = \widehat{\mathcal{Z}}_\Sigma(D^1, S^0)$ from (40) by $\widehat{\mathcal{Z}}_\Sigma(D^2, S^1)$. In this case, the distinction between disjoint and non-disjoint index sets α and β in Baskakov's formula is not necessary for the corresponding $*$-product because two monomials of the form (42) with overlapping index sets always multiply to 0 in $\mathbf{B}(\Sigma)$.

We finally consider another description of $H^*(\mathcal{Z}(\Sigma))$ in the polytopal case. Let P be a simple polytope with m facets, and let Σ be the boundary complex of the dual simplicial polytope. For any subset $\alpha \subset [m]$, let $P_\alpha \subset P$ be the union of the corresponding facets.

Lemma 4.1 *There is a ring isomorphism*

$$\Theta_\alpha \colon H^*(P, P_\alpha) \to H^*_c(C\,\Sigma_\alpha)$$

for any $\alpha \subset [m]$. Moreover, the diagram

$$\begin{array}{ccc} H^*(P,P_\alpha)\otimes H^*(P,P_\beta) & \xrightarrow{\ \cup\ } & H^*(P,P_{\alpha\cup\beta}) \\ \downarrow{\scriptstyle \Theta_\alpha\otimes\Theta_\beta} & & \downarrow{\scriptstyle \Theta_{\alpha\cup\beta}} \\ H^*_c(C\,\Sigma_\alpha)\otimes H^*_c(C\,\Sigma_\beta) & \xrightarrow{\ *\ } & H^*_c(C\,\Sigma_{\alpha\cup\beta}) \end{array}$$

commutes for all $\alpha, \beta \subset [m]$.

Proof Let Σ' be the barycentric subdivision of Σ, considered as a triangulation of ∂P. As a topological space, Σ_α can be identified with a subcomplex of Σ', hence $C\,\Sigma_\alpha$ with a subcomplex of $C\,\Sigma' \approx P$. We can also identify P_α with the union of the closed blocks (or cells) in Σ' dual to the vertices in α, *cf.* [20, §64].

We claim that the canonical inclusion of pairs

$$(C\,\Sigma_\alpha, \Sigma_\alpha) \to (C\,\Sigma', P_\alpha) \tag{56}$$

is a strong deformation retract. Similar to the proof of [20, Lemma 70.1], we can define a strong deformation retraction that moves the vertex $v_\sigma \in C\,\Sigma'$ corresponding to a simplex $\sigma \in \Sigma$ to the vertex $v_{\sigma\cap\alpha} \in C\,\Sigma_\alpha$ along a straight line, which is inside σ if $\sigma \cap \alpha \neq \varnothing$. If σ has no vertex in α, then v_σ is moved to the apex $v_\varnothing$ of the cone, and $v_\varnothing$ is mapped to itself. We extend the map linearly to each simplex $\tau \in C\,\Sigma'$. If τ is contained in $\sigma \in \Sigma$, then it is mapped to the cone over the simplex $\sigma \cap \alpha \in \Sigma_\alpha$ (with the empty simplex $\varnothing$ giving the apex). The deformation retraction restricts to one from P_α onto Σ_α. We therefore get an isomorphism

$$\Theta_\alpha \colon H^*(P, P_\alpha) \to H^*(C\,\Sigma_\alpha, \Sigma_\alpha) = H^*_c(C\,\Sigma_\alpha) \tag{57}$$

in cohomology.

To show that the above diagram commutes, we work on the chain level. We use simplicial chains for the left-hand side of (56), which canonically map to singular chains on the right. We choose a vertex ordering for $C\,\Sigma_{\alpha\cup\beta}$ such that all vertices smaller than the apex $v_\varnothing$ are in α and all greater ones in β. (Some may be in both.) To a simplex $\sigma \in C\,\Sigma_{\alpha\cup\beta}$, we have to apply the Alexander–Whitney diagonal and possibly the projections from $C\,\Sigma_{\alpha\cup\beta}$ to $C\,\Sigma_\alpha$ and $C\,\Sigma_\beta$, which send "superfluous" vertices to $v_\varnothing$. Afterwards we evaluate the resulting tensor product on $a \otimes b$ where $a, b \in C(P)$ are cocycles vanishing on P_α and P_β, respectively.

Because of the way we have ordered the simplices, the following happens: If σ does not contain $v_\varnothing$, then the result is 0 for both ways of going through the diagram. Otherwise we obtain $(-1)^{|b||\sigma'|}\,a(\sigma')\,b(\sigma'')$ for both ways where σ' is the front face of σ ending in $v_\varnothing$ and σ'' the back face starting there. Hence the diagram commutes in either case. □

As a consequence, we get a ring isomorphism

$$H^*(\mathcal{Z}_{\mathbb{R}}(P)) = \bigoplus_{\alpha\subset[m]} H^*(P, P_\alpha) \tag{58}$$

where the multiplication on the right-hand side is given by the cup products

$$H^*(P, P_\alpha) \otimes H^*(P, P_\beta) \to H^*(P, P_{\alpha\cup\beta}) \tag{59}$$

for all $\alpha, \beta \subset [m]$. This description of the cohomology ring of a real moment-angle manifold was stated without proof by Gitler and López de Medrano [14, p. 1526].[4]

The Alexander-dual description for moment-angle manifolds,

$$H^*(\mathcal{Z}(\Sigma)) = \bigoplus_{\alpha\subset[m]} \tilde{H}_{d+m-|\alpha|-*}(P_\alpha) \tag{60}$$

where $d = \dim P - 1$, has been provided by Bosio–Meersseman [6, Thm. 10.1], with the product given up to sign by the intersection products

[4] In [16, p. 489], López de Medrano writes that "at the end [of [14]] we announced, prematurely, a formula for the cohomology ring of any [$\mathcal{Z}_{\mathbb{R}}(P)$], but the proof ran into some technical problems".

$$\tilde{H}_{d-k}(P_\alpha) \otimes \tilde{H}_{d-l}(P_\beta) \to \tilde{H}_{d-(k+l)}(P_{\alpha\cap\beta}) \tag{61}$$

for $\alpha, \beta \subset [m]$ with $\alpha \cup \beta = [m]$ and $k, l \geq 0$.

Acknowledgments I thank Don Stanley for his questions about dga models and Santiago López de Medrano for stimulating discussions. I also thank the organizers of the "Thematic Program on Toric Topology and Polyhedral Products" at the Fields Institute for creating an environment favourable to research and an anonymous referee for comments that helped to improve the presentation of the paper.

The author was supported by an NSERC Discovery Grant.

References

1. A. Bahri, M. Bendersky, F.R. Cohen, Polyhedral products and features of their homotopy theory, in Handbook of Homotopy Theory, ed. by H. Miller (CRC Press, Boca Raton, 2020), pp. 103–144
2. A. Bahri, M. Bendersky, F.R. Cohen, S. Gitler, Cup-products for the polyhedral product functor. Math. Proc. Cambridge Philos. Soc. **153**, 457–469 (2012)
3. I.V. Baskakov, Cohomology of K-powers of spaces and the combinatorics of simplicial divisions (Russian). Uspekhi Mat. Nauk **57**, 147–148 (2002); English translation: Russian Math. Surveys **57**, 989–990 (2002)
4. I.V. Baskakov, V.M. Buchstaber, T.E. Panov, Cellular cochain algebras and torus actions (Russian). Uspekhi Mat. Nauk **59**, 159–160 (2004); English translation: Russian Math. Surveys **59**, 562–563 (2004)
5. A. Berglund, Homotopy invariants of Davis–Januszkiewicz spaces and moment-angle complexes. unpublished notes (2010)
6. F. Bosio, L. Meersseman, Real quadrics in $\mathbf{C}^n$, complex manifolds and convex polytopes. Acta Math. **197**, 53–127 (2006)
7. V.M. Buchstaber, T.E. Panov, Torus Actions and Their Applications in Topology and Combinatorics (American Mathematical Society, Providence, RI, 2002)
8. V.M. Buchstaber, T.E. Panov, Toric Topology (American Mathematical Society, Providence, RI, 2015)
9. L. Cai, On products in real moment-angle manifolds. J. Math. Soc. Japan **69**, 503–528 (2017)
10. S. Eilenberg, J.C. Moore, Homology and fibrations I: Coalgebras, cotensor product and its derived functors. Comment. Math. Helv. **40**, 199–236 (1966)
11. M. Franz, On the integral cohomology of smooth toric varieties. arXiv:math/0308253v1
12. M. Franz, The cohomology rings of smooth toric varieties and quotients of moment-angle complexes. Geom. Topol. **25**, 2109–2144 (2021)
13. M. Franz, The cohomology rings of real toric spaces and smooth real toric varieties. Proc. Roy. Soc. Edinburgh Sect. A **152**(3), 720–737 (2021)
14. S. Gitler, S. López de Medrano, Intersections of quadrics, moment-angle manifolds and connected sums. Geom. Topol. **17**, 1497–1534 (2013)
15. M. de Longueville, The ring structure on the cohomology of coordinate subspace arrangements. Math. Z. **233**, 553–577 (1999)
16. S. López de Medrano, From singularities to polyhedral products, in Handbook of Geometry and Topology of Singularities II, ed. by C.-J.L. Molina, et al. (Springer, Cham, 2021), pp. 463–499
17. S. Mac Lane, Homology (Springer, New York, 1975)
18. W.S. Massey, Homology and Cohomology Theory (Dekker, New York, 1978)
19. W.S. Massey, Singular Homology Theory (Springer, New York, 1980)

20. J.R. Munkres, Elements of Algebraic Topology (Addison-Wesley, Menlo Park, CA, 1984)
21. T.E. Panov, N. Ray, Categorical aspects of toric topology, in Toric Topology (Osaka, 2006), ed. by M. Harada et al. (American Mathematical Society, Providence, RI, 2008), pp. 293–322
22. Q. Zheng, The cohomology algebra of polyhedral product spaces. J. Pure Appl. Algebra **220**, 3752–3776 (2016)

Duality in Toric Topology

Jelena Grbić and Matthew Staniforth

1 Introduction

The polyhedral product $(\mathbf{X}, \mathbf{A})^{\mathcal{K}}$ of topological pairs (X_i, A_i) is a subspace of the Cartesian product $\prod X_i$, which is specified by the face category of a simplicial complex $\mathcal{K}$. The homotopy theory of polyhedral product spaces is a rapidly evolving area of algebraic topology, and the tools of homotopy theory can often be enhanced using both algebraic and combinatorial techniques when being brought to bear on the study of polyhedral products.

A polyhedral product of particular interest in toric topology is the moment-angle complex, where $(X_i, A_i) = (D^2, S^1)$, which comes readily equipped with the action of a torus. A study of moment-angle complexes and related polyhedral products not only allows us to gain insight into these spaces themselves but also provides us with a framework within which we can investigate an interplay of homotopy theoretic, algebraic and combinatorial phenomena. In this paper, we investigate the interaction of duality phenomena in these areas.

An integral Poincaré duality space X is one whose action of its integral cohomology algebra on its integral homology satisfies Poincaré duality, that is, there exists $n \in \mathbb{N}$ and $[\mu] \in H_n(X)$ such that the cap product

$$[\mu] \frown (-)\colon H^l(X) \longrightarrow H_{n-l}(X)$$

is an isomorphism for all l. Any manifold satisfies Poincaré duality, but not every Poincaré duality space is a manifold. For example, an integral homology n-manifold, a space with the same integral local homology groups as $\mathbb{R}^n$, satisfies

J. Grbić (✉) · M. Staniforth
School of Mathematical Sciences, University of Southampton, Southampton, UK
e-mail: j.grbic@soton.ac.uk; m.staniforth@soton.ac.uk

A. Bahri et al. (eds.), *Toric Topology and Polyhedral Products*, Fields Institute Communications 89, https://doi.org/10.1007/978-3-031-57204-3_8

Poincaré duality. We characterise Poincaré duality moment-angle complexes $\mathcal{Z}_{\mathcal{K}}$ in terms of a duality of the underlying simplicial complex $\mathcal{K}$.

A generalised homology n-sphere (GHS^n) is a homology n-manifold with the homology of S^n. In 2015, Fan and Wang [6] characterised the simplicial complexes $\mathcal{K}$ for which the geometric realisation $|\mathcal{K}|$ is a GHS^n in terms of a duality condition on the homology and cohomology groups of full subcomplexes of $\mathcal{K}$, which we refer to as combinatorial Alexander duality. In Theorem 2.3, we show that the moment-angle complex $\mathcal{Z}_{\mathcal{K}}$ being a Poincaré duality space is equivalent to the condition that $\mathcal{K}$ exhibits combinatorial Alexander duality. As a corollary of our result and Cai's [5] classification of moment-angle manifolds, we obtain that there are no Poincaré duality moment-angle complexes, which are not manifolds.

Duality phenomena are ubiquitous across mathematics and are not limited to topology and combinatorics. A duality that appears in commutative algebra is Gorenstein duality: a property for a d-dimensional Noetherian ring R that is measured by the functor $\mathrm{Ext}^{d-t}(-, R)$. The Avramov-Golod theorem [3, Theorem 3.4.5] equivocates Gorenstein duality of the Stanley-Reisner ring $\mathbb{K}[\mathcal{K}]$, where $\mathbb{K}$ is a field, with Poincaré of its Tor-algebra. The framework of toric topology allows us to investigate an interplay of Gorenstein duality in the integral Stanley-Reisner ring $\mathbb{Z}[\mathcal{K}]$ with topological and combinatorial dualities, in the simplicial complex $\mathcal{K}$ and the moment-angle complex $\mathcal{Z}_{\mathcal{K}}$, respectively. Paraphrasing Stanley's result [7] on Gorenstein Stanley-Reisner rings $\mathbb{Z}[\mathcal{K}]$, in Theorem 2.5, we show that Poincaré duality in $\mathcal{Z}_{\mathcal{K}}$ is equivalent to Gorenstein duality in $\mathbb{Z}[\mathcal{K}]$, realising an interplay of algebraic, combinatorial and topological dualities. This complements the result proven by Buchstaber and Panov [4, Theorem 4.6.8] in the case of coefficients over a field.

As a Cartesian product of simplicial complexes is not a simplicial complex, a polyhedral product $(\mathcal{K}_{\mathbf{i}}, \mathcal{L}_{\mathbf{i}})^{\mathcal{K}}$ of simplicial pairs $(\mathcal{K}_i, \mathcal{L}_i)$ is not a simplicial complex. A related notion to the polyhedral product exists, where a simplicial complex, known as the polyhedral join product [8], is constructed as a union of join products of simplicial complexes. The special cases known as substitution complexes and composition complexes were studied by Abramyan-Panov [1] and Ayzenberg [2], respectively. In Theorem 2.7, using the polyhedral join product, we specify a family of polyhedral products, which satisfy Poincaré duality.

2 Duality of $\mathcal{Z}_{\mathcal{K}}$ and $\mathbb{Z}[\mathcal{K}]$

2.1 Preliminaries: The Cohomology of $\mathcal{Z}_{\mathcal{K}}$

For a positive integer m, a simplicial complex on the vertex set $[m] = \{1, ..., m\}$ is a subset of $2^{[m]}$, which is closed under taking subsets, and contains the empty set. We allow a simplicial complex $\mathcal{K}$ to contain ghost vertices, that is, we allow that there might exist $i \in [m]$ such that $\{i\} \notin \mathcal{K}$.

Definition 2.1 Let $\mathcal{K}$ be a simplicial complex on vertex set $[m]$, and denote by $(\mathbf{X}, \mathbf{A}) = \{(X_i, A_i)\}_{i=1}^m$ an m-tuple of CW pairs. The *polyhedral product* is defined as

$$(\mathbf{X}, \mathbf{A})^{\mathcal{K}} = \bigcup_{\sigma\in\mathcal{K}} (\mathbf{X}, \mathbf{A})^{\sigma} \subseteq \prod_{i=1}^{m} X_i, \text{ where } (\mathbf{X}, \mathbf{A})^{\sigma} = \prod_{i=1}^{m} Y_i,\ Y_i = \begin{cases} X_i & \text{for } i \in \sigma \\ A_i & \text{for } i \notin \sigma. \end{cases}$$

If $(X_i, A_i) = (X, A)$ for all i, we denote the polyhedral product by $(X, A)^{\mathcal{K}}$. When $(X_i, A_i) = (D^2, S^1)$ for all i, the polyhedral product is denoted by $\mathcal{Z}_{\mathcal{K}}$ and referred to as the moment-angle complex on $\mathcal{K}$.

We begin with a description of the integral cellular cochain complex $C^*(\mathcal{Z}_{\mathcal{K}}; \mathbb{Z})$ due to Panov and Buchstaber [4, Section 4.4].

Let d^m denote the m-dimensional unit ball in $\mathbb{C}^m$. The disk d^1 admits a decomposition into *three* cells: the basepoint $*$, the boundary circle S and the 2-cell D.

Taking products, we obtain a cellular decomposition of d^m. A cell e of d^m is a product of cells of d of the form $\prod_{i=1}^m Y_i$, where, for each i, Y_i is either the basepoint $*$, the 1-cell S or the 2-cell D. This can be phrased in terms of subsets of $[m]$. Each cell of d^m corresponds to exactly one pair (J, I) of subsets $J, I \subseteq [m]$, with $J \cap I = \emptyset$, and this pair characterises unique a cell of d^m. The subset J corresponds to the 1-cells, I corresponds to the 2-cells, and $[m]\backslash(J \cup I)$ corresponds to the 0-cells. We denote the cell corresponding to the pair (J, I) by $\kappa(J, I)$. The dimension of such a cell is

$$\dim \kappa(J, I) = |J| + 2|I|.$$

This CW structure on d^m induces a sub-CW structure on $\mathcal{Z}_{\mathcal{K}} \subseteq \mathrm{d}^m$. For $J \subseteq [m]$, we denote by $\mathcal{K}_J = \{\sigma \in \mathcal{K} \mid \sigma \subseteq J\}$ the full subcomplex of $\mathcal{K}$ on J. Then, $\kappa(J\backslash\sigma, \sigma) \in \mathcal{Z}_{\mathcal{K}}$ if and only if $\sigma \in \mathcal{K}_J$.

We denote by $C_*(\mathcal{K})$ and $C_*(\mathcal{Z}_{\mathcal{K}})$ the simplicial and cellular chain complexes of $\mathcal{K}$ and $\mathcal{Z}_{\mathcal{K}}$, respectively. Here and throughout, coefficients are taken to be in $\mathbb{Z}$, and we observe the convention that $C_{-1}(\mathcal{K}) \cong C^{-1}(\mathcal{K}) \cong \langle \emptyset_* \rangle = \mathbb{Z}$. There is the isomorphism of graded modules

$$h : \bigoplus_{J\subseteq[m]} C_*(\mathcal{K}_J) \xrightarrow{\cong} C_*(\mathcal{Z}_{\mathcal{K}}), \quad \sigma \mapsto \kappa(J\backslash\sigma, \sigma)$$

where the grading on the left-hand side is given by $|\sigma| = 2|\sigma| + |J|$ for $\sigma \in \mathcal{K}_J$.

By theorems of Hochster, Baskakov, Panov and Buchstaber, the map h induces the isomorphism of cohomology rings [4, Theorem 4.5.7]

$$h^* : H^*(\mathcal{Z}_{\mathcal{K}}) \cong \bigoplus_{J\subseteq[m]} \tilde{H}^*(\mathcal{K}_J) \tag{1}$$

where the ring structure on the right-hand side is induced by the cochain-level Baskakov product

$$C^{p-1}(\mathcal{K}_I)\otimes C^{q-1}(\mathcal{K}_J) \to C^{p+q-1}(\mathcal{K}_{I\cup J}),\ \sigma^*\otimes\tau^* \mapsto \begin{cases}(\sigma\cup\tau)^* & \text{if } I\cap J=\emptyset\\ 0 & \text{otherwise}\end{cases}$$

where $(\sigma\cup\tau)^*$ is zero if $\sigma\cup\tau\notin\mathcal{K}_{I\cup J}$ and otherwise denotes the cochain dual to $(\sigma\cup\tau)\in C_*(\mathcal{K}_{I\cup J})$.

2.2 *Poincaré Duality of $\mathcal{Z}_\mathcal{K}$*

For $d\in\mathbb{N}$, a CW complex X is an *n-Poincaré duality space* if there exists a class $[\mu]\in H_n(X)$ such that the cap product

$$[\mu]\frown(-)\colon H^l(X)\to H_{n-l}(X)$$

is an isomorphism for all l. Here, n is referred to as the Poincaré duality dimension of X, and $[\mu]$ is referred to as the fundamental class.

The characterisation of the structure of the cellular chains and cochains of $\mathcal{Z}_\mathcal{K}$ in terms of the simplicial chains of full subcomplexes of $\mathcal{K}$ allows us to reframe statements about Poincaré duality of $\mathcal{Z}_\mathcal{K}$ as statements about duality of the simplicial chains and cochains of full subcomplexes of $\mathcal{K}$.

We start with a description of the cap product in $\mathcal{Z}_\mathcal{K}$, on the cellular level, in terms of the combinatorics of the simplicial complex $\mathcal{K}$. We then exploit this combinatorial description to obtain a characterisation of Poincaré duality of $\mathcal{Z}_\mathcal{K}$ in terms of $\mathcal{K}$ and also in terms of the Stanley-Reisner ring $\mathbb{Z}[\mathcal{K}]$.

Proposition 2.2 *Let $\kappa(J\backslash\sigma,\sigma)\in C_*(\mathcal{Z}_\mathcal{K})$ and $\kappa(\widehat{J}\backslash\widehat{\sigma},\widehat{\sigma})^*\in C^*(\mathcal{Z}_\mathcal{K})$, corresponding to simplices $\sigma\in\mathcal{K}_J$ and $\widehat{\sigma}\in\mathcal{K}_{\widehat{J}}$, respectively. Then the cap product is given by*

$$\kappa(J\backslash\sigma,\sigma)\frown\kappa(\widehat{J}\backslash\widehat{\sigma},\widehat{\sigma})^* = \begin{cases}0 & \widehat{J}\nsubseteq J\\ 0 & \widehat{\sigma}\nsubseteq\sigma\\ \kappa\left((J\backslash\sigma)\backslash(\widehat{J}\backslash\widehat{\sigma}),\sigma\backslash\widehat{\sigma}\right) & \textit{otherwise.}\end{cases}$$

Proof For any CW complex, the cellular chain-level cup $\smile$ and cap $\frown$ products satisfy

$$\langle\alpha,\phi\smile\psi\rangle=\langle\alpha\frown\phi,\psi\rangle$$

for $\alpha\in C_{k+l}(\mathcal{Z}_\mathcal{K}),\phi\in C^l(\mathcal{Z}_\mathcal{K}),\psi\in C^k(\mathcal{Z}_\mathcal{K})$, where $\langle-,-\rangle$ denotes the evaluation pairing.

Let $\kappa(J\backslash\sigma,\sigma) \in C_*(\mathcal{Z_K})$ and $\kappa(\widehat{J}\backslash\widehat{\sigma},\widehat{\sigma})^* \in C^*(\mathcal{Z_K})$. We write the cap product $\kappa(J\backslash\sigma,\sigma) \frown \kappa(\widehat{J}\backslash\widehat{\sigma},\widehat{\sigma})^*$ in terms of generators $C_*(\mathcal{Z_K})$. For $L \subseteq [m]$ and $\tau \in \mathcal{K}_L$, the coefficient of a generator $\kappa(L\backslash\tau,\tau) \in C_*(\mathcal{Z_K})$ in $\kappa(J\backslash\sigma,\sigma) \frown \kappa(\widehat{J}\backslash\widehat{\sigma},\widehat{\sigma})^*$ is given by

$$\langle(\kappa(J\backslash\sigma,\sigma) \frown \kappa(\widehat{J}\backslash\widehat{\sigma},\widehat{\sigma})^*), \kappa(L\backslash\tau,\tau)^*\rangle = \langle(\kappa(J\backslash\sigma,\sigma), \kappa(\widehat{J}\backslash\widehat{\sigma},\widehat{\sigma})^* \smile \kappa(L\backslash\tau,\tau)^*\rangle.$$

Now,

$$\langle(\kappa(J\backslash\sigma,\sigma), \kappa(\widehat{J}\backslash\widehat{\sigma},\widehat{\sigma})^* \smile \kappa(L\backslash\tau,\tau)^*\rangle \neq 0$$

is equivalent to

$$(J\backslash\sigma,\sigma) = ((\widehat{J}\backslash\widehat{\sigma}) \cup (L\backslash\tau), \widehat{\sigma} \cup \tau).$$

Thus, $\kappa(J\backslash\sigma,\sigma) \frown \kappa(\widehat{J}\backslash\widehat{\sigma},\widehat{\sigma})^*$ is non-trivial if and only if $\widehat{J}\backslash\widehat{\sigma} \subseteq J\backslash\sigma$ and $\widehat{\sigma} \subseteq \sigma$, whence $\kappa(J\backslash\sigma,\sigma) \frown \kappa(\widehat{J}\backslash\widehat{\sigma},\widehat{\sigma})^* = \kappa((J\backslash\sigma)\backslash(\widehat{J}\backslash\widehat{\sigma}), \sigma\backslash\widehat{\sigma})$.

We show that n-Poincaré duality spaces $\mathcal{Z_K}$ are characterised by a duality in $\mathcal{K}$ referred to as combinatorial Alexander duality. A space X is a GHS^n if it is a homology n-manifold with the homology of S^n. Fan and Wang [6, Theorem 3.4] showed that for $\mathcal{K}$ a simplicial complex of dimension n on vertex set $[m]$, $\mathcal{K}$ is a GHS^n if and only if

$$\tilde{H}^l(\mathcal{K}_J) \cong \tilde{H}_{n-l-1}(\mathcal{K}_{[m]\backslash J})$$

for all $J \subseteq [m]$, $0 \leq l \leq n$. In this case, we say that $\mathcal{K}$ has n-dimensional combinatorial Alexander duality.

We are now ready to give our combinatorial classification of Poincaré duality moment-angle complexes.

Theorem 2.3 *Let $\mathcal{K}$ be a simplicial complex on $[m]$ with non-trivial cohomology. Then $\mathcal{Z_K}$ is an $(n+m)$-Poincaré duality space if and only if $\mathcal{K}$ satisfies $(n-1)$-dimensional combinatorial Alexander duality.*

Proof The sufficient implication is settled by a result of Cai [5, Corollary 2.10]; if $\mathcal{K}$ satisfies $(n-1)$-dimensional combinatorial Alexander duality and has non-trivial cohomology, then $\mathcal{Z_K}$ is an $(n+m)$-dimensional manifold.

We show the necessary implication. Let $\mathcal{K}$ be a simplicial complex on $[m]$, with non-trivial cohomology, and suppose that $\mathcal{Z_K}$ is an $(n+m)$-Poincaré duality space. We show first that $\mathcal{K}$ has the homology of S^{n-1}. We subsequently utilise this fact in showing that a certain chain is a representative of the fundamental class $[\mu] \in H_{n+m}(\mathcal{Z_K})$.

As the simplicial complex $\mathcal{K}$ has non-trivial cohomology, for some l, there exists $0 \neq [\tau] \in \tilde{H}^l(\mathcal{K})$. Let $\tau = \sum_j \alpha_j \tau_j^*$, where $\tau_j^* \in C^l(\mathcal{K})$ are basis cochains, corresponding to simplices τ_j. We show that l must equal $n-1$.

The image of $[\tau]$ under isomorphism (1) is the class

$$h^*([\tau]) = \left[\sum_j A_j \kappa([m]\backslash\tau_j, \tau_j)^*\right] \in H^{l+m+1}(\mathcal{Z}_\mathcal{K})$$

where $A_j = \text{sgn}(\tau_j, [m])\alpha_j$.

Let $0 \neq [\mu] \in H_{n+m}(\mathcal{Z}_\mathcal{K})$ denote the fundamental class, represented by $\mu = \sum_i a_i \kappa(J_i\backslash\sigma_i, \sigma_i)$. We evaluate the product

$$0 \neq [\mu] \frown h^*([\tau]) = \left[\sum_{i,j} a_i A_j \left(\kappa(J_i\backslash\sigma_i, \sigma_i) \frown \kappa([m]\backslash\tau_j, \tau_j)^*\right)\right] \in H_{n-(l+1)}(\mathcal{Z}_\mathcal{K}).$$

By Proposition 2.2,

$$\kappa(J_i\backslash\sigma_i, \sigma_i) \frown \kappa([m]\backslash\tau_j, \tau_j)^* \neq 0$$

implies that

$$\tau_j \subseteq \sigma_i, \text{ and } [m]\backslash\tau_j \subseteq J_i\backslash\sigma_i$$

and therefore

$$J_i = [m] \text{ and } \tau_j = \sigma_i.$$

Thus, the non-triviality of $\kappa(J_i\backslash\sigma_i, \sigma_i) \frown \kappa([m]\backslash\tau_j, \tau_j)^*$ implies that

$$\kappa(J_i\backslash\sigma_i, \sigma_i) \frown \kappa([m]\backslash\tau_j, \tau_j)^* = \kappa(\emptyset, \emptyset).$$

Therefore,

$$\begin{aligned}[\mu] \frown h^*([\tau]) &= \left[\sum_{i,j} \text{sgn}(\tau_j, [m]) a_i \alpha_{\tau_j} \left(\kappa(J_i\backslash\sigma_i, \sigma_i) \frown \kappa([m]\backslash\tau_j, \tau_j)^*\right)\right] \\ &= [A\kappa(\emptyset, \emptyset)] \in H_0(\mathcal{Z}_\mathcal{K})\end{aligned}$$

where $A \neq 0$. It follows that $h^*([\tau]) \in H^{n+m}(\mathcal{Z}_\mathcal{K})$, so that $[\tau] \in \tilde{H}^{n-1}(\mathcal{K})$ by the definition of the isomorphism h^*.

We have obtained that the $(n-1)$-st cohomology group of $\mathcal{K}$ is the only non-trivial cohomology group. It remains to show that $\tilde{H}^{n-1}(\mathcal{K}) \cong \mathbb{Z}$. By Poincaré duality, we have that $H^{n+m}(\mathcal{Z}_\mathcal{K}) \cong H_0(\mathcal{Z}_\mathcal{K}) \cong \mathbb{Z}$. By (1), $\tilde{H}^{n-1}(\mathcal{K})$ includes into $H^{n+m}(\mathcal{Z}_\mathcal{K}) \cong \mathbb{Z}$ as a component of a direct sum. It follows that $h^*: H^{n+m}(\mathcal{Z}_\mathcal{K}) \to \tilde{H}^{n-1}(\mathcal{K}) \cong \mathbb{Z}$ is an isomorphism. Therefore, $\mathcal{K}$ has the homology of S^{n-1}, as claimed. It follows that the fundamental class $[\mu] \in H_{n+m}(\mathcal{Z_K})$ can be represented by a cellular chain of the form $\mu = \sum_i a_i \kappa([m]\backslash\sigma_i, \sigma_i)$.

We now show that $\mathcal{K}$ has combinatorial Alexander duality, that is, for any $J \subseteq [m]$, and $0 \leq l \leq n+m$,

$$\tilde{H}^l(\mathcal{K}_J) \cong \tilde{H}_{n-l-2}(\mathcal{K}_{[m]\backslash J}).$$

By Poincaré duality, we have the sequence of isomorphisms

$$\bigoplus_{J\subseteq[m]} \tilde{H}^{\hat{l}-|J|-1}(\mathcal{K}_J) \cong H^{\hat{l}}(\mathcal{Z_K}) \cong H_{n+m-\hat{l}}(\mathcal{Z_K}) \cong \bigoplus_{L\subseteq[m]} \tilde{H}_{(n+m-\hat{l})-|L|-1}(\mathcal{K}_L)$$

given by the composition $(h_*)^{-1} \circ ([\mu] \frown (-)) \circ h^*$, where $[\mu] \in H_{n+m}(\mathcal{Z_K})$ denotes the fundamental class of $\mathcal{Z_K}$. Substituting $\hat{l} = l + |J| + 1$,

$$\begin{aligned}\bigoplus_{J\subseteq[m]} \tilde{H}^l(\mathcal{K}_J) &\cong H^{l+|J|+1}(\mathcal{Z_K}) \cong H_{n+m-(l+|J|+1)}(\mathcal{Z_K})\\ &\cong \bigoplus_{L\subseteq[m]} \tilde{H}_{n+m-l-|J|-|L|-2}(\mathcal{K}_L).\end{aligned}$$

Denoting the composite isomorphism by Φ, we show that Φ respects the direct sum decompositions. In particular, for all $J \subseteq [m]$,

$$\Phi(\tilde{H}^l(\mathcal{K}_J)) \subseteq \tilde{H}_{n+m-l-|J|-([m]-|J|)-2}(\mathcal{K}_{[m]\backslash J}) = \tilde{H}_{n-l-2}(\mathcal{K}_{[m]\backslash J}).$$

Suppose that for $J \subseteq [m]$, $\mathcal{K}_J$ has non-trivial cohomology. Otherwise, the statement follows vacuously. Let $0 \neq [\tau] \in \tilde{H}^l(\mathcal{K}_J)$ with representative cochain $\tau = \sum_j \alpha_j \tau_j^*$. Then

$$h^*([\tau]) = \left[\sum_j \mathrm{sgn}(\tau_j, J)\alpha_j \kappa(J\backslash\tau_j, \tau_j)^*\right] \in H^{l+|J|+1}(\mathcal{Z_K}).$$

Let $[\mu]$ denote the fundamental class of $\mathcal{Z_K}$ with representative chain $\mu = \sum_i a_i \kappa([m]\backslash\sigma_i, \sigma_i)$. Evaluating the cap product gives

$$\begin{aligned}[\mu] \frown h_c([\tau]) &= \left[\sum_{i,j} \mathrm{sgn}(\tau_j, J) a_i \alpha_j \kappa([m]\backslash\sigma_i, \sigma_i) \frown \kappa(J\backslash\tau_j, \tau_j)^*\right]\\ &= \left[\sum_{\hat{i},\hat{j}} A_{\hat{i},\hat{j}} \kappa([m]\backslash\sigma_{\hat{i}}, \sigma_{\hat{i}}) \frown \kappa(J\backslash\tau_{\hat{j}}, \tau_{\hat{j}})^*\right]\end{aligned}$$

$$= \left[\sum_{\hat{i},\hat{j}} A_{\hat{i},\hat{j}} \kappa(([m]\backslash\sigma_{\hat{i}})\backslash(J\backslash\tau_{\hat{j}}), \sigma_{\hat{i}}\backslash\tau_j) \right]$$

$$= \left[\sum_{\hat{i},\hat{j}} A_{\hat{i},\hat{j}} \kappa(([m]\backslash J)\backslash(\sigma_{\hat{i}}\backslash\tau_{\hat{j}}), \sigma_{\hat{i}}\backslash\tau_j) \right] \in H_{n+m-(l+|J|+1)}(\mathcal{Z_K})$$

where $\hat{i}, \hat{j}$ are the pairs for which the cap product $\kappa([m]\backslash\sigma_{\hat{i}}, \sigma_{\hat{i}}) \frown \kappa(J\backslash\tau_{\hat{j}}, \tau_{\hat{j}})^*$ is non-trivial and $A_{\hat{i},\hat{j}} = \mathrm{sgn}(\tau_j, J)a_i\alpha_j \neq 0$. The last equality follows since both J and σ_i contain τ_j, and $J \cap \sigma_i = \tau_j$ since the cap product is non-trivial.

The image of $[\mu] \frown h^*([\tau])$ under the inverse of the homology isomorphism of (1) is

$$(h_*)^{-1}([\mu] \frown h_c([\tau]) = \left[\sum_{\hat{i},\hat{j}} A_{\hat{i},\hat{j}} \sigma_{\hat{i}}\backslash\tau_{\hat{j}} \right] \in \tilde{H}_{n-l-2}(\mathcal{K}_{[m]\backslash J}).$$

We therefore have that under the composition of isomorphisms

$$(h_h)^{-1} \circ ([\mu] \frown (-)) \circ h_c \colon \bigoplus_{J \subseteq [m]} \tilde{H}^{l-|J|-1}(\mathcal{K}_J) \to \bigoplus_{L \subseteq [m]} \tilde{H}_{(n+m-l)-|L|-1}(\mathcal{K}_L)$$

the image of each of the groups $\tilde{H}^l(\mathcal{K}_J)$ is contained in $\tilde{H}_{n-l-2}(\mathcal{K}_{[m]\backslash J})$. These groups are therefore isomorphic, and $\mathcal{K}$ therefore satisfies $(n-1)$-dimensional combinatorial Alexander duality.

We obtain a relation between Poincaré duality of moment-angle complexes and combinatorial Alexander duality of simplicial complexes and conclude that there are no Poincaré duality moment-angle complexes that are not manifolds.

Corollary 2.4 *Let $\mathcal{K}$ be a simplicial complex on $[m]$ with non-trivial cohomology. Then the following are equivalent:*

1. *$\mathcal{Z_K}$ is an $(n+m)$-Poincaré duality space over $\mathbb{Z}$.*
2. *$\mathcal{K}$ has $(n-1)$-dimensional combinatorial Alexander duality.*
3. *$\mathcal{Z_K}$ is an $(n+m)$-dimension manifold.*

2.3 Gorenstein Duality of $\mathbb{Z}[\mathcal{K}]$

A Noetherian ring satisfies Gorenstein duality if its localisation at every maximal ideal exhibits a certain form of self-duality. In this paper, we relate the Gorenstein duality of Stanley-Reisner rings of simplicial complexes to Poincare duality of moment-angle complexes $\mathcal{Z_K}$.

Let $\mathcal{K}$ be a simplicial complex on vertex set $[m]$ and R a commutative ring. The Stanley-Reisner ring is

$$R[\mathcal{K}] = R[v_1, ..., v_m]/\mathcal{I}_{\mathcal{K}}$$

where

$$\mathcal{I}_{\mathcal{K}} = (v_{i_1}...v_{i_j} \mid \{i_1, ..., i_j\} \notin \mathcal{K})$$

is the Stanley-Reisner ideal, that is, the ideal generated by monomials corresponding to missing faces of $\mathcal{K}$.

By a result of Stanley [7, Theorem 5.1], the Stanley-Reisner ring $\mathbb{Z}[\mathcal{K}]$ having Gorenstein duality is equivalent to $\mathcal{K}^*$ being an integral generalised homology d-sphere, where $\mathcal{K}^* = \mathcal{K}_{\{v \in [m] \mid st_{\mathcal{K}}(v) \neq \mathcal{K}\}}$ is the core of $\mathcal{K}$ and d is the dimension of $\mathcal{K}^*$.

Notice that if $\mathcal{K}$ has non-trivial cohomology, then $\mathcal{K} = \mathcal{K}^*$. Theorem 2.3 together with Stanley's [7, Theorem 5.1] relates Poincaré duality of moment-angle complexes, Gorenstein duality of Stanley-Reisner rings and combinatorial Alexander duality of simplicial complexes. We obtain an interplay between algebraic, combinatorial and topological dualities.

Theorem 2.5 *Let $\mathcal{K}$ be a simplicial complex on $[m]$ with non-trivial cohomology. The following are equivalent:*

1. *$\mathcal{Z}_{\mathcal{K}}$ is an $(n+m)$-Poincaré duality space.*
2. *$\mathcal{K}$ has $(n-1)$-dimensional combinatorial Alexander duality.*
3. *$\mathbb{Z}[\mathcal{K}]$ has Gorenstein duality.*

2.4 The Polyhedral Join Product

We extend our characterisation of Poincaré duality in $\mathcal{Z}_{\mathcal{K}}$ by utilising the polyhedral join product of simplicial complexes.

Definition 2.6 Let $\mathcal{K}$ be a simplicial complex on $[m]$, and for $1 \leq i \leq m$, let $(\mathcal{K}_i, \mathcal{L}_i)$ be a simplicial pair on $[l_i]$, where the sets $[l_i]$ are pairwise disjoint. The *polyhedral join product* is the simplicial complex on vertex set $[l_1] \sqcup ... \sqcup [l_m]$, defined as

$$(\mathcal{K}_{\mathbf{i}}, \mathcal{L}_{\mathbf{i}})^{*\mathcal{K}} = \bigcup_{\sigma \in \mathcal{K}} (\mathcal{K}_{\mathbf{i}}, \mathcal{L}_{\mathbf{i}})^{*\sigma} \text{ where } (\mathcal{K}_{\mathbf{i}}, \mathcal{L}_{\mathbf{i}})^{*\sigma} = \mathop{*}_{i=1}^{m} \mathcal{Y}_i, \; \mathcal{Y}_i = \begin{cases} \mathcal{K}_i & i \in \sigma \\ \mathcal{L}_i & \text{otherwise.} \end{cases}$$

Let $l = \sum_{i=1}^{m} l_i$ where $l_i \geq 1 \; \forall i$, and let $(\mathbf{X}, \mathbf{A})$ be an l-tuple of CW complexes, partitioned into m distinct l_i tuples with $\mathbf{X_i} = \{X_{ij}\}_{j=1}^{l_i}$ and $\mathbf{A_i} = \{A_{ij}\}_{j=1}^{l_i}$. It was proven by Vidaurre [8, Theorem 2.9] that the polyhedral join product and the polyhedral product interact in the following way:

$$(\mathbf{X}, \mathbf{A})^{(\mathcal{K}_\mathbf{i}, \mathcal{L}_\mathbf{i})^{*\mathcal{K}}} = \left((\mathbf{X}_\mathbf{i}, \mathbf{A}_\mathbf{i})^{\mathcal{K}_\mathbf{i}}, (\mathbf{X}_\mathbf{i}, \mathbf{A}_\mathbf{i})^{\mathcal{L}_\mathbf{i}}\right)^{\mathcal{K}}. \tag{2}$$

We make use of this fact in extending our classification of Poincaré duality moment-angle complexes to polyhedral products whose entries are themselves moment-angle complexes.

Proposition 2.7 *Let $\mathcal{K}$ be a simplicial complex on $[m]$ and, for $1 \leq i \leq m$, $(\mathcal{K}_i, \mathcal{L}_i)$ a simplicial pair on $[l_i]$. Suppose that the polyhedral join $(\mathcal{K}_\mathbf{i}, \mathcal{L}_\mathbf{i})^{*\mathcal{K}}$ has non-trivial cohomology. Then the following are equivalent:*

1. *The polyhedral product $(\mathcal{Z}_{\mathcal{K}_\mathbf{i}}, \mathcal{Z}_{\mathcal{L}_\mathbf{i}})^{\mathcal{K}}$ is a Poincaré duality space.*
2. *The polyhedral join product $(\mathcal{K}_\mathbf{i}, \mathcal{L}_\mathbf{i})^{*\mathcal{K}}$ has combinatorial Alexander duality.*
3. *The Stanley-Reisner ring $\mathbb{Z}[(\mathcal{K}_\mathbf{i}, \mathcal{L}_\mathbf{i})^{*\mathcal{K}})]$ has Gorenstein duality.*

Proof The equivalence of (i) and (ii) follows from (2) together with Theorem 2.5. The equivalence of (ii) and (iii) follows from Theorem 2.5. □

Example 2.8

1. Let $\mathcal{K} = \partial\Delta^1$, and let $(\mathcal{K}_1, \mathcal{L}_1) = (\mathcal{K}_2, \mathcal{L}_2) =($  $)$, where $\circ$ denotes a ghost vertex. Then $(\mathcal{K}_\mathbf{i}, \mathcal{L}_\mathbf{i})^{*\mathcal{K}}$ is a 6-vertex triangulation of S^2 and in particular is a generalised homology sphere, so that

$$(\mathcal{Z}_{\mathcal{K}_\mathbf{i}}, \mathcal{Z}_{\mathcal{L}_\mathbf{i}})^{\mathcal{K}} = (S^3 \times D^2, S^3 \times S^1)^{\partial\Delta^1}$$

is a Poincaré duality space. Indeed, applying (2), and realising this 6-vertex triangulation of S^2 as $\partial\Delta^1 * \partial\Delta^1 * \partial\Delta^1$, we have

$$(S^3 \times D^2, S^3 \times S^1)^{\partial\Delta^1} = \mathcal{Z}_{\partial\Delta^1} \times \mathcal{Z}_{\partial\Delta^1} \times \mathcal{Z}_{\partial\Delta^1} = S^3 \times S^3 \times S^3.$$

2. Generalising the previous example, let $\mathcal{K} = \partial\Delta^1$ and let $(\mathcal{K}_1, \mathcal{L}_1) = (\mathcal{K}_2, \mathcal{L}_2) = (\partial\Delta^{n-1} * \{v\}, \partial\Delta^n * \{\circ\})$, where $\circ$ denotes a ghost vertex. Then $(\mathcal{K}_\mathbf{i}, \mathcal{L}_\mathbf{i})^{*\mathcal{K}}$ is a $(2n+2)$-vertex triangulation of the $(n+1)$-sphere, and thus

$$(\mathcal{Z}_{\mathcal{K}_\mathbf{i}}, \mathcal{Z}_{\mathcal{L}_\mathbf{i}})^{\mathcal{K}} = (S^{2n-1} \times D^2, S^{2n-1} \times S^1)^{\partial\Delta^1}$$

is a Poincaré duality space.

3. Let $\mathcal{K} = \partial\Delta^1 * \{j\}$, $(\mathcal{K}_1, \mathcal{L}_1) = (\mathcal{K}_2, \mathcal{L}_2) = (\{v\}, \{\emptyset\})$ and $(\mathcal{K}_j, \mathcal{L}_j) = ($ $, \emptyset)$. Then $(\mathcal{K}_\mathbf{i}, \mathcal{L}_\mathbf{i})^{*\mathcal{K}}$ is a 7-vertex triangulation of S^2. Here, $(\mathcal{Z}_{\mathcal{K}_1}, \mathcal{Z}_{\mathcal{L}_1}) = (S^3, T^2)$ and $(\mathcal{Z}_{\mathcal{K}_2}, \mathcal{Z}_{\mathcal{L}_2}) = ((S^3 \times S^4)^{\#5}, T^5)$, so that

$$\left((S^3, T^2), ((S^3 \times S^4)^{\#5}, T^5)\right)^{\Delta^1}$$

is a Poincaré duality space.

These examples demonstrate that there are a variety of polyhedral join products that give rise to Poincaré duality spaces $\mathcal{Z}_{\mathcal{K}}$. The classification of polyhedral join products that are generalised homology spheres in the special case of composition complexes enables us to extend our results on duality. Recall that composition complexes are the special case of the polyhedral join product where for all i, $\mathcal{K}_i = \Delta^{n_i}$, $n_i \geq 1$. Ayzenberg [2, Theorem 6.6] proved that the composition complex $\mathcal{K}(\mathcal{K}_1, ..., \mathcal{K}_m) = (\Delta^{n_i}, \mathcal{K}_i)^{*\mathcal{K}}$ is a generalised homology sphere if and only if $\mathcal{K}$ is a generalised homology sphere; for any non-ghost vertex i of $\mathcal{K}$, $\mathcal{K}_i = \partial\Delta^{l_i-1}$, and for any ghost vertex i of $\mathcal{K}$, $\mathcal{K}_i$ is a generalised homology sphere.

Utilising this result together with (2), and the fact that the polyhedral product is a homotopy functor [4, Proposition 8.1.1], we obtain the following corollary.

Corollary 2.9 *Let $\mathcal{K}$ be a complex on $[m]$ with no ghost vertices, and let $\mathcal{K}_1, ..., \mathcal{K}_m$ be complexes on $[l_1], ..., [l_m]$, respectively. Then, $(\mathbf{C}\mathcal{Z}_{\mathcal{K}_i}, \mathcal{Z}_{\mathcal{K}_i})^{\mathcal{K}}$ is a Poincaré duality space if and only if $\mathcal{K}$ is a generalised homology sphere, and for all i, $\mathcal{K}_i = \partial\Delta^{[l_i]}$.*

Example 2.10

1. Let $\mathcal{K}$ be a simplicial complex on $[m]$. For $1 \leq i \leq m$, let $l_i \geq 2$. Then $(CT^{l_i}, T^{l_i})^{\mathcal{K}}$ is a Poincaré duality space if and only if $\mathcal{K}$ consists solely of ghost vertices.
2. Let $\mathcal{K}$ be a simplicial complex on $[m]$, and for $1 \leq i \leq m$, let $l_i \geq 1$. Then $(D^{2l_i}, S^{2l_i-1})^{\mathcal{K}}$ is a Poincaré duality space if and only if $\mathcal{K}$ is a generalised homology sphere.

References

1. S.A. Abramyan, T.E. Panov, Higher Whitehead products for moment-angle complexes and substitutions of simplicial complexes. Tr. Mat. Inst. Steklova **305**(Algebraicheskaya Topologiya Kombinatorika i Matematicheskaya Fizika), 7–28 (2019)
2. A.A. Ayzenberg, Substitutions of polytopes and of simplicial complexes, and multigraded Betti numbers. Trans. Moscow Math. Soc. **74**, 175–202 (2013)
3. W. Bruns, H.J. Herzog, Cohen-Macaulay rings, in *Cambridge Studies in Advanced Mathematics*, 2nd edn. (Cambridge University, Cambridge, 1998)
4. V.M. Buchstaber, T.E. Panov, Toric topology, in *Mathematical Surveys and Monographs*, vol. 204 (American Mathematical Society, Providence, RI, 2015)
5. L. Cai, On products in a real moment-angle manifold. J. Math. Soc. Japan **69**(2), 503–528 (2017)
6. F. Fan, X. Wang, On the cohomology of moment-angle complexes associated to Gorenstein* complexes. arXiv:1508.00159 (2015), pp. 1–49
7. R.P. Stanley, Combinatorics and commutative algebra, in *Progress in Mathematics*, vol. 41, 2nd edn. (Birkhäuser Boston, Inc., Boston, MA, 1996)
8. E.Vidaurre, On polyhedral product spaces over polyhedral joins. Homology Homotopy Appl. **20**(2), 259–280 (2018)

Bundles over Connected Sums

Lisa Jeffrey and Paul Selick

1 Introduction

Let A be a connected sum $A \cong B\#C$ of n-manifolds. See, for example, Hatcher [1] for the definition of connected sum. Let $F \to L \to C$ be a bundle over C where F is a manifold.

Using the definition, we get a map $A \to C$. Let $F \to M \to A$ be the pullback of the bundle $F \to L \to C$ to A.

Letting B' denote the complement of a chart in B and setting $X' := (B' \times F)/(* \times F)$, we prove the following. There is a cofibration $M \to L \to \Sigma X'$ for which the corresponding long exact homology/cohomology sequences split to give $H_*(M) \cong H_*(X') \oplus H_*(L)$ and $H^*(M) \cong H^*(X') \oplus H^*(L)$. (See Theorems 3.1 and 3.3.)

These results suggest the possibility that M is the connected sum of L and some manifold X whose $(n-1)$ skeleton is homotopy equivalent to X', but we give an example to show that this is not necessarily the case. (See Example 3.4.) As we shall see, if $M \simeq X\#L$, then the cofibration sequence $X' \to M \to L$ would have to split to give $M \simeq X' \vee L$, but this fails in Example 3.4.

In the final section, we consider bundles over some 6-manifolds including the case where A is a symplectic manifold and the M is the total space of its associated prequantum line bundle. We find that in that case, $M \simeq \#^{2r}(S^3 \times S^4)\#L$ where L is a 7-manifold whose nonzero cohomology groups are $\mathbb{Z}$ in degrees $0, 7$ and $\mathbb{Z}/k$ in degree 4, where r and k are determined by the cohomology of A. (See Theorem 4.1.)

The authors would like to thank Sebastian Chenery who pointed out an error in an earlier version of this paper.

L. Jeffrey (✉) · P. Selick
Department of Mathematics, University of Toronto, Toronto, ON, Canada
e-mail: jeffrey@math.toronto.edu; selick@math.toronto.edu

A. Bahri et al. (eds.), *Toric Topology and Polyhedral Products*, Fields Institute Communications 89, https://doi.org/10.1007/978-3-031-57204-3_9

For topological spaces X and Y, let $X \cong Y$ denote "X is homeomorphic to Y", and let $X \simeq Y$ denote "X is homotopy equivalent to Y".

2 Connected Sums

Let D^n denote the closed disk $D^n := \{x \in \mathbb{R}^n \mid \|x\| \leq 1\}$.

Lemma 2.1 *For any points a, b in the interior of D^n, there exists a self-diffeomorphism $f : D^n \to D^n$ such that $f(a) = b$ and $f|_{\partial D^n}$ is the identity.*

Proof Set $f(a) = b$. For $x \neq a$, let X_x be the point at which the production of the line segment joining a to x meets ∂D^n. Then $x = ta + (1 - t)X_x$ for some t. Set $f(x) = tb + (1 - t)X_x$. □

More generally, we have:

Lemma 2.2 *Let U_p, V_q be subcharts of D^n. Then there exists a self-diffeomorphism $f : D^n \to D^n$ such that the restriction of f to U_p is the standard diffeomorphism on open balls and such that $f|_{\partial D^n}$ is the identity.*

For a point p in an n-manifold X, define a subchart around p to be an open neighbourhood U_p of p, which is diffeomorphic to an open ball in $\mathbb{R}^n$ within some chart of X.

For a connected n-manifold X, let $X' = X \setminus D^n$ denote the complement of a subchart of X.

Lemma 2.3 *Up to diffeomorphism, X' is independent of the choice of the subchart removed.*

Proof Let U_p, U_q be subcharts of X. In the special case where there exists a chart W containing both $\bar{U}_p$ and $\bar{U}_q$, this follows from the earlier lemma. Then for arbitrary U_p, U_q, find a finite (by compactness) chain of charts connecting U_p to V_q, using connectivity. □

After removal of the subchart, there is a deformation retraction $X' \simeq X^{(n-1)}$ to the $(n-1)$-skeleton of X. Let $f_X : S^{n-1} \to X'$ denote the attaching map of the top cell of X.

Suppose that X, Y are simply connected oriented n-manifolds.

In a connected sum, $X\#Y = X' \cup_{S^{n-1}\times I} Y'$ (where the orientation on one of the inclusions $S^{n-1} \times \{0\} \hookrightarrow X'$ or $S^{n-1} \times \{1\}$ is reversed so that $X\#Y$ inherits an orientation), there is a canonical projection $X\#Y \to X$. Similarly, we have $X\#Y \to Y$. The canonical projections $X\#Y \to X$ and $X\#Y \to Y$ preserve the orientation class. That is, they induce isomorphisms on $H_n(\,)$.

Collapsing the centre of the tube $S^{n-1} \times I$ within $X\#Y$ gives a map $X\#Y \to X' \vee Y'$. If we form $(X\#Y)'$ by choosing the subchart to be removed to be within the centre of the tube, then collapsing to produce $X' \vee Y'$ has collapsed a contractible subset of $(X\#Y)'$ giving a homotopy equivalence $(X\#Y)' \xrightarrow{\simeq} X' \vee Y'$.

By writing $S^n = S^n\#S^n$ and considering naturality of the pinch, we see that the homotopy class of the attaching map of the top cell in $X\#Y$ is given by $f_{X\#Y} = f_X + f_Y$ within $\pi_n(X) \oplus \pi_n(Y) \subset \pi_n(X' \vee Y')$.

Choosing the subchart to be removed from $X\#Y$ to be within Y' gives a (non-canonical) inclusion $X' \hookrightarrow (X\#Y)'$ with $(X\#Y)'/X' \cong Y'$. The composite $X' \hookrightarrow (X\#Y)' \to X$ with the canonical projection is an injective map from a compact Hausdorff space, so it is a homeomorphism to its image. Composing with the inverse of this homeomorphism is a left splitting of the inclusion $X' \hookrightarrow (X\#Y)'$. Similarly, there is a left splitting of the inclusion $Y' \hookrightarrow (X\#Y)'$.

Lemma 2.4 *Let M be a closed n-manifold, and let $A \subset M'$ be a closed n-dim subset of M with $\partial \bar{A} \cong S^{n-1}$. Then $M \cong N\#X$ for some manifolds N and X with $N' = A$. Furthermore, the canonical projection $M' \to N' = A$ is a left splitting of the inclusion $A \hookrightarrow M'$.*

Proof Set $\hat{X} := M \setminus A$. Then $\hat{X}$ is a manifold-with-boundary with $\partial \hat{X} = \partial \bar{A}$. Let $T \cong S^{n-1} \times I$ be a tubular neighbourhood of $\partial \hat{X}$ in $\hat{X}$ and set $X' := \hat{X} \setminus T$. Then

$$M = \bar{A} \cup_{S^{n-1}\times\{0\}} T \cup_{S^{n-1}\times\{1\}} \overline{X'}$$

so $M = N\#X$ where $N = A \cup_{(T\times\{0\})} D^n$ and $X = X' \cup_{(T\times\{1\})} D^n$. By construction, $A \hookrightarrow M' \to N' = A$ is the identity on A. □

3 The Cofibration Sequence Associated with a Bundle over a Connected Sum

For definitions and properties of principal cofibrations used in this section, see pp. 56–61 of [3].

Let B, C be closed n-manifolds and let $A := B\#C$. Suppose

$$F \to L \to C$$

is a (locally trivial) fibre bundle whose fibre F is a manifold. Then L is a manifold of dimension $n + \dim(F)$, which we will denote by m.

Let $F \to M \to A$ be the pullback of the bundle under the canonical projection $A \to C$. The total space M is a manifold of dimension m.

Let $\hat{L}$ be the total space of the restriction of the bundle to $C' := C \setminus \text{chart}$.

By definition,

$$A = B' \cup_{S^{n-1}\times I} C'$$

where, by construction, the restriction of the bundle to B' is trivial.

Taking inverse images under the bundle projection $M \twoheadrightarrow A$ gives

$$M = (B' \times F) \cup_{(S^{n-1} \times I \times F)} \hat{L}.$$

In other words, we have

$$\begin{array}{ccccc} S^{n-1} \times I \times F & \hookrightarrow & (B' \times F) & \longrightarrow & (B' \times F)/(S^{n-1} \times I \times F) \\ \downarrow & & \downarrow & & \parallel \\ \hat{L} & \hookrightarrow & M & \longrightarrow & M/\hat{L} \end{array}$$

where the left square is a pushout.

The space

$$\begin{aligned} M/\hat{L} &= (B' \times F)/(S^{n-1} \times I \times F) \\ &= \big(B'/(S^{n-1} \times I) \times F\big)/(* \times F) \\ &= (B \times F)/(* \times F). \end{aligned}$$

has the same homology as $B \vee (B \wedge F)$. In fact, if F is a suspension, then $(B \times F)/(* \times F) \simeq B \vee (B \wedge F)$ (Selick, [3] Prop 7.7.8).

Set $X' := (B' \times F)/(* \times F)$.

Theorem 3.1 *There is a cofibration diagram*

$$\begin{array}{ccccc} M' & \longrightarrow & L' & \longrightarrow & \Sigma X' \\ \downarrow & & \downarrow & & \parallel \\ M & \longrightarrow & L & \longrightarrow & \Sigma X' \end{array}$$

(i.e. the rows are cofibrations and the right square is a pushout.)

Proof We had

$$M/\hat{L} = (B' \times F)/(S^{n-1} \times I \times F)$$

Also, since $L = \hat{L} \cup_{S^{n-1} \times I \times F} F$, we have

$$L/\hat{L} = (S^n \times F)/(* \times F)$$

(which can be regarded as the special case of the preceding with $B = S^n$). Thus, we have a diagram

$$\begin{array}{ccccc} & & & & X' \\ & & & & \downarrow \\ \hat{L} & \longrightarrow & M & \longrightarrow & M/\hat{L} = (B\times F)/(*\times F) \\ \| & & \downarrow & & \downarrow \\ \hat{L} & \longrightarrow & L & \longrightarrow & L/\hat{L} = (S^n\times F)/(*\times F) \\ & & \downarrow & & \downarrow \\ & & \Sigma X' & = \!= \!= \!= & \Sigma X' \end{array}$$

in which the bottom right square is a pushout and the rows and right columns are cofibrations, which yields the cofibration $M \to L \to \Sigma X'$. Deleting a chart from L and deleting its preimage from M gives the first row of the theorem. □

From the long exact homology sequence of the cofibration, we get:

Corollary 3.2 *The lift $M \to L$ of the canonical projection preserves the orientation class. That is, it induces isomorphisms on $H_m(\,)$, where $m = \dim L - \dim M$.*

This corollary can be proved in other ways such as naturality of the Serre spectral sequence.

Let $f : X \to Y$ be a differentiable map between compact oriented m-manifolds. Let $D_X : H^k(X) \cong H_{n-k}(X)$ and $D_Y : H^k(Y) \cong H_{n-k}(Y)$ be the Poincaré duality isomorphisms. Suppose f has degree λ (multiplies by λ on $H_n(\,)$). Then $f_* \circ D_X \circ f^* = \lambda D_Y$. In particular, if f preserves the orientation class (that is, has degree 1), then f^* is injective and f_* is surjective. Applying this to $M \to L$ shows:

Theorem 3.3 (Decomposition Theorem) *In the long exact homology sequence of the cofibration, the connecting map $\partial : H_q(L) \to H_{q-1}(X')$ is zero. Likewise, in the long exact cohomology sequence, the map $\delta : H^{q-1}(X') \to H^q(L)$ is zero. Thus, for $0 < q < m$, we have $H_q(M) \cong H_q(X') \oplus H_q(L)$ and $H^q(M) \cong H^q(X') \oplus H^q(L)$.*

This suggests that perhaps there is a manifold X such that $M \simeq X\#L$ where X is homotopy equivalent to the one-point compactification of X', but this is not necessarily true.

Example 3.4 Consider $A = \mathbb{C}P^2$ and write $A = B\#C$ where $B = \mathbb{C}P^2$ and $C = S^4$. Consider the trivial bundle $S^7 \times S^4 \to S^4$. Then $M = S^7 \times \mathbb{C}P^2$; $B' = S^2$; $C' = *$; $A' = B' \vee C' = S^2$, while

$$M' = (F \times A)' = (F \times A') \cup_{F' \times A'} (F' \times A)$$

$$= (S^7 \times S^2) \cup_{* \times S^2} (* \times \mathbb{C}P^2) = \mathbb{C}P^2 \vee S^7 \vee S^9$$

and $L = S^4 \times S^7$ so $L' = S^4 \vee S^7$. Our cofibration is

$$(S^2 \times S^7)/(* \times S^7) \to M' \to S^4 \vee S^7$$

which becomes $S^2 \vee S^9 \to \mathbb{C}P^2 \vee S^7 \vee S^9 \to S^4 \vee S^7$. This does not split, so in this example, M does not become homotopy equivalent to $X\#L$ for any X.

4 Bundles over 6-Manifolds

Let A be a 6-manifold such that $H^*(A)$ is simply connected and torsion-free. Suppose $H^2(A) = \mathbb{Z}$.

Let $x \in H^2(A)$ be a generator and let $V \in H^6(A)$ be the volume form. Then $x^3 = kV$ for some integer k.

By Wall [4], we can write $A = B\#C$ where $B = (S^3 \times S^3)^{\#r}$ for some r and C is a simply connected torsion-free 6-manifold with $H^3(C) = 0$ and $H^2(C) = \mathbb{Z}$.

Although M is a S^1 bundle over A, it does not immediately follow from Wall's result that M also admits a decomposition as a connected sum. We shall see that this is in fact true. This is the content of our Theorem 4.1 below.

Associated with x, there are complex line bundles over A and C classified by x. Let M and L denote the sphere bundles of these line bundles. Then there are S^1-bundles $S^1 \to M \to A$ and $S^1 \to L \to C$. Note that the long exact homotopy sequence tells us that $\pi_1(M) = \mathbb{Z}$ and $\pi_q(M) = \pi_q(A)$ for $q \neq 1$.

As in Ho-Jeffrey-Selick-Xia [2], we calculate that the cohomology of the 7-manifold L is given by $H^q(L) = \begin{cases} \mathbb{Z} & q = 0, 7; \\ \mathbb{Z}/k & q = 4. \\ 0 & otherwise. \end{cases}$

Theorem 4.1 *We have*

$$M \simeq \#^{2r}(S^3 \times S^4)\#L,$$

where the homology of the space L is specified above.

Proof In the notation of the preceding section applied to $S^1 \to M \to A$, we have $B' = \vee_{2r} S^3$, $L' = P^4(k)$ and

$$X' := (B' \times S^1)/(* \times S^1) \simeq B' \vee (B' \wedge S^1) \vee_{2r} (S^3 \vee \Sigma^3 S^1)$$

where $P^n(k)$ denotes the Moore space $S^{n-1} \cup_k e^n$. Thus, our cofibration sequence becomes

$$\vee_{2r} (S^3 \vee \Sigma^3 S^1) \to M' \to P^4(k)$$

or equivalently

$$\vee_{2r}(S^3 \vee S^4) \to M' \to P^4(k).$$

The composition of the bundle map $M' \to A'$ with the canonical projection $A' \to B'$ provides a splitting of the restriction of

$$\vee_{2r}(S^3 \vee S^4) \to M'$$

to $\vee_{2r} S^3$.

For degree reasons, the cofibration

$$\vee_{2r}(S^3 \vee S^4) \to M' \to P^4(k)$$

is principal, induced from some attaching map $P^3(k) \to \vee_{2r}(S^3 \vee S^4)$ whose image (for degree reasons) lands in $\vee_{2r} S^3$. Since the restriction of $\vee_{2r}(S^3 \vee S^4) \to M'$ to $\vee_{2r} S^3$ splits, this implies that this attaching map is trivial. Thus, the cofibration splits to give

$$M' \simeq \vee_{2r}(S^3 \vee S^4) \vee_{2r} P^4(k).$$

To obtain M from M', we attach the top cell giving

$$H^q(M) = H^q(M') \oplus H^q(S^7) = \begin{cases} \mathbb{Z} & q = 0, 7 \\ \mathbb{Z}^{2r} & q = 3 \\ \mathbb{Z}^{2r} \oplus \mathbb{Z}/k & q = 4 \\ 0 & \text{otherwise.} \end{cases}$$

Letting $\tilde{V}$ denote the generator of $H^7(M)$, using Poincaré duality, we can pair the generators $\langle u_1, u_2, \ldots u_{2r} \rangle$ of $\mathbb{Z}$ in degrees 3 with the generators $\langle v_1, v_2, \ldots v_{2r} \rangle$ of $\mathbb{Z}$ in degrees 4 so that $u_i v_j = \delta_{ij} \tilde{V}$. If we reduce to $\mathbb{Z}/k$ coefficients, there is also a nonzero cup product ab where a, b are generators of $H^3(M; \mathbb{Z}/k)$ and $H^4(M; \mathbb{Z}/k)$, respectively.

Examining the cohomology of M, we see that

$$H^*(M) = H^*\big(\#^{2r}(S^3 \times S^4)\#L\big)$$

where $H^q(L) = \begin{cases} \mathbb{Z} & q = 0, 7 \\ \mathbb{Z}/k & q = 4 \\ 0 & \text{otherwise.} \end{cases}$

The attaching maps f_M and $f_{\#^{2r}(S^3 \times S^4)\#L}$ are both

$$[\iota_1^3, \iota_1^4] + [\iota_2^3, \iota_2^4] + \ldots + [\iota_r^3, \iota_r^4] + f_L$$

where the Whitehead product $[\iota^3, \iota^4]$ is the attaching map

$$f_{S^3 \times S^4},$$

and so

$$M \simeq \#^{2r}(S^3 \times S^4)\#L.$$

$\square$

References

1. A. Hatcher, *Algebraic Topology* (Cambridge University, Cambridge, 2002)
2. N. Ho, L. Jeffrey, P. Selick, E. Xia, Flat connections and the commutator map for $SU(2)$. Q. J. Math. **72**, 163–197 (2021)
3. P. Selick, *Introduction to Homotopy Theory*, vol. 9 (AMS (Fields Institute Monographs), New York, 1997)
4. C.T.C. Wall, Classification problems in differential topology. V. On certain 6-manifolds. Invent. Math. **1**, 355–374 (1966); corrigendum, ibid. **2**, 306 (1966)

The $SO(4)$ Verlinde Formula Using Real Polarizations

Lisa Jeffrey and Kaidi Ye

1 Introduction

The Verlinde formula [9] is a formula for the dimension of the geometric quantization of the space $\mathcal{M}_g(G)$ of conjugacy classes of representations of the fundamental group of an oriented compact, closed 2-manifold of genus g into a compact Lie group G. The quantization is the dimension of the space of holomorphic sections of the kth power of a holomorphic line bundle L over $\mathcal{M}_g(G)$. This formula may be expressed as a trigonometric polynomial (see [12] and [13]) but may also be defined as the number of integer labellings of a collection of non-intersecting simple closed curves on Σ^g. The Verlinde formula is important because a basis for the quantization is the major ingredient in any physical calculations.

A system of integer labellings of curves corresponds to integer values of the moment maps for torus actions on the space $\mathcal{M}_g(G)$. In toric geometry, the integer values of the moment map for the torus action enumerate a basis for the holomorphic sections of the prequantum line bundle. A prequantum line bundle L is a complex line bundle over a symplectic manifold (M, ω) with a connection whose curvature is $2\pi i \omega$. For the definition of a prequantum line bundle, see, for example, Chapter 11 of [3].

The paper of Jeffrey and Weitsman [5] exhibited a real polarization on the moduli space $\mathcal{M}_g(SU(2))$ of gauge equivalence classes of flat $SU(2)$ connections on a compact oriented 2-manifold Σ^g without boundary. The topology of this space was studied in the fundamental paper [1]. If k is a nonnegative integer, these authors then used this real polarization to interpret the $SU(2)$ Verlinde formula (a formula specifying the dimension of the level k quantization) using a real polarization

L. Jeffrey (✉) · K. Ye
Department of Mathematics, University of Toronto, Toronto, ON, Canada
e-mail: jeffrey@math.toronto.edu; kaidi.ye@mail.utoronto.ca

A. Bahri et al. (eds.), *Toric Topology and Polyhedral Products*, Fields Institute Communications 89, https://doi.org/10.1007/978-3-031-57204-3_10

(e.g., as described by Sniatycki [8]). A basis for the level k quantization (which for a complex polarization is defined as the space of holomorphic sections of L^k, where L is the prequantum line bundle over $\mathcal{M}_g(SU(2))$ is parametrized by labelling the closed curves in a trinion decomposition for the surface by integers between 0 and k subject to certain conditions. The construction of [5] interprets the geometric quantization of the $SU(2)$ moduli space as an analogue of the geometric quantization of a toric manifold. See also [7] and [11].

In this article, we give a similar interpretation for the Verlinde formula for $SO(4)$. Our construction below extends the analogy with toric geometry to this case. We make use of the fact that $SU(2) \times SU(2)$ is a double cover of $SO(4)$.

Suppose G is a compact Lie group with maximal torus T. For a closed surface of genus g, the choice of a trinion decomposition[1] provides $(3g - 3)\dim(T)$ independent variables. The dimension of the moduli space is $(2g - 2)\dim(G)$. This means the number of independent variables is equal to half the dimension of the moduli space of gauge equivalence classes of flat connections if and only if $\dim(G) = 3\dim(T)$. This is true for $G = SU(2)$ or $SU(2) \times SU(2)$ and also $SO(4)$. These are examples of semisimple groups for which this relation holds.

Goldman [4] proved that if two simple closed curves in the surface Σ^g do not intersect, then the functions determined by the holonomies around these curves Poisson commute. See Sect. 5 for detailed discussion. Hence, the variables given by the holonomies along these curves provide the structure of a toric manifold. The Hamiltonian flows of these functions are only well defined on an open dense set, namely, the set where none of the simple closed curves in the trinion decomposition are sent to the center $Z(G)$ under the representation. On this open dense set, the integrable system specified by the holonomies around the curves in the trinion decomposition has the structure of a toric manifold.

For every circle C_j $(j = 1, \ldots, 3g - 3)$ in the trinion decomposition of Σ^g, we label it with two nonnegative integers l_j, l'_j where $l_j, l'_j \in \{0, 1, \ldots, k\}$ for k a nonnegative integer, which is the power of the prequantum line bundle over the moduli space.

For a complex polarization, this characterization of the $SO(4)$ Verlinde formula was already known—see, for example, [2] (p. 711) and [6] for the $SO(3)$ case. Our contribution is to recover this formula using a real polarization.

Definition 1.1 A *trinion* (also called a pair of pants) is a two-dimensional disc with two holes. We can denote a pair of pants as

$$D = \{z \in \mathbb{C} : |z| \leq 2\} - (\{z : |z - 1| < \frac{1}{2}\} \cup \{z : |z + 1| < \frac{1}{2}\}).$$

Definition 1.2 A *trinion decomposition* of Σ^g is the choice of $3g-3$ disjoint simple closed curves $\gamma_1, \ldots, \gamma_{3g-3}$ in Σ^g so that Σ^g is formed as the union of a collection of trinions with pairs of boundary components joined along the γ_j.

[1] The terms "trinion" and "trinion decomposition" are defined at the end of this section.

The Verlinde formula for $SO(4)$ tells us that these labellings must satisfy the following conditions. Choose an ordering of all the boundary circles of the trinion decomposition (in other words, parametrize these by integers between 1 and $3g-3$). Suppose m, n, p are the values of this index for the boundary components of one of the trinions in the trinion decomposition. Then

$$l_m + l'_m \in 2\mathbb{Z}$$

$$l_n + l'_n \in 2\mathbb{Z}$$

$$l_p + l'_p \in 2\mathbb{Z}$$

$$|l_m - l_n| \le l_p \le \min\{2k - l_m - l_n, l_m + l_n\},$$

$$|l'_m - l'_n| \le l'_p \le \min\{2k - l'_m - l'_n, l'_m + l'_n\},$$

$$l_m + l_n + l_p \in 2\mathbb{Z},$$

$$l'_m + l'_n + l'_p \in 2\mathbb{Z}.$$

A choice of trinion decomposition for Σ^g may be represented by a trivalent graph, where each vertex represents a trinion. Two vertices are joined by an edge if the corresponding trinions share a common boundary circle. A labelling is an assignment of an integer in $[0, k]$ to each boundary circle of the trinion decomposition, equivalently to each edge in the trivalent graph. We impose a collection of inequalities for each trinion. Equivalently we impose inequalities on the labellings of any three edges of the trivalent graph, which have a common vertex.

For example, consider the case $g = 3$. Figure 1 shows the integer labelling. The integer labelling of the boundary loops of every trinion D_i, $i = 1, 2, 3, 4$ must satisfy the above constraints.

This paper is organized as follows. In Sect. 2, we describe real polarizations. In Sect. 3, we describe relations between several moduli spaces. In Sect. 4, we describe the moduli space of a trinion. In Sect. 5, we describe the period of a Hamiltonian function. In Sect. 6, we describe the Verlinde formula.

2 Real Polarization

Let $G = SO(4)$ and let Σ^g be a compact, oriented 2-manifold with genus g. Let $P = \Sigma^g \times G$ be a fixed trivial G-bundle on Σ^g. We express the space of flat connections as

$$\mathcal{A}_{\mathcal{F}} = \{A \in \Omega^1(\Sigma^g) \otimes \mathfrak{so}(4) | F_A = 0\}$$

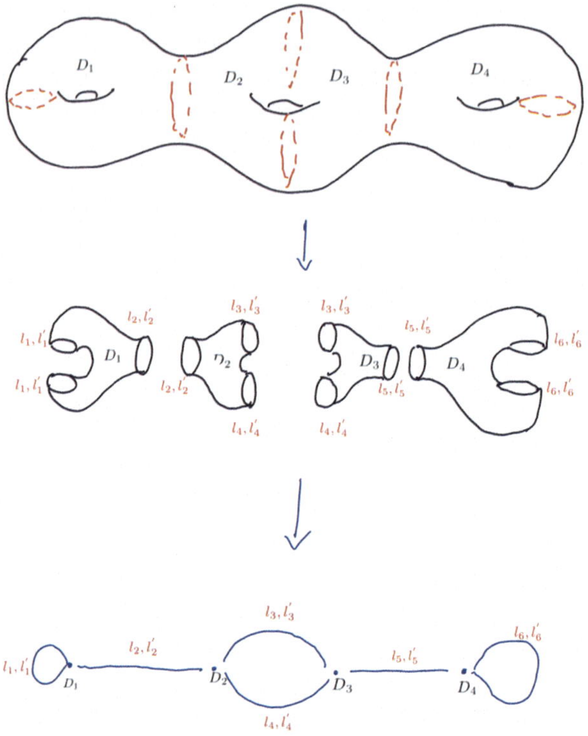

Fig. 1 Trinion decomposition and its integer labelling, with the corresponding labelled dual graph

where the curvature of the connection A is defined as

$$F_A = dA + A \wedge A.$$

This is a subspace of the space of all connections, denoted

$$\mathcal{A} = \{(a, b)\} = \Omega^1(\Sigma^g) \otimes \mathfrak{su}(2) \oplus \Omega^1(\Sigma^g) \otimes \mathfrak{su}(2)$$

We are using the fact $\mathfrak{so}(4) \cong \mathfrak{su}(2) \oplus \mathfrak{su}(2)$ for the above, where a and b are Lie algebra valued 1-forms.

Definition 2.1 Let $C_i \subset \Sigma^g$ be a closed, oriented simple curve. Define

$$\tilde{f} : \mathcal{A}_{\mathcal{F}} \longrightarrow \mathbb{R}^2$$

$$\tilde{f} : (a, b) \mapsto (\tilde{f}^a_{C_i}, \tilde{f}^b_{C_i})$$

where

$$\tilde{f}^a_{C_i} = \frac{1}{2}\mathrm{tr}(\mathrm{Hol}_{C_i}(a)) \tag{1}$$

$$\tilde{f}^b_{C_i} = \frac{1}{2}\mathrm{tr}(\mathrm{Hol}_{C_i}(b)). \tag{2}$$

These functions denote the holonomies of the connection a and b about the closed curve C_i.

Then we may define the polarization map based on [5] and [10] via the double covering map $\Phi : SU(2) \times SU(2) \to SO(4)$ as follows:

$$\tilde{\mathcal{F}} : \mathcal{A}_F \to \mathbb{R}^{6g-6}$$

$$\tilde{\mathcal{F}} : (a, b) \mapsto (\tilde{f}^a_{C_1}, \tilde{f}^b_{C_1}, ..., \tilde{f}^a_{C_{3g-g}}, \tilde{f}^b_{C_{3g-3}}).$$

$\tilde{\mathcal{F}}$ descends to

$$\mathcal{F} : \mathcal{M}_g(SO(4)) \to \mathbb{R}^{6g-6}$$

$$\mathcal{F} : [(a, b)] \mapsto (f^a_{C_1}, f^b_{C_1}, ..., f^a_{C_{3g-3}}, f^b_{C_{3g-3}}). \tag{3}$$

Here a and b are $\mathfrak{su}(2)$ valued 1-forms on Σ^g, and $[(a, b)]$ denotes the equivalence class of (a, b) in $\mathcal{M}_g(SO(4))$ (the equivalence class under the action of the gauge group).

Here:

1. Both $f^a_{C_i}$ and $f^b_{C_i}$ descend from $\tilde{f}^a_{C_i}$ and $\tilde{f}^b_{C_i}$, and we define $\tilde{f}^a_{C_i}, \tilde{f}^b_{C_i}$ in Eqs. (1) and (2). This is because the trace of the holonomy of a flat connection is unchanged under a gauge transformation.
2. We define the map by

$$\Phi : SU(2) \times SU(2) \longrightarrow SO(4)$$

$$\Phi : (u, v) \mapsto R_{u,v}(\cdot)$$

where $u, v \in SU(2)$ and

$$R_{u,v} : \mathbf{q} \mapsto u \cdot \mathbf{q} \cdot v^{-1}, \mathbf{q} \in \mathbb{H}. \tag{4}$$

Note that we identify $SU(2)$ as the collection of unit quaternions, in other words $SU(2) = \{\mathbf{q} : \|\mathbf{q}\| = 1, \mathbf{q} \in \mathbb{H}\}$. Here $\mathbb{H}$ denotes the space of quaternions, which is isomorphic to $\mathbb{R}^4$.

We have the following properties:

(a) The map $R_{u,v}(\cdot)$ represents all possible rotations in the space $\mathbb{H}$. If we identify $\mathbb{H} = \{a + b\hat{i} + c\hat{j} + d\hat{k}\}$ with $\mathbb{R}^4$, then $R_{u,v}(\cdot)$ represents all possible rotations in $\mathbb{R}^4$.

(b) The map Φ is a group homomorphism from the product group $SU(2) \times SU(2)$ to $SO(4)$ with kernel $\{(1, 1), (-1 - 1)\}$.
(c) The first group isomorphism theorem implies that $\frac{SU(2)\times SU(2)}{\{(1,1),(-1-1)\}} \cong SO(4)$.
(d) By knowing the kernel of Φ, we may represent G as the following:

$$G = \{[u : v]\} \tag{5}$$

where $u, v \in SU(2)$ and the notation $[u : v]$ denotes the equivalence class of the ordered pair (u, v) under the equivalence relation

$$(u, v) \sim (-u, -v). \tag{6}$$

(The notation is analogous to the usual notation for real projective space $\mathbb{R}P^n$ as a quotient of S^n by the diagonal action of multiplication by -1.)

3. The set $\{C_i\}_{i=1}^{i=3g-3}$ is a collection of closed oriented curves in Σ^g, which specifies a trinion decomposition of Σ^g (or equivalently a trivalent graph, as we explained above).

1. From now on, let $G = SO(4)$ and let Σ^g be a compact, oriented 2-manifold with genus g. Let $P = \Sigma^g \times G$ be a fixed trivial G-bundle on Σ^g. We express the moduli space of flat $SO(4)$ connections on Σ^g by

$$\mathcal{M}_g(SO(4)) = \mathcal{A}_F/\mathcal{G}$$

where $\mathcal{G} = C^\infty(\Sigma^g, G)$ and $\mathcal{G}$ acts on $\mathcal{A}_F$ by $A^g = g^{-1}Ag + g^{-1}dg, g \in \mathcal{G}$. Since the bundle is trivial, the connection A may be written as a Lie algebra valued 1-form on Σ^g so that the curvature F_A vanishes.
2. We can also represent $\mathcal{M}_g(SO(4))$ as

$$\text{Hom}(\pi_1(\Sigma^g), G)/G$$

where $\text{Hom}(\pi_1(\Sigma^g), G) = \{(a_1, b_1, ..., a_g, b_g) \in G^{2g} : \prod_{i=1}^g a_i b_i {a_i}^{-1} {b_i}^{-1} = 1\}$ and G acts on $\text{Hom}(\pi_1(\Sigma^g), G)$ by simultaneous conjugation. Here the 1 on the right denotes the rank 4 identity matrix. We can use Φ and the notation in (5) to rewrite $\text{Hom}(\pi_1(\Sigma^g), G)/G$ as follows:

$$\mathcal{M}_g(SO(4)) = \Big\{\Big([u_1 : v_1], [p_1 : q_1], \ldots, [u_g : v_g], [p_g : q_g]\Big) \in G^{2g} \mid \prod_{j=1}^{g} [u_j : v_j][p_j : q_j][u_j : v_j]^{-1}[p_j : q_j]^{-1} = 1\Big\}/G. \tag{7}$$

If we use the description $\mathcal{M}_g(SO(4)) = \mathrm{Hom}(\pi_1(\Sigma^g), G)/G$, then we can rewrite the map $\tilde{\mathcal{F}}$ as

$$\tilde{\mathcal{F}} : \mathrm{Hom}(\pi_1(\Sigma^g), G) \longrightarrow \mathbb{R}^{6g-6}$$

$$\tilde{\mathcal{F}} = (\tilde{\theta_1}, \tilde{\theta}_1' ... \tilde{\theta}_{3g-3}, \tilde{\theta}_{3g-3}')$$

where the pair

$$(\tilde{\theta}_i, \tilde{\theta}_i') : \mathrm{Hom}(\pi_1(\Sigma^g), G) \longrightarrow \mathbb{R}^2$$

$$(\tilde{\theta}_i, \tilde{\theta}_i') : \rho([C_i]) \mapsto (\frac{1}{2}\mathrm{Tr}(u_i), \frac{1}{2}\mathrm{Tr}(v_i))$$

and

$$\rho([C_i]) \sim \begin{bmatrix} u_i & 0 \\ 0 & v_i . \end{bmatrix},$$

Here $\sim$ means "is conjugate to." Here $u_j, v_j \in SU(2)$.

The map $(\tilde{\theta}_i, \tilde{\theta}_i')$ descends to $(\theta_i, \theta_i') : \mathcal{M}_g(SO(4)) \longrightarrow \mathbb{R}^2$ and $\tilde{\mathcal{F}}$ descends to $\mathcal{F} : \mathcal{M}_g(SO(4)) \longrightarrow \mathbb{R}^{6g-6}$.

Now we can define the holonomy angle as:

Definition 2.2 *The holonomy angle pair* (θ_i, θ_i') of curve C_i from a trinion decomposition is

$$\theta_i = \cos^{-1}(\frac{1}{2}\mathrm{tr}(u_i))$$

and

$$\theta_i' = \cos^{-1}(\frac{1}{2}\mathrm{tr}(v_i))$$

where $[u_i : v_i] \in G$, and we make $\cos^{-1}(x)$ to take values in $[0, \pi]$ with $x \in [-1, 1]$. This is the choice made in the paper [5]. It is the choice of a connected interval U for which the function $\cos : \theta \mapsto \cos(\theta)$ gives a bijection from U to $[-1, 1]$.

$$\cos(\pi - \theta_i) = -\cos(\theta_i)$$

and

$$\cos(\pi - \theta_i') = -\cos(\theta_i')$$

with condition $\theta_i + \theta_i^{'} \leq \pi$.

We can define a map $\bar{\Theta}$ by using the holonomy angle pairs.

Definition 2.3 The map $\bar{\Theta} : \mathcal{M}_g(SO(4)) \to \mathbb{R}^{6g-6}$ is defined by

$$\bar{\Theta} = (\theta_1, \theta_1^{'}, ..., \theta_{3g-3}, \theta_{3g-g}^{'}) : \mathcal{M}_g(SO(4)) \to \mathbb{R}^{6g-6}. \tag{8}$$

The components of this map have Hamiltonian flows with constant periods. In other words, these components form moment maps for circle actions. They are only defined where the component functions are differentiable (so that their Hamiltonian flows can be defined).

3 The Relation Among Moduli Spaces for $G = SU(2), SU(2) \times SU(2), SO(4)$ over Σ^g

Let the moduli space of conjugacy classes of representations of the fundamental group of Σ^g be

$$\mathcal{M}_g(SU(2)) = \left\{(u_1, v_1, \ldots, u_g, v_g) \in SU(2)^{2g} | \Pi_{i=1}^g u_i v_i {u_i}^{-1} {v_i}^{-1} = 1\right\} / SU(2),$$

and for $G = SU(2) \times SU(2)$, we define

$$\begin{aligned} &\mathcal{M}_g(SU(2) \times SU(2)) \\ &= \left\{ \left(\begin{bmatrix} u_1 & 0 \\ 0 & v_1 \end{bmatrix}, \begin{bmatrix} u_1' & 0 \\ 0 & v_1' \end{bmatrix}, \ldots, \begin{bmatrix} u_g & 0 \\ 0 & v_g \end{bmatrix}, \begin{bmatrix} u_g' & 0 \\ 0 & v_g' \end{bmatrix} \right) | \right. \\ &\times \left. \begin{bmatrix} \Pi_{i=1}^g u_i u_i' {u_i}^{-1} {u_i'}^{-1} = 1 & 0 \\ 0 & \Pi_{i=1}^g v_i v_i' {v_i}^{-1} {v_i'}^{-1} = 1 \end{bmatrix} u_i, v_i, u_i', v_i' \in SU(2) \right\} \\ &/(SU(2) \times SU(2)) \end{aligned}$$

Recall that for $G = SO(4)$, we can define the moduli space in terms of $SU(2)$ by using the double covering map Φ and (5) as follows:

$$\mathcal{M}_g(SO(4)) = \left\{ \left([u_1 : v_1], [p_1 : q_1], \ldots, [u_g : v_g], [p_g : q_g] \right) : \right.$$

$$\left. \Pi_{i=1}^g u_i p_i {u_i}^{-1} {p_i}^{-1} = \Pi_{i=1}^g v_i q_i {v_i}^{-1} {q_i}^{-1} = 1 \right\} / G.$$

It is clear that

$$\mathcal{M}_g(SU(2) \times SU(2)) = \mathcal{M}_g(SU(2)) \times \mathcal{M}_g(SU(2))$$

and the relation between $\mathcal{M}_g(SU(2) \times SU(2))$ and $\mathcal{M}_g(SO(4))$ is described as follows:

Proposition 3.1 *There is a surjective map ϕ from the moduli space $\mathcal{M}_g(SU(2) \times SU(2))(\epsilon = 1) \coprod \mathcal{M}_g(SU(2) \times SU(2))(\epsilon = -1)$ to $\mathcal{M}_g(SO(4))$. In other words, the map $\phi : \mathcal{M}_g(SU(2) \times SU(2))(\epsilon = 1) \coprod \mathcal{M}_g(SU(2) \times SU(2))(\epsilon = -1) \longrightarrow \mathcal{M}_g(SO(4))$ is surjective.*

Proof Let $\mathfrak{P} = SU(2) \times SU(2)$ and $G = SO(4)$. Let Φ again be the double covering map from $\mathfrak{P}$ to G. By the first isomorphism theorem, we know $\mathfrak{P}/\ker(\Phi) \cong G$. We use $\mathfrak{P}$, G to rewrite $\mathcal{M}_g(SU(2) \times SU(2))$ as follows:

$$\mathcal{M}_g(SU(2) \times SU(2)) = \mathrm{Hom}(\pi_1(\Sigma^g), \mathfrak{P})/\mathfrak{P}. \tag{9}$$

Let $\epsilon = \pm 1$. Let $\widetilde{\mathcal{M}_g}(SU(2) \times SU(2))(\epsilon) = \{\rho \in \mathrm{Hom}(\mathrm{F}_{2g}, \mathfrak{P}) | \rho(\mathrm{c}) = \epsilon \mathrm{I}\}$. Here c is the product of commutators of a chosen collection of generators $\{a_j, b_j, j = 1, \ldots, g\}$ for the free group F_{2g} on $2g$ generators, which is the fundamental group of $\Sigma^g \setminus \{\mathrm{pt}\}$. Also I denotes the identity element of $SU(2) \times SU(2)$. Let $\widetilde{\mathcal{M}_g}(SO(4)) = \mathrm{Hom}(\pi_1(\Sigma^g), \mathrm{G})$ and x be a point in $\widetilde{\mathcal{M}_g}(SO(4))$. For every $g \in G$, there is $q \in \mathfrak{P}$ such that $\Phi(q) = g$, since Φ is a surjective map.

$$g \cdot x = gxg^{-1} = \Phi(q)x\Phi(q)^{-1}.$$

Now, we want to define the following maps:

$$\tilde{\phi} : \widetilde{\mathcal{M}_g}(SU(2) \times SU(2))(\epsilon) \to \widetilde{\mathcal{M}_g}(SO(4))$$

by

$$\tilde{\phi} : \psi \mapsto \Phi^{2g}(\psi) \tag{10}$$

where $\Phi^{2g}(\psi) = (\Phi(a_1), \Phi(b_1), ..., \Phi(a_g), \Phi(b_g))$.

The image of this map is in $\widetilde{\mathcal{M}_g}(SO(4))$, since

$$(a_1, b_1, ..., a_g, b_g) \to (\Phi(a_1), \Phi(b_1), ..., \Phi(a_g), \Phi(b_g))$$

so that

$$\Pi_{i=1}^g \Phi(a_i)\Phi(b_i)\Phi(a_i)^{-1}\Phi(b_i)^{-1} = \Phi(\Pi_{i=1}^g a_i b_i {a_i}^{-1} {b_i}^{-1}) = \Phi(\epsilon) = 1. \tag{11}$$

We have

$$\phi : \widetilde{\mathcal{M}_g}(SU(2) \times SU(2))(\epsilon)/\mathfrak{P} \to \widetilde{\mathcal{M}_g}(SO(4))/\{\mathfrak{P}/ker(\Phi)\} \tag{12}$$

$$\phi : [\psi] \mapsto [\tilde{\phi}(\psi)]$$

where $[\psi]$, $[\tilde{\phi}(\psi)]$ are the orbits with representative elements ψ, $\tilde{\phi}(\psi)$.

Note that

$$\tilde{\phi}^{-1}(\widetilde{\mathcal{M}_g}\Big(SO(4)\Big)) = \widetilde{\mathcal{M}_g}(SU(2) \times SU(2))(1) \sqcup \widetilde{\mathcal{M}_g}(SU(2) \times SU(2))(-1).$$

If we use $\tilde{\phi}(\psi) = \Phi^{2g}(\psi)$ for the map in (12), we can rewrite (12) as

$$\phi : [\psi] \mapsto [\Phi^{2g}(\psi)]. \tag{13}$$

We want to show (a) ϕ is well defined and (b) it is a surjective map.

Part (a): Proof that ϕ is well defined

Let $[\psi] \in \mathcal{M}_g(SU(2) \times SU(2))(\epsilon)$ and ψ, ψ' are both in the orbit $[\psi]$. In other words, $\exists q \in \mathfrak{P}$, such that $q\psi q^{-1} = \psi'$. We want to show $\phi([\psi]) = \phi([\psi'])$.

The fact that $\phi([\psi]) = \phi([\psi'])$ means they are in the same orbit in $\mathcal{M}_g(SO(4))$. We will show that $\exists \delta \in G$ such that $\delta\tilde{\phi}(\psi)\delta^{-1} = \tilde{\phi}(\psi')$.

Let $\delta = \Phi(q)$. We check the following:

$$\delta\tilde{\phi}(\psi)\delta^{-1} = \Phi(q)\tilde{\phi}(\psi)\Phi(q)^{-1} = \Phi(q)\tilde{\phi}(\psi)\Phi(q^{-1}). \tag{14}$$

Recall that $\tilde{\phi}(\psi) = \Phi^{2g}(\psi)$, and then we have that the RHS of (14) is

$$\Phi(q)\Phi^{2g}(\psi)\Phi(q^{-1})$$

or

$$\Phi^{2g}(q\psi q^{-1})$$

Equation (14) shows

$$\Phi^{2g}(q\psi q^{-1}) = \Phi^{2g}(\psi') = \tilde{\phi}(\psi').$$

Now we can confirm that $\phi([\psi])$, $\phi([\psi'])$ are in the same orbit.

Part (b): Proof of surjectivity

Define $\mathcal{M}_g(SU(2) \times SU(2))(\epsilon)$ to be the quotient of $\widetilde{\mathcal{M}_g}(SU(2) \times SU(2))(\epsilon)$ under the action of G by conjugation. The Eq. (11) shows that the product of commutators is I as long as the product of the preimages of commutators is equal to ϵI where $\epsilon \in \{\pm 1\}$ and I is the identity element. Hence the map $\tilde{\phi}$ is surjective. We know Φ^{2g} is surjective as well. We have clearly defined the map ϕ, and we have proved ϕ is well defined. Thus, the surjectivity of ϕ can be seen as follows. Pick any orbit in $\mathcal{M}_g(SO(4))$ with representative element h (in other words, $[h] \in \mathcal{M}_g(SO(4))$. By the surjectivity of $\tilde{\phi}$, we know $\exists \psi \in \coprod_\epsilon \widetilde{\mathcal{M}_g}(SU(2) \times SU(2))(\epsilon)$ such that $\tilde{\phi}(\psi) = h$, where $\epsilon = \pm 1$. So the orbit of ψ will be mapped to the orbit of h under the map ϕ. □

4 Moduli Space of a Trinion for $G = SO(4)$

Let $(\theta_i, \theta_i^{'})$ be the holonomy angle pair that corresponds to the boundary circle C_i of a trinion D, where $i = 1, 2, 3$.

The moduli space of representations of the fundamental group of a trinion D can be represented as follows:

$$\mathcal{M}(D) = \mathrm{Hom}(\pi_1(D), G)/G = \left\{(s_1, s_2, s_3)|s_i \in G, s_1 s_2 s_3 = 1\right\}/G,$$

where G acts on $\mathrm{Hom}(\pi, G)$ by simultaneous conjugation.

The map $\Theta : \mathcal{M}(D) \to [0, \pi]^6 \subset \mathbb{R}^6$ can be expressed as

$$\Theta = (\theta_1, \theta_1^{'}, \theta_2, \theta_2^{'}, \theta_3, \theta_3^{'}) : \mathcal{M}(D) \to [0, \pi]^6 \subset \mathbb{R}^6.$$

Theorem 4.1 *Let $\theta_j, \theta_j^{'} \in [0, \pi]$. The map Θ is a surjective map from $\mathcal{M}(D)$ to its image. Its image is the product of two polytopes*

$$\Delta_1 = \{\theta_2, \theta_2, \theta_3 : |\theta_1 - \theta_2| \le \theta_3 \le \min(\theta_1 + \theta_2, 2\pi - (\theta_1 + \theta_2)), \theta_i \in [0, \pi]\} \tag{15}$$

and

$$\Delta_2 = \{\theta_1', \theta_2', \theta_3' : |\theta_1' - \theta_2'| \le \theta_3' \le \min(\theta_1' + \theta_2', 2\pi - (\theta_1' + \theta_2')), \theta_i' \in [0, \pi]\}. \tag{16}$$

Proof For the image of $\overline{\Theta}$, we need to find the conditions on conjugacy classes $[s_i]$ so that $[s_1][s_2][s_3] = 1$ where $s_i \in G$. We can represent $s_1 s_2 s_3 = 1$ as

$$R(u_1, v_1)R(u_2, v_2)R(u_3, v_3) = 1$$

where R was defined in (4) above and $u_{i=1,2,3}, v_{i=1,2,3} \in SU(2)$. We denote the rank 4 identity matrix by 1. This implies

$$u_1 u_2 u_3 = 1 = v_1 v_2 v_3 \tag{17}$$

or

$$u_1 u_2 u_3 = -1 = v_1 v_2 v_3. \tag{18}$$

For Eq. (17), the situation is exactly the same as in the $SU(2)$ case in [5], which gives us (15). The proof of Eq. (15) in [5] applies to prove Eq. (17). It also proves Eq. (18) by replacing θ_3 by $\pi - \theta_3$ (in the notation of that article).

This means the solutions of Eq. (17) are in bijective correspondence with (15), while those of (18) are also in bijective correspondence with (15) via $\theta_3 \mapsto \pi - \theta_3$.

These inequalities define a tetrahedron with the following *four* vertices.

$$V_1 = (0, 0, \pi),\ V_2 = (\pi, 0, 0)$$

$$V_3 = (\pi, \pi, \pi),\ V_4 = (0, \pi, 0).$$

Compare the tetrahedron Δ_2 defined by V_1, V_2, V_3, V_4 with the tetrahedron Δ_1 defined by Proposition 3.1 in [JW] with vertices

$$Q_1 = (0, 0, 0),\ Q_2 = (0, \pi, \pi),$$

$$Q_3 = (\pi, 0, \pi),\ Q_4 = (\pi, \pi, 0).$$

We found they are clearly isomorphic as one can transform into the other by a composition of rotations and translations, in other words, a rigid motion.

We can always obtain Δ_2 by applying a sequence of cubic lattice preserving isometries on Δ_1 as follows:

1. Reflecting Δ_1 with respect to the $\theta_1\theta_{u_2}$ plane to get Δ_1'
2. Reflecting Δ_1' with respect to the $\theta_1\theta_{u_3}$ plane to get Δ_1''
3. Reflecting Δ_1'' with respect to the $\theta_1\theta_{u_2}$ plane to get Δ_1'''
4. Translating Δ_1''' along the θ_{u_2} axis toward the positive direction by π unit to get Δ_1''''

Then Δ_1'''' is the same as Δ_2.

If we re-scale everything by $\frac{k}{\pi}$ and $k \in \mathbb{Z}_{>0}$, then isometry (d) is just translation by k units, and it preserves our lattice points. After this scaling, our lattice points are $\mathbb{Z}^3$.

The above equation gives one V_1, V_2, V_3, V_4.

After the quotient, Eq. (18) gives us (15) and (16) as well. □

5 Period of the Hamiltonian Flow

The period of the Hamiltonian flow associated with each loop C (in other words, C is a simple closed curve) can be obtained by studying the period of the Goldman flow for each loop C.

We may define the Goldman flow for $\mathrm{Hom}(\pi_1(\Sigma^g), \mathrm{SO}(4))$ as follows.

In the paper [4, Section 1.1], Goldman assumes we have an invariant function $f : G \to \mathbb{R}$ where f is a C^1 function such that $f(gAg^{-1}) = f(A)$ for all $g, A \in G$ and G is a matrix Lie group with finitely many components. An associated function F for f is a function $F : G \longrightarrow \mathfrak{g}$ with the following properties. For X in the Lie algebra of G, and $< \cdot, \cdot >$ an inner product on this Lie algebra invariant under the adjoint action, we have

$$< X, F(A) >= df_A(X) = \frac{d}{dt}|_{t=0} f(A\exp(tX)).$$

It follows that if $f(g) = \mathrm{tr(g)}$, then

$$F(A) = \frac{1}{4}(A - A^{-1}).$$

The computation of the above F is in [4] Cor 1.9.

For our case, we use $\cos^{-1}$ composed with the trace function for the holonomy angle pair instead of just the trace function for our invariant function. The details are as follows. Let $\theta_G^C : G \to \mathbb{R}$ denote the invariant function that is associated with the loop C. Let $\Gamma : [-1, 1] \times [-1, 1] \to [0, \pi] \times [0, \pi]$ be the map

$$\Gamma(x, y) = \left(\cos^{-1}(x), \cos^{-1}(y)\right),$$

and

$$\delta : \mathbb{R}^2 \to \mathbb{R}$$

$$\delta : (x, y) \mapsto x + y$$

We pick $\mathcal{S} \in T$ (where T is the maximal torus of G) and denote $\mathcal{S}$ in the same fashion as we did in (5). That is

$$\mathcal{S} = [s : s'] \in G$$

and $s = \begin{bmatrix} e^{i\theta_i} & 0 \\ 0 & e^{-i\theta_i} \end{bmatrix}, s' = \begin{bmatrix} e^{i\theta_i'} & 0 \\ 0 & e^{-i\theta_i'} \end{bmatrix} \in T \subset SU(2).$

We can write θ_G^C as follows:

$$\begin{aligned} \theta_G^C(\mathcal{S}) &= \delta \circ \Gamma([\frac{1}{2}\mathrm{tr}(s), \frac{1}{2}\mathrm{tr}(s')]) \\ &= \delta \circ \Gamma \circ (\cos(\theta_i), \cos(\theta_i')) \\ &= \theta_i + \theta_i' \end{aligned} \tag{19}$$

As a result, we should modify its associated function accordingly.

$$< X, F_c(\mathcal{S}) > = \frac{d}{dt}|_{t=0}(\cos(\theta_i(s)), \cos(\theta_i^{'}(s^{'})))$$
$$= \begin{bmatrix} -\sin(\theta_i(s)) & 0 \\ 0 & -\sin(\theta_i^{'}(s^{'})) \end{bmatrix} \begin{bmatrix} \frac{d}{dt}|_{t=0}(\theta_i(s)) & 0 \\ 0 & \frac{d}{dt}|_{t=0}(\theta_i^{'}(s^{'})) \end{bmatrix} \tag{20}$$

This implies

$$F_{\theta_G^C} = \begin{bmatrix} \Lambda & 0 \\ 0 & \Lambda' \end{bmatrix} F_c(\mathcal{S}) \tag{21}$$

where Λ acts on a 2×2 diagonal block matrix as

$$\Lambda \begin{bmatrix} a & 0 \\ 0 & b \end{bmatrix} = \frac{-1}{\sin(\theta_i)} \begin{bmatrix} a & 0 \\ 0 & b \end{bmatrix}$$

and

$$\Lambda' \begin{bmatrix} a' & 0 \\ 0 & b' \end{bmatrix} = \frac{-1}{\sin(\theta_i')} \begin{bmatrix} a' & 0 \\ 0 & b' \end{bmatrix}.$$

So $\begin{bmatrix} \Lambda & 0 \\ 0 & \Lambda' \end{bmatrix}$ acts on $\begin{bmatrix} a & 0 & 0 & 0 \\ 0 & b & 0 & 0 \\ 0 & 0 & a' & 0 \\ 0 & 0 & 0 & b' \end{bmatrix}$ as

$$\begin{bmatrix} \frac{-1}{\sin(\theta_i)} \begin{bmatrix} a & 0 \\ 0 & b \end{bmatrix} & O \\ O & \frac{-1}{\sin(\theta_i')} \begin{bmatrix} a' & 0 \\ 0 & b' \end{bmatrix} \end{bmatrix}.$$

We compute $F_{\theta_G^C}$ below.

$$
\begin{aligned}
&F_{\theta^C_{SO(4)}}(\mathcal{S}) \\
&= \begin{bmatrix} \Lambda & 0 \\ 0 & \Lambda' \end{bmatrix} \frac{1}{4}(\mathcal{S} - \mathcal{S}^{-1}) \\
&= \begin{bmatrix} \Lambda & 0 \\ 0 & \Lambda' \end{bmatrix} \begin{bmatrix} \frac{1}{4}(s - s^{-1}) & 0 \\ 0 & \frac{1}{4}(s' - (s')^{-1}) \end{bmatrix} \\
&= \frac{1}{4} \begin{bmatrix} \Lambda & 0 \\ 0 & \Lambda' \end{bmatrix} \begin{bmatrix} \mathrm{diag}(e^{i\theta_i} - e^{-i\theta_i}, e^{-i\theta_i} - e^{i\theta_i}) & 0 \\ 0 & \mathrm{diag}(e^{i\theta_i'} - e^{-i\theta_i'}, e^{-i\theta_i'} - e^{i\theta_i'}) \end{bmatrix} \\
&= \frac{i}{2} \begin{bmatrix} \Lambda & 0 \\ 0 & \Lambda' \end{bmatrix} \begin{bmatrix} \mathrm{diag}(\sin(\theta_i), -\sin(\theta_i)) & 0 \\ 0 & \mathrm{diag}(\sin(\theta_i'), -\sin(\theta_i')) \end{bmatrix} \\
&= \frac{1}{2} \begin{bmatrix} -i & 0 & 0 & 0 \\ 0 & i & 0 & 0 \\ 0 & 0 & -i & 0 \\ 0 & 0 & 0 & i \end{bmatrix}.
\end{aligned}
\tag{22}
$$

(Above, $\mathrm{diag}(a_1, \ldots, a_n)$ is the $n \times n$ diagonal matrix where the jth entry is a_j.) Now we are ready to construct the flow as follows:

$$\tilde{\Xi}_t^C : \tilde{U}_C \to \tilde{U}_C$$

Here the subset $\tilde{U}_C$ of $\widetilde{\mathcal{M}}_g(SO(4))$ is defined as

$$\tilde{U}_C = \{\phi \in \mathrm{Hom}(\pi_1(\Sigma)^g, G) | \phi[C] \neq \pm 1\}.$$

These flows were constructed in [4].

1. If the curve C is a non-separating curve (i.e., $\Sigma^g \setminus C$ is connected):
   Let $B \subset \Sigma^g$ be another oriented simple closed curve that intersects C once transversely. Here $< [B] >$ and $\pi_1(\Sigma^g - C)$ are two subgroups of $\pi_1(\Sigma^g)$.
   If $\alpha \in \pi_1(\Sigma^g - C)$, then

$$\tilde{\Xi}_t^C(\phi)(\alpha) = \phi(\alpha)$$

$$\tilde{\Xi}_t^C([B]) = \phi([B])\zeta_t{}^C(\phi) \tag{23}$$

   Here the element ζ_t^C is defined in (24) below.

Definition 5.1 The map ${\zeta_t}^C : \tilde{U}_C \to G$ can be defined as follows:

$$
\begin{aligned}
{\zeta_t}^C(\phi) &= \exp(4\pi^2 t F_{\theta_G^C}(\phi([C]))) \\
&= \begin{bmatrix} e^{-2\pi^2 ti} & 0 & 0 & 0 \\ 0 & e^{2\pi^2 ti} & 0 & 0 \\ 0 & 0 & e^{-2\pi^2 ti} & 0 \\ 0 & 0 & 0 & e^{2\pi^2 ti} \end{bmatrix}.
\end{aligned} \tag{24}
$$

2. If the curve C is a separating curve (suppose C separates Σ^g into two components Σ_1 and Σ_2): For $\alpha \in \pi_1(\Sigma_1)$, we have

$$\tilde{\Xi}_t^C(\phi)(\alpha) = \phi(\alpha)$$

For $\alpha \in \pi_1(\Sigma_2)$, we have

$$\tilde{\Xi}_t^C(\phi)(\alpha) = {\zeta_t}^C(\phi)\phi(\alpha){\zeta_t}^C(\phi)^{-1} \tag{25}$$

Recall that in G, we have the following equivalence relation. The element $[u : u']$ of G denotes the equivalence class of $(u, u') \in SU(2) \times SU(2)$ under the equivalence relation $(u, u') \sim (-u, -u')$. It follows that

$$e^{i\theta_*} \backsim -e^{i\theta_*}$$

and

$$\Rightarrow e^{i\theta_*} \backsim e^{i(\theta_* + \pi)}.$$

This implies the period of ${\zeta_t}^C(\phi)$ is $\frac{1}{2\pi}$. Note that the period of ${\zeta_t}^C$ is the same as the period of the corresponding Goldman flow $\tilde{\Xi}_t^{\,C}$ if C is a non-separating curve. Its period will be two times the period of the corresponding Goldman flow $\tilde{\Xi}_t^{\,C}$ if it is a separating curve.
Now we can confirm that:

1. For the non-separating case, the period of $\tilde{\Xi}_t^C$ is $\frac{1}{2\pi}$.
2. For the separating case, the period of $\tilde{\Xi}_t^C$ is $\frac{1}{4\pi}$.

Theorem 5.2 *Let* $U_C = (\theta_i{}^{-1}(0, \pi), \theta_i'^{-1}(0, \pi))$. *The flow* $\tilde{\Xi}_t^C$ *on* $\tilde{U}_C$ *covers the Hamiltonian flow on* U_c *associated with the function* θ_G^C.

Here U_C is the region in $\mathcal{M}_g(SO(4))$ such that the Goldman flow is properly defined (i.e., $\sin(\theta_i') \neq 0$ and $\sin(\theta_i) \neq 0$), and (θ_i, θ_i') is the holonomy angle pair for curve $C \subset \Sigma^g$.

As a consequence of the above, the period of $\tilde{\Xi}_t^C$ determines the period of the Hamiltonian flow on U_c associated with the function θ_G^C. It has period $\frac{1}{2\pi}$ if C is a non-separating curve and $\frac{1}{4\pi}$ if C is a separating curve. As the result of [5], all the Hamiltonian flows are periodic with constant period and so induce a torus action on an open dense set of $(\mathcal{M}_g(SO(4)), \omega)$. If we define

$$H_i = \frac{k}{\pi}\theta_G^C \tag{26}$$

where k comes from the symplectic variety $(\mathcal{M}_g(SO(4)), k\omega)$, then the period of H_i is the following:

1. $\frac{k}{2}$ if C is a non-separating curve
2. $\frac{k}{4}$ if C is a separating curve

Remarks

1. For C a non-separating curve:
   If we require the value of θ_G^C to be a nonnegative even integer, then H_i gives a period 1 Hamiltonian flow.
   For C a separating curve:
   If we require the value of θ_G^C to be a nonnegative integer divisible by 4, then H_i gives a period 1 Hamiltonian flow.
2. The Hamiltonian flow associated with a holonomy angle pair is two times faster than the $G = SU(2)$ case in [5]. Therefore, for $G = SO(4)$ case, the period 1 flow requires 2 as its additional factor with respect to the $G = SU(2)$ case in [5].

6 Verlinde Formula

From Theorem 4.4, Proposition 5.5 of [5], and the flow speed associated with H_i, we deduce the following result.

Let $x \in B_g^{ind}$, where B_g^{ind} is the set of points x in the image of $\mathcal{M}_g(SO(4))$ under the moment map for which each holonomy angle pair gives us two linearly independent Hamiltonian vector fields. Then x is a Bohr-Sommerfeld point (defined in [5]), if and only if for any $y \in \mathcal{F}^{-1}(x)$ where $\mathcal{F}$ defined in (3) satisfies the below conditions:

$$H_i(y) \in 2\mathbb{Z}.$$

Here H_i was defined in (26).

Let $(\theta_i, \theta_i^{'})$ be a holonomy angle pair of the ith boundary circle C_i of one trinion. We define

$$l_i = \frac{k\theta_i}{\pi} \tag{27}$$

and

$$l_i' = \frac{k\theta_i'}{\pi} \tag{28}$$

where $k \in \mathbb{Z}$ is a chosen constant and l_i, l_i' are also required to be integers. Equation (19) implies that

$$l_i + l_i' = H_i \in 2\mathbb{Z}, \tag{29}$$

where H_i was introduced in the previous section. From Theorem 4.4 [5] and the remark in Sect. 4, we know the value of H_i needs to be an even number.

The definition (2.2) gives you the following condition regarding the relation between l_i and l_i' :

$$l_i + l_i' \leq k \tag{30}$$

Let $\mathcal{H}(M)$ be the Hilbert space that is obtained from quantization of space M via real polarization. Using notations from (3.1), we have

$$\dim(\mathcal{H}(\widetilde{\mathcal{M}_g}(\mathrm{SU}(2)\times\mathrm{SU}(2))(\epsilon = -1)))+\dim(\mathcal{H}(\widetilde{\mathcal{M}_g}(\mathrm{SU}(2)\times\mathrm{SU}(2))(\epsilon = 1))) =$$

$$2^{3g-3}\dim(\mathcal{H}(\mathcal{M}_g(\mathrm{SO}(4)))).$$

We impose Eq. (30) for $i = 1, \ldots, 3g-3$ to count representations of the fundamental group into $SO(4)$ instead of into $SU(2)\times SU(2)$. This follows because of properties (b,c) (p. 156) and the equivalence relation stated in (6).

We may also apply (27) and (28) to (15) and (16) for each trinion. Then we have the following:

$$|l_1 - l_2| \leq l_3 \leq \min\{2k - l_1 - l_2, l_1 + l_2\} \tag{31}$$

and

$$|l_1' - l_2'| \leq l_3' \leq \min\{2k - l_1' - l_2', l_1' + l_2'\}. \tag{32}$$

From [5, Proposition 5.4], we know there exists another set of period 1 generators of the period lattice. (See [5] Definition 4.1.)

$$\gamma = \frac{k}{\pi}(\theta_1 + \theta_2 + \theta_3)$$

$$\gamma' = \frac{k}{\pi}(\theta_1' + \theta_2' + \theta_3')$$

We apply Theorem 4.4 and Proposition 5.5 of [5] for $\gamma, \gamma^{'}$. We will have the following for each trinion:

$$l_1 + l_2 + l_3 \in 2\mathbb{Z} \tag{33}$$

and

$$l_1^{'} + l_2^{'} + l_3^{'} \in 2\mathbb{Z} \tag{34}$$

Equations (29), (31), (32), (33), and (34) give us the Verlinde formula for $G = SO(4)$. These equations must be imposed for every 3-tuples of indices (l_1, l_2, l_3), (l_1', l_2', l_3'), which correspond to the three edges coming from a vertex in the trivalent graph dual to a trinion decomposition. Let us consider the following example when genus $g = 2$. We can find the number of Bohr-Sommerfeld fibers as follows:

1. We first apply a trinion decomposition for the surface $\Sigma^{g=2}$ as Fig. 2. It should give two trinions D_1, $D_2$and a trivalent graph.
2. Each edge of the trivalent graph corresponds to the boundary circles C_1, C_2, and $C_3 (D_1 \cap D_2 = \{C_1, C_2, C_3\})$. We should label each edge E_i with a pair of integers $l_i, l_i^{'} \in [0, k]$. By (33), (34), (32), (31), and (29), we have the following system of equations:

$$|l_1 - l_2| \leq l_3 \leq l_1 + l_2,$$

$$l_1 + l_2 + l_3 \leq 2k,$$

$$|l_1^{'} - l_2^{'}| \leq l_3^{'} \leq l_1^{'} + l_2^{'},$$

$$l_1^{'} + l_2^{'} + l_3^{'} \leq 2k.$$

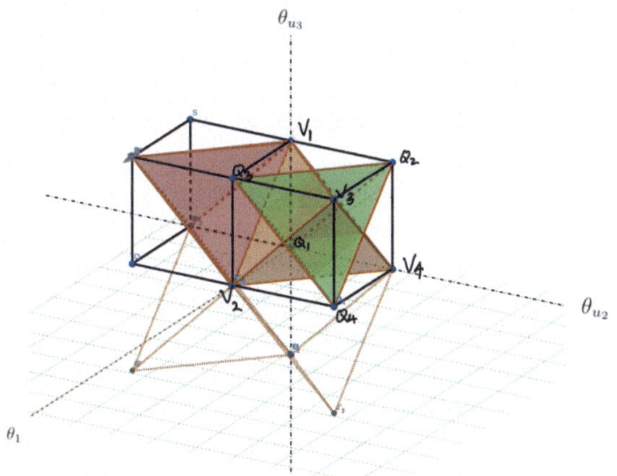

Fig. 2 The green tetrahedron is Δ_1, the red tetrahedron is Δ_1''', and the orange tetrahedron is Δ_2. We created this diagram by using https://www.geogebra.org/3d

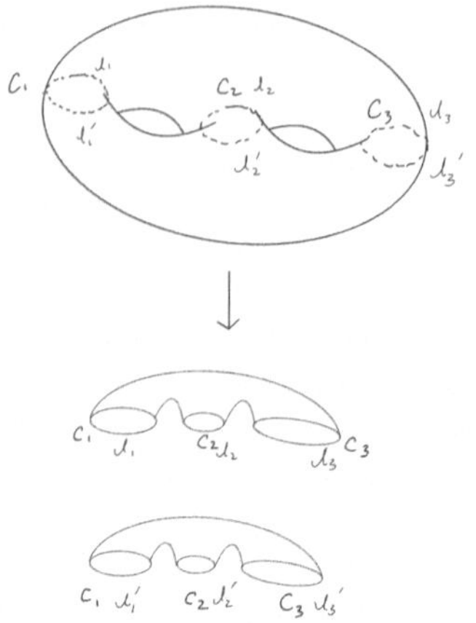

Together with the following conditions:

$$l_1 + l_2 + l_3 \in 2\mathbb{Z},$$

$$l_1^{'} + l_2^{'} + l_3^{'} \in 2\mathbb{Z},$$

$$l_i + l_i^{'} \in 2\mathbb{Z}, i = 1, 2, 3.$$

$$l_i + l_i^{'} \leq k$$

These labellings enumerate representations of the free group on $2g$ generators into $SU(2) \times SU(2)$, which come from homomorphisms into $SO(4)$. Each representation of the free group on $2g$ generators into $SO(4)$ gives rise to two homomorphisms into $SU(2) \times SU(2)$ (because if one multiplies the image elements by -1, the equation will still hold).

The integer solutions that are from the above equations and satisfy the above conditions are located in the polyhedra as well as on its surface inscribed in the cube with side length k. Section 4 gives us the details about this polyhedron. The number of desired solutions, as described above, will give us the number of Bohr-Sommerfeld fibers.

The image of the moment map for $\mathcal{M}_g(SU(2) \times SU(2))$ is obtained by imposing Eqs. (15) and (16) at every trinion arising from a trinion decomposition or every vertex of the trivalent graph. By Eqs. (27) and (28), they will give us a collection of integer labels $(l_i, l_i^{'})$ for each edge E_i of the trivalent graph. The set of Bohr-Sommerfeld points is in bijective correspondence with the set of integer labels $(l_i, l_i^{'})$ that meet the conditions (31), (32), and (33).

For $\mathcal{M}_g(SO(4))$ case, we need to consider the following:

Let $T_{SU(2)\times SU(2)}$ and $T_{SO(4)}$ be fixed maximal tori of $SU(2) \times SU(2)$ and $SO(4)$, respectively. The double covering map restricts on these maximal tori to a map from

$T_{SU(2)\times SU(2)}$ to $T_{SO(4)}$. This map is also surjective with kernel $\{I, -I\}$. Thus, the restriction to the covering map gives us the following:

$$\begin{aligned}
\Phi : (u, v) \mapsto R_{u,v}(\cdot) &= uqv^{-1} \\
&= \begin{bmatrix} e^{-2(\pi)^2(t_s)i} & 0 \\ 0 & e^{2(\pi)^2(t_s)i} \end{bmatrix} (q) \begin{bmatrix} e^{-2(\pi)^2(t_h)i} & 0 \\ 0 & e^{2(\pi)^2(t_h)i} \end{bmatrix} \\
&= \begin{bmatrix} e^{-2(\pi)^2(t_s)i} & 0 \\ 0 & e^{2(\pi)^2(t_s)i} \end{bmatrix} \begin{bmatrix} Z_q & 0 \\ 0 & \bar{Z}_q \end{bmatrix} \begin{bmatrix} e^{-2(\pi)^2(t_h)i} & 0 \\ 0 & e^{(2\pi)^2(t_h)i} \end{bmatrix}^{-1} \\
&= \begin{bmatrix} Z_q e^{-2(\pi)^2(t_s-t_h)i} & 0 \\ 0 & -\bar{Z}_q e^{2(\pi)^2(t_h-t_s)i} \end{bmatrix}.
\end{aligned} \tag{35}$$

Here $u, v, q \in T_{SU(2)\times SU(2)}$. We represent them by quaternionic notation.

The period of the flow depends on the period of $\begin{bmatrix} Z_q e^{-2(\pi)^2(t_s-t_h)i} & 0 \\ 0 & -\bar{Z}_q e^{2(\pi)^2(t_h-t_s)i} \end{bmatrix}$. Because of the kernel$\{I, -I\}$of the covering map, the period is half of the period of the corresponding flow in the $SU(2)$ case in [5]. So, for integer lattices, we require

$$\frac{k(t_s - t_h)}{2\pi} \in \mathbb{Z}. \tag{36}$$

If we let $l_s = \frac{k(t_s)}{\pi}$ and $l_h = \frac{k(t_h)}{\pi}$, (36) is equivalent with

$$l_s + l_h \in 2\mathbb{Z}.$$

This is exactly the condition listed in (29). This explains why, in the $\mathcal{M}_g(SO(4))$ case, we need to impose one additional condition (29).

Acknowledgments This question was originally suggested to the authors by Jonathan Weitsman, and he provided key suggestions during the project. The authors would like to acknowledge their gratitude to Jonathan Weitsman for his guidance and insight. The first author is partially supported by an NSERC Discovery Grant.

References

1. M. Atiyah, R. Bott, The Yang-Mills equations over a Riemann surface. Phil. Trans. Roy. Soc. **A308**, 523–615 (1982).
2. P. Francesco, P. Mathieu, D. Senechal, *Conformal Field Theory* (Springer, Berlin, 2012)
3. S. Dwivedi, J. Herman, L. Jeffrey, T. van den Hurk, *Hamiltonian Group Actions and Equivariant Cohomology* (Springer (Springer Briefs), Berlin, 2019)
4. W. Goldman, Invariant functions on Lie groups and Hamiltonian flows of surface group representations. Invent.Math. **85**, 263–302 (1986)

5. L.C. Jeffrey, J. Weitsman, Bohr-Sommerfeld orbits in the moduli space of flat connections and the Verlinde dimension formula. Commun. Math. Phys. **150**, 593–630 (1992)
6. D. Krepski, E. Meinrenken, On the Verlinde formulas for $SO(3)$ bundles. Quart. J. Math. **64**(1), 235–252 (2013)
7. T. Ramadas, I.M. Singer, J. Weitsman, Some comments on Chern-Simons gauge theory. Commun. Math. Phys. **126**, 409–420 (1989)
8. J. Sniatycki, On cohomology groups appearing in geometric quantization, in *Differential Geometric Methods in Mathematical Physics*. Springer Lecture Notes in Mathematics, vol. 570 (1977), pp. 46–66
9. E. Verlinde, Fusion rules and modular transformations in $2d$ conformal field theory. Nucl. Phys. **B300**, 351 (1988)
10. J. Weitsman, Real polarization of the moduli space of flat connections on a Riemann surface. Commun. Math. Phys. **145**, 425–433 (1992)
11. J. Weitsman, Quantization via real polarization of the moduli space of flat connections and Chern-Simons gauge theory in genus one. Commun. Math. Phys. **137**, 175–190 (1991)
12. M. Thaddeus, Conformal field theory and the cohomology of the moduli space of stable bundles J. Differential Geom. **35**(1), 131–149 (1992)
13. A. Szenes, Verification of Verlinde's formulas for $SU(2)$. Internat. Math. Res. Notices **7**, 93–98 (1991)

GKM Graph Locally Modeled by $T^n \times S^1$-Action on $T^*\mathbb{C}^n$ and Its Graph Equivariant Cohomology

Shintarô Kuroki and Vikraman Uma

1 Introduction

A *GKM graph* is a labeled graph defined by the special but wide class of manifolds with torus actions, called *GKM manifolds*. From the torus action on a GKM manifold, a GKM graph is defined by its zero and one-dimensional orbits together with the labels on edges defined by the tangential representations around fixed points. Goresky-Kottwicz-MacPherson in [8] show that if a GKM manifold satisfies a certain condition, called *equivariant formality*, then its equivariant cohomology is isomorphic to an algebra defined from its GKM graph. We call this algebra a *graph equivariant cohomology* in this paper. Motivated by the work of Goresky-Kottwicz-MacPherson, Guillemin-Zara in [12] introduce the abstract GKM graph without considering any GKM manifolds, and they translate some geometric properties of GKM manifolds into combinatorial ones of GKM graphs. After the works of Guillemin-Zara, a GKM graph can be regarded as a combinatorial approximation of space with torus action, and it has been studied by some mathematicians, e.g., see [4, 6, 7, 9, 10, 19, 20]. In this paper, we introduce a certain class of GKM graphs with legs and attempt to unify two slightly different classes of manifolds from the GKM theoretical point of view, i.e., *toric hyperKähler manifolds* and *cotangent bundles of toric manifolds*, where a *leg* is a half-line whose boundary corresponds to the initial vertex. We briefly recall toric hyperKähler manifolds and introduce the motivation of the present paper. We shall then state our main results and outline the organization of this paper.

S. Kuroki (✉)
Okayama University of Science, Okayama, Japan
e-mail: kuroki@ous.ac.jp

V. Uma
Department of Mathematics, Indian Institute of Technology, Chennai, India
e-mail: vuma@iitm.ac.in

A. Bahri et al. (eds.), *Toric Topology and Polyhedral Products*, Fields Institute Communications 89, https://doi.org/10.1007/978-3-031-57204-3_11

A *toric hyperKähler variety* is defined by the hyperKähler quotient of a torus action on the cotangent bundle $T^*\mathbb{C}^m$. This space is introduced by Goto and Bielwasky-Dancer in [1, 11] as the hyperKähler analogue of the symplectic toric manifolds. The non-singular toric hyperKähler varieties are $4n$-dimensional non-compact manifolds with T^n-action. They are completely determined by some class of hyperplane arrangements in $\mathbb{R}^n$ (see [1]), like symplectic toric manifolds are completely determined by Delzant polytopes in $\mathbb{R}^n$ (see [5]). The equivariant topology and geometry of toric hyperKähler manifolds are studied by some mathematicians, e.g., [13–16, 18, 22]. In particular, Harada-Proudfoot show that every toric hyperKähler manifold admits the residual S^1-action and the equivariant cohomology of a toric hyperKähler manifold with $T^n \times S^1$-action is determined by the half-space arrangements in $\mathbb{R}^n$. They also show that the toric hyperKähler manifolds with $T^n \times S^1$-actions satisfy the *GKM condition*, i.e., its zero and one-dimensional orbits have the structure of a graph. Note that the T^n-action on a toric hyperKähler manifold does not satisfy the GKM condition. Therefore, we can define the labeled graph (with legs) from toric hyperKähler manifolds with $T^n \times S^1$-actions. The GKM graph of a toric hyperKähler manifold is obtained from the one-dimensional intersections of hyperplanes like the GKM graph of a symplectic toric manifold is obtained from the one-skeleton of a moment-polytope. By definition, the tangential representations of $T^n \times S^1$-actions on the fixed points are isomorphic (up to automorphism on $T^n \times S^1$) to the standard T^n-action on $T^*\mathbb{C}^n$ together with the scalar multiplication of S^1 on the fiber, i.e.,

$$(t_1, \ldots, t_n, r) \cdot (z_1, \ldots, z_n, w_1, \ldots, w_n) \mapsto (t_1 z_1, \ldots, t_n z_n, r t_1^{-1} w_1, \ldots, r t_n^{-1} w_n), \tag{1}$$

where

$$(t_1, \ldots, t_n) \in T^n, \quad r \in S^1, \quad (z_1, \ldots, z_n, w_1, \ldots, w_n) \in T^*\mathbb{C}^n (\simeq \mathbb{C}^n \times \mathbb{C}^n).$$

We call the action defined by (1) is the *standard* $T^n \times S^1$*-action on* $T^*\mathbb{C}^n$.

On the other hand, the cotangent bundle T^*M of a $2n$-dimensional toric manifold M also has the $T^n \times S^1$-action. More precisely, because T acts on M smoothly, each element $t \in T$ induces the diffeomorphism $t : M \to M$, say $t : p \mapsto t \cdot p$ for $p \in M$. By taking its differential $dt : TM \to TM$, we have the lift of the T-action on the tangent bundle TM. Note that $(dt)_p : T_pM \to T_{t \cdot p}M$ is the linear isomorphism, because the cotangent bundle T^*M is defined by the bundle over M whose fibres are $T_p^*M := \mathrm{Hom}(T_pM, \mathbb{R})$. Therefore, for an element $f \in T_p^*M$, we can define the $t \in T$ action by $t \cdot f := f \circ (dt)_p^{-1} : T_{t \cdot p}M \to \mathbb{R}$. Together with the scalar multiplication by S^1 on each fiber T_p^*M, we have the $T^n \times S^1$-action on T^*M. It follows from the definition that this also satisfies the GKM conditions and the tangential representation around every fixed point in T^*M is isomorphic to the standard $T^n \times S^1$-action on $T^*\mathbb{C}^n$.

Note that T^*M of a toric manifold M is not a toric hyperKähler manifold except in the case when M is a product of some projective spaces; see [1]. So the cotangent bundles of toric manifolds and the toric hyperKähler manifolds are different classes of manifolds. However, it is known that their equivariant cohomologies are quite similar. The equivariant cohomology $H^*_{T^n}(M)$ of a symplectic toric manifold M with T^n-action is isomorphic to the Stanley-Reisner ring of the moment-polytope (see e.g. [3, Lemma 7.4.34] for more general class of manifolds with T^n-actions). Because there is an equivariant deformation retract from T^*M to M, we see that $H^*_{T^n}(T^*M)$ is also isomorphic to the Stanley-Reisner ring of a polytope. On the other hand, by Konno's theorem [14], the equivariant cohomology of a toric hyperKähler manifold with T^n-action is isomorphic to the Stanley-Reisner ring of hyperplane arrangements. Therefore, these distinct classes of manifolds have similar equivariant cohomology ring structures. So it may be natural to ask whether we can unify these classes of manifolds. One answer is that there exists an embedding from a symplectic toric manifold M to a toric hyperKähler manifold (see [1, 13]). In this paper, we answer this question from a different direction, namely, we unify the equivariant cohomologies of these classes by using GKM graphs.

To achieve that, we introduce the class of GKM graphs whose axial functions around vertices are modeled by the standard $T^n \times S^1$-action on $T^*\mathbb{C}^n$, called a *GKM graph locally modeled by* $T^n \times S^1$*-action on* $T^*\mathbb{C}^n$ or $T^*\mathbb{C}^n$*-modeled GKM graph* for short in Definition 2.4. This GKM graph behaves like the hyperplane arrangements but does not always come from the hyperplane arrangements; see Sect. 3. We study the graph equivariant cohomology of $T^*\mathbb{C}^n$-modeled GKM graphs. The first main theorem of this paper is as follows (the technical notions will be introduced in Sects. 3 and 4):

Theorem 1.1 (Theorem 4.1) *Let $\mathcal{G}$ be a $2n$-valent $T^*\mathbb{C}^n$-modeled GKM graph and $\mathbf{L} = \{L_1, \cdots, L_m\}$ be the set of all hyperplanes in $\mathcal{G}$. Assume that $\mathcal{G}$ satisfies the following two assumptions:*

1. *For each $L \in \mathbf{L}$, there exist the unique pair of the halfspace H and its opposite side $\overline{H}$ such that $H \cap \overline{H} = L$;*
2. *For every subset $\mathbf{L}' \subset \mathbf{L}$, its intersection $\bigcap_{L \in \mathbf{L}'} L$ is empty or connected.*

Then, the following ring isomorphism holds:

$$H^*(\mathcal{G}) \simeq \mathbb{Z}[\mathcal{G}].$$

To prove this, we introduce an *x-forgetful graph* $\widetilde{\mathcal{G}}$ from a $T^*\mathbb{C}^n$-modeled GKM graph $\mathcal{G}$ in Sect. 5. An x-forgetful graph $\widetilde{\mathcal{G}}$ is a labeled graph but not a GKM graph, and it may be regarded as the combinatorial counterpart of the T^n-actions on toric hyperKähler manifolds and cotangent bundles over toric manifolds. We define its graph equivariant cohomology $H^*(\widetilde{\mathcal{G}})$ and prove its ring structure in Theorem 5.1. In Sect. 6, we give a proof of Theorem 4.1 by using Theorem 5.1. In Sect. 7, we also study the $H^*(BT^n)$-module structure of $H^*(\widetilde{\mathcal{G}})$. As the second main result of this paper, in Theorem 7.6, we exhibit an $H^*(BT^n)$-module basis of $H^*(\widetilde{\mathcal{G}})$ by using the shellablility of a simplicial complex $\Delta_{\mathbf{L}}$ associated with $\mathbf{L}$. Dividing $H^*(\widetilde{\mathcal{G}})$ by $H^{>0}(BT^n)$, we also introduce $H^*_{ord}(\widetilde{\mathcal{G}})$, which corresponds to the ordinary cohomology of the usual equivariant cohomology. Then, we show that the $H^*(BT^n)$-module basis of $H^*(\widetilde{\mathcal{G}})$ induces a $\mathbb{Z}$-module basis for $H^*_{ord}(\widetilde{\mathcal{G}})$. Finally, in the case when $\widetilde{\mathcal{G}}$ corresponds to the line arrangements in $\mathbb{R}^2$ (which corresponds geometrically to the 8-dimensional toric hyperKähler manifolds), we describe the structure constants of $H^*_{ord}(\widetilde{\mathcal{G}})$ with respect to this basis.

2 GKM Graph Locally Modeled by $T^n \times S^1$-Action on $T^*\mathbb{C}^n$

This section aims to define a GKM graph with legs and its graph equivariant cohomology. In particular, we introduce *GKM graphs locally modeled by $T^n \times S^1$-action on $T^*\mathbb{C}^n$* as the special class of GKM graphs with legs.

2.1 Notations

We first prepare some notations. In this paper, Γ is a connected graph which possibly has legs, where a *leg* means an outgoing half-line from one vertex (see the left graph in the Fig. 1).

We define a graph with legs more precisely. Let $\mathcal{V}$ be a set of vertices, E be a set of edges and Leg be a set of legs in Γ, and $\mathcal{E} = E \cup Leg$. The graph Γ is denoted by

$$\Gamma = (\mathcal{V},\ \mathcal{E}).$$

Fig. 1 These are examples of regular graphs with legs and orientations. The left 2-valent graph has two legs; on the other hand, the right 3-valent graph has no legs. Note that all edges have two orientations and all legs have only one orientation

In this paper, we assume that $\mathcal{V}$ and $\mathcal{E}$ are finite sets. We also assume that Γ is an oriented graph. For $\epsilon \in E$, we denote by $i(\epsilon)$ and $t(\epsilon)$ the initial vertex and the terminal vertex of ϵ, respectively. We denote the opposite directed edge of ϵ as $\overline{\epsilon}$, i.e., $i(\overline{\epsilon}) = t(\epsilon)$ and $t(\overline{\epsilon}) = i(\epsilon)$. For $\ell \in Leg$, there is no terminal vertex, but there exists an initial vertex $i(\ell)$. Note that the leg in Γ can be characterized by the element ϵ in $\mathcal{E}$ such that there is no $\overline{\epsilon}$. For a vertex $p \in \mathcal{V}$, we put the set of all outgoing edges and legs from $p \in \mathcal{V}$ by

$$\mathcal{E}_p = \{\epsilon \in \mathcal{E} \mid i(\epsilon) = p\}.$$

Assume that $|\mathcal{E}_p| = m$ for all $p \in \mathcal{V}$, where the symbol $|X|$ represents the cardinality of the finite set X. We call such a graph a *(regular) m-valent graph*; see Fig. 1.

Let $\Gamma = (\mathcal{V}, \mathcal{E})$ be a graph with legs. We denote a subgraph of Γ by $G = (\mathcal{V}^G, \mathcal{E}^G)$, that is, G satisfies $\mathcal{V}^G \subset \mathcal{V}$ and $\mathcal{E}^G \subset \mathcal{E}$. We use the following symbols.

- $\mathcal{E}_p^G$: the set of all outgoing edges and legs in $\mathcal{E}^G$ from $p \in \mathcal{V}^G$.
- $E^G \subset \mathcal{E}^G$ (resp. $E_p^G \subset \mathcal{E}_p^G$): the set of all edges (resp. out going from p) in G, i.e., if $\epsilon \in E^G$, then the both $i(\epsilon), t(\epsilon) \in \mathcal{V}^G$.
- $Leg^G \subset \mathcal{E}^G$ (resp. $(Leg)_p^G \subset \mathcal{E}_p^G$): the set of all legs (resp. out going from p) in G.

Note the following remark about legs in G.

Remark 2.1 Because Γ is an oriented graph, we may consider the subgraph $G = (\mathcal{V}^G, \mathcal{E}^G)$ of $\Gamma = (\mathcal{V}, \mathcal{E})$ such that there exists a leg $\epsilon \in Leg^G$ in G which is an edge $\epsilon \in \mathcal{E}$ in Γ. In other words, $t(\epsilon) \notin \mathcal{V}^G$ but $t(\epsilon) \in \mathcal{V}$; or $\overline{\epsilon} \notin \mathcal{E}^G$ but $\overline{\epsilon} \in \mathcal{E}$.

2.2 GKM Graph with Legs and Its Graph Equivariant Cohomology

In this section, we shall define a *GKM graph* (*(possibly) with legs*) and its *graph equivariant cohomology*.

Let $\Gamma = (\mathcal{V}, \mathcal{E})$ be an m-valent graph. We first prepare the following notations. Let $\nabla = \{\nabla_\epsilon \mid \epsilon \in E\}$ be a collection of bijective maps

$$\nabla_\epsilon : \mathcal{E}_{i(\epsilon)} \to \mathcal{E}_{t(\epsilon)}$$

for all edges $\epsilon \in E$. A *connection* on Γ is a set $\nabla = \{\nabla_\epsilon | \epsilon \in E\}$, which satisfies the following two conditions:

- $\nabla_{\overline{\epsilon}} = \nabla_\epsilon^{-1}$.
- $\nabla_\epsilon(\epsilon) = \overline{\epsilon}$.

We can easily check that an m-valent graph Γ admits different $((m-1)!)^g$ connections, where g is the number of (unoriented) edges E.

Let T^n be an n-dimensional torus. In particular, we often denote a 1-dimensional torus by S^1. If we do not emphasis the dimension of T^n, then we denote it by T. Let $\mathfrak{t}$ be a Lie algebra of T, $\mathfrak{t}_{\mathbb{Z}}$ be the lattice of $\mathfrak{t}$, and $\mathfrak{t}^*$ (resp. $\mathfrak{t}_{\mathbb{Z}}^*$) be the dual of $\mathfrak{t}$ (resp. $\mathfrak{t}_{\mathbb{Z}}$). The symbol $\mathrm{Hom}(T, S^1)$ represents a set of all homomorphisms from the torus T to S^1. It is well-known that $\mathrm{Hom}(T^n, S^1) \simeq \mathbb{Z}^n$. Moreover, it may be regarded as $\mathfrak{t}_{\mathbb{Z}}^*$ and $H^1(T) \simeq H^2(BT)$, where BT is the classifying space of T. In this paper, if we omit the coefficient of the cohomology, then it means the cohomology with integer coefficients. Therefore, we have the identification

$$\mathrm{Hom}(T, S^1) \simeq \mathfrak{t}_{\mathbb{Z}}^* \simeq H^2(BT).$$

Define an *axial function* by the function

$$\alpha : \mathcal{E} \longrightarrow H^2(BT)$$

such that it satisfies the following three conditions:

- $\alpha(\bar{\epsilon}) = \pm\alpha(\epsilon)$ for all edges $\epsilon \in E$.
- $\alpha(\mathcal{E}_p) = \{\alpha(\epsilon) \mid \epsilon \in \mathcal{E}_p\}$ are *pairwise linearly independent* for all $p \in \mathcal{V}$, that is, for every two distinct elements $\epsilon_1, \epsilon_2 \in \mathcal{E}_p$, $\alpha(\epsilon_1), \alpha(\epsilon_2)$ are linearly independent in $H^2(BT)$.
- There is a connection ∇ which satisfies the following *congruence relation* for all edges $\epsilon \in E$:

$$\alpha(\epsilon') - \alpha(\nabla_\epsilon(\epsilon')) \equiv 0 (\mathrm{mod}\, \alpha(\epsilon))$$

  for all $\epsilon' \in \mathcal{E}_{i(\epsilon)}$.

Definition 2.2 (GKM graph with legs) Let $\mathcal{G} = (\Gamma, \alpha, \nabla)$ be a collection of an m-valent graph $\Gamma = (\mathcal{V}, \mathcal{E})$, where the map

$$\alpha : \mathcal{E} \longrightarrow H^2(BT^n)$$

is an axial function ($n \leq m$) and ∇ is a connection on Γ. We call $\mathcal{G} = (\Gamma, \alpha, \nabla)$ a *GKM graph (with legs)*.

Remark 2.3 Suppose that $\mathcal{E}_p$ satisfies the 3-linearly independent condition for all $p \in \mathcal{V}$, i.e., for every distinct three elements $\epsilon_1, \epsilon_2, \epsilon_3 \in \mathcal{E}_p$, the axial function $\alpha(\epsilon_1), \alpha(\epsilon_2), \alpha(\epsilon_3)$ are linearly independent. Then, by the similar proof for the cases of GKM graphs without legs in [12], if there exists a connection ∇, then the connection ∇ is unique. In particular, if the GKM graph $\mathcal{G}$ satisfies the 3-linearly independent condition, then for any two edges (or legs) ϵ, ϵ' in E_p we can determine the 2-valent GKM subgraph which contains ϵ, ϵ'. Hence, we often omit the connection ∇ in the GKM graph $\mathcal{G}$. Namely, we often denote the GKM graph by $\mathcal{G} = (\Gamma, \alpha)$ if the connection is obviously determined by the context.

Due to the theory of toric hyperKähler varieties (see [1, 13, 14]), the tangential representation on each fixed point is isomorphic to the T^n-action on $T^*\mathbb{C}^n$ ($\simeq \mathbb{H}^n$, i.e., the n-dimensional quaternionic space), which is defined by the strandard T^n-action on $\mathbb{C}^n$, i.e.,

$$(t_1, \ldots, t_n) \cdot (z_1, \ldots, z_n, w_1, \ldots, w_n) := (t_1 z_1, \ldots, t_n z_n, t_1^{-1} w_1, \ldots, t_n^{-1} w_n),$$

where $(t_1, \ldots, t_n) \in T^n$, $z = (z_1, \ldots, z_n) \in \mathbb{C}^n$, and $(w_1, \ldots, w_n) \in T_z^*(\mathbb{C}^n)$. On the other hand, Harada-Proudfoot [13] found that there exists the residual S^1-action on the toric hyperKähler varieties, and this action fits into the GKM theory. In this case, the tangential representation on each fixed point may be regarded as the $T^n \times S^1$-action on $T^*\mathbb{C}^n$, i.e.,

$$(t_1, \ldots, t_n, r) \cdot (z_1, \ldots, z_n, w_1, \ldots, w_n) := (t_1 z_1, \ldots, t_n z_n, r t_1^{-1} w_1, \ldots, r t_n^{-1} w_n),$$

where $r \in S^1$. Therefore, the toric hyperKähler manifolds with $T^n \times S^1$-actions induce the GKM graphs with legs whose axial functions around every vertex are isomorphic to

$$\{e_1^*, \ldots, e_n^*, -e_1^* + x, \ldots, -e_n^* + x\},$$

where $e_1^*, \ldots, e_n^*$ is a generator of a rank n subspace in $H^2(BT^n \times BS^1) \simeq \mathbb{Z}^n \oplus \mathbb{Z}$ and x is a generator of $H^2(BS^1) \simeq \mathbb{Z}$. By defining this abstractly, we introduce the following notion:

Definition 2.4 (GKM graph locally modeled by $T^n \times S^1$-action on $T^*\mathbb{C}^n$) Let $\mathcal{G} = (\Gamma,\ \alpha,\ \nabla)$ be a $2n$-valent GKM graph with legs with an axial function

$$\alpha : \mathcal{E} \longrightarrow H^2(BT^n \times BS^1) \simeq \mathfrak{t}_{\mathbb{Z}}^* \oplus \mathbb{Z}x,$$

where x is a generator of the dual of the Lie algebra of S^1. We call $\mathcal{G} = (\Gamma,\ \alpha,\ \nabla)$ a *GKM graph modeled by the $T^n \times S^1$-action on $T^*\mathbb{C}^n$* (or simply a *$T^*\mathbb{C}^n$-modeled GKM graph*) if it satisfies the following conditions for all $p \in \mathcal{V}$:

1. We can divide $\mathcal{E}_p$ into $\{\epsilon_1^+,\ \cdots,\ \epsilon_n^+,\ \epsilon_1^-,\ \cdots,\ \epsilon_n^-\}$ such that

$$\alpha(\epsilon_j^+) + \alpha(\epsilon_j^-) = x$$

for all $j = 1,\ \cdots,\ n$.

2. The set $\{\alpha(\epsilon_j^+),\ x \mid j = 1, \cdots, n\}$ spans $\mathfrak{t}_{\mathbb{Z}}^* \oplus \mathbb{Z}x$, i.e.,

$$\langle \alpha(\epsilon_1^+),\ \cdots,\ \alpha(\epsilon_n^+),\ x \rangle = \mathfrak{t}_{\mathbb{Z}}^* \oplus \mathbb{Z}x.$$

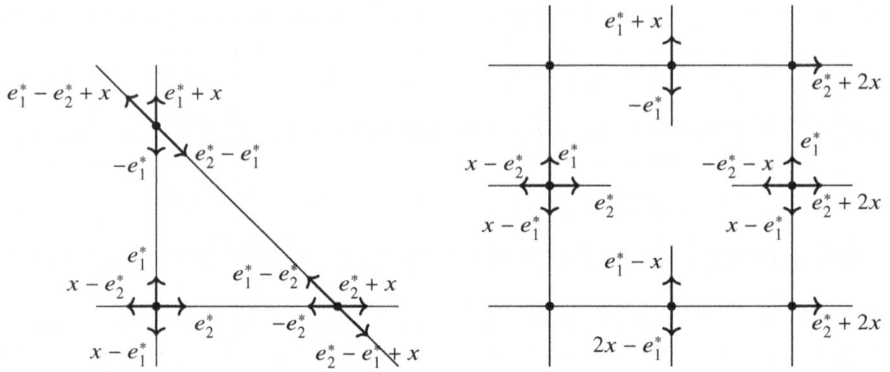

Fig. 2 $T^*\mathbb{C}^2$-modeled GKM graphs, where $\langle \mathrm{e}_1^*, \mathrm{e}_2^* \rangle \simeq (\mathfrak{t}^2)^*_{\mathbb{Z}}$. In the left and the right figures, we assume $\alpha(\epsilon) = -\alpha(\overline{\epsilon})$ and omit some axial functions, which are automatically determined by the definition

We call $\{\epsilon_j^+, \epsilon_j^-\}$ such that $\alpha(\epsilon_j^+) + \alpha(\epsilon_j^-) = x$ a 1-*dimensional pair* in $\mathcal{E}_p$. Furthermore, we call an element x a *residual basis*.

Figure 2 shows some examples of $T^*\mathbb{C}^n$-modeled GKM graphs.

Remark 2.5 Note that the axial function on $T^*\mathbb{C}^n$-modeled GKM graphs satisfies the 3-linearly independent condition for all vertices. Therefore, the connection on a $T^*\mathbb{C}^n$-modeled GKM graph $\mathcal{G}$ is uniquely determined, and we may denote it by

$$\mathcal{G} = (\Gamma, \alpha).$$

We also have the following lemma for the 1-dimensional pair $\{\epsilon^+, \epsilon^-\}$.

Lemma 2.6 *Let* $\{\epsilon^+, \epsilon^-\}$ *be a* 1*-dimensional pair in* $\mathcal{E}_{i(\epsilon')}$ *for some edge* $\epsilon' \in E$. *Then,* $\{\nabla_{\epsilon'}(\epsilon^+), \nabla_{\epsilon'}(\epsilon^-)\}$ *is also a* 1*-dimensional pair in* $\mathcal{E}_{t(\epsilon')}$.

Proof We first divide the edges and legs $\mathcal{E}_{i(\epsilon')}$ by the 1-dimensional pairs as follows:

$$\mathcal{E}_{i(\epsilon')} = \{\epsilon_1^+, \epsilon_1^-\} \cup \cdots \cup \{\epsilon_n^+, \epsilon_n^-\}$$

Because the axial function α satisfies the congruence relation, there are integers k_j^+ and k_j^-, $j = 1, \ldots, n$, such that

$$\alpha(\nabla_{\epsilon'}(\epsilon_j^+)) - \alpha(\epsilon_j^+) = k_j^+ \alpha(\epsilon'),$$

$$\alpha(\nabla_{\epsilon'}(\epsilon_j^-)) - \alpha(\epsilon_j^-) = k_j^- \alpha(\epsilon').$$

Since $\{\epsilon_j^+, \epsilon_j^-\}$ is a 1-dimensional pair, we also have

$$\begin{aligned}&(\alpha(\nabla_{\epsilon'}(\epsilon_j^+)) - \alpha(\epsilon_j^+)) + (\alpha(\nabla_{\epsilon'}(\epsilon_j^-)) - \alpha(\epsilon_j^-)) \\ =&\alpha(\nabla_{\epsilon'}(\epsilon_j^+)) + \alpha(\nabla_{\epsilon'}(\epsilon_j^-)) - x \\ =&(k_j^+ + k_j^-)\alpha(\epsilon').\end{aligned} \tag{2}$$

In order to show the statement, it is enough to show that the following equation holds:

$$k_j^- = -k_j^+.$$

Suppose on the contrary that $k_j^- \neq -k_j^+$. Then, by (2), we have

$$\alpha(\nabla_{\epsilon'}(\epsilon_j^-)) = -\alpha(\nabla_{\epsilon'}(\epsilon_j^+)) + x + (k_j^+ + k_j^-)\alpha(\epsilon') \neq -\alpha(\nabla_{\epsilon'}(\epsilon_j^+)) + x.$$

This implies that $\{\nabla_{\epsilon'}(\epsilon_j^+),\ \nabla_{\epsilon'}(\epsilon_j^-)\}$ is not a 1-dimensional pair. Therefore, there is another element $\epsilon (\neq \nabla_{\epsilon'}(\epsilon_j^-))$ in $\mathcal{E}_{t(\epsilon')}$ such that

$$\alpha(\epsilon) = -\alpha(\nabla_{\epsilon'}(\epsilon_j^+)) + x.$$

This gives that

$$\alpha(\nabla_{\epsilon'}(\epsilon_j^-)) = \alpha(\epsilon) + (k_j^+ + k_j^-)\alpha(\epsilon').$$

However, since $\{\nabla_{\epsilon'}(\epsilon_j^-), \epsilon, \overline{\epsilon'}\} \subset \mathcal{E}_{t(\epsilon')}$, this is a contradiction to the fact that $T^*\mathbb{C}^n$-modeled GKM graph is always 3-linearly independent. Hence, we must have $k_j^- = -k_j^+$. This establishes the statement. □

Finally, in this section, we also define the notion of *graph equivariant cohomology*. Let $\mathcal{G} = (\Gamma,\ \alpha,\ \nabla)$ be a GKM graph (with legs) such that $\alpha : \mathcal{E} \to H^2(BT)$. With the definition similar to that of the GKM graph without legs, the graph equivariant cohomology is defined as follows.

Definition 2.7 (graph equivariant cohomology) The following ring is called a *graph equivariant cohomology* of $\mathcal{G}$:

$$H^*(\mathcal{G}) = \{\varphi : \mathcal{V} \to H^*(BT) \mid \varphi(i(\epsilon)) - \varphi(t(\epsilon)) \equiv 0 \ (\mathrm{mod}\ \alpha(\epsilon))\}.$$

We call the relation $\varphi(i(\epsilon)) - \varphi(t(\epsilon)) \equiv 0 \ (\mathrm{mod}\ \alpha(\epsilon))$ a *congruence relation* of φ on an edge $\epsilon \in E$.

3 Some Notions of $T^*\mathbb{C}^n$-Modeled GKM Graphs

Let $\mathcal{G} = (\Gamma, \alpha, \nabla)$ be a $2n$-valent $T^*\mathbb{C}^n$-modeled GKM graph such that $\alpha : \mathcal{E} \to H^2(BT^n) \oplus \mathbb{Z}x$, where x is a residual basis. The goal of this paper is to compute $H^*(\mathcal{G})$ for a certain class of $T^*\mathbb{C}^n$-modeled GKM graphs. To do that, we prepare some notions and properties of $T^*\mathbb{C}^n$-modeled GKM graphs. We first introduce the following notion.

We define a *GKM subgraph* $\mathcal{H} = (H, \alpha^H, \nabla^H)$ by a k-valent subgraph H of Γ such that the axial function is defined by

$$\alpha^H := \alpha|_{\mathcal{E}^H}$$

and the connection is defined by

$$\nabla^H := \{\nabla_\epsilon|_{\mathcal{E}^H} \mid \epsilon \in E^H\}.$$

We also denote $\nabla^H_\epsilon := \nabla_\epsilon|_{\mathcal{E}^H}$ for an edge $\epsilon \in E^H$.

3.1 Hyperplane

In this section, we introduce the notion of a hyperplane in $\mathcal{G}$ and show a key property Lemma 3.3, which will be used to show the main theorem of this paper.

Definition 3.1 (hyperplane) Let $\mathcal{G}$ be a $T^*\mathbb{C}^n$-modeled GKM graph. Assume that a GKM subgraph $\mathcal{L} = (L, \alpha^L, \nabla^L)$ of $\mathcal{G}$ is a $(2n-2)$-valent subgraph of Γ and it is a $T^*\mathbb{C}^{n-1}$-modeled GKM graph with the residual basis x, i.e., there are 1-dimensional pairs $\{\epsilon_j^+, \epsilon_j^-\} \subset \mathcal{E}_p^L$, $j = 1, \dots, n-1$, on each vertex such that

$$\alpha^L(\epsilon_j^+) + \alpha^L(\epsilon_j^-) = x.$$

Such a GKM subgraph $\mathcal{L}$ is said to be a *hyperplane* if L is a maximal $(2n-2)$-valent connected subgraph in Γ, i.e., if L' is a $(2n-2)$-valent connected subgraph in Γ such that $L \subset L'$, then $L = L'$.

Example 3.2 The following two figures show an example and a non-example of hyperplanes (Fig. 3).

For the hyperplane $\mathcal{L}$, we have the following property (this may be regarded as the analogue of the property of facets in a torus graph [20]).

Lemma 3.3 *Let $\mathcal{G} = (\Gamma, \alpha, \nabla)$ be a $2n$-valent $T^*\mathbb{C}^n$-modeled GKM graph. Take a vertex $p \in \mathcal{V}$. Then, for every 1-dimensional pair $\{\epsilon^+, \epsilon^-\} \subset \mathcal{E}_p$, there exists a unique hyperplane $\mathcal{L} = (L, \alpha^L, \nabla^L)$ such that $\mathcal{E}_p^L = \mathcal{E}_p \setminus \{\epsilon^+, \epsilon^-\}$.*

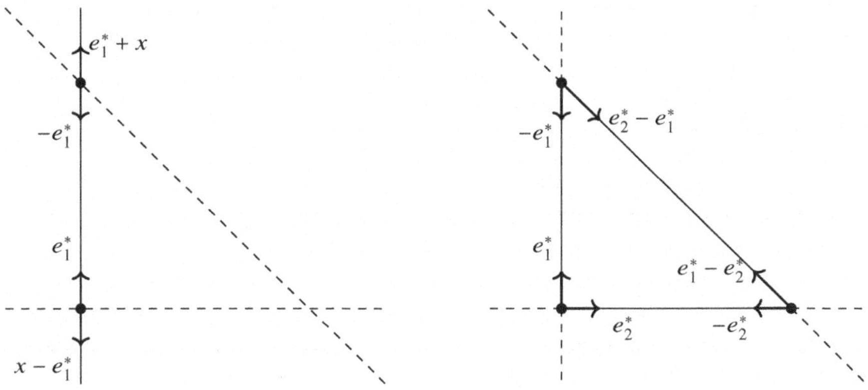

Fig. 3 The left figure shows a hyperplane of the left GKM graph in Fig. 2. The right figure is a 2-valent GKM subgraph of the left GKM graph in Fig. 2, but it is not a hyperplane

Proof We first prove the existence of the hyperplane, which satisfies the statement. Put $\mathcal{E}_p^L = \mathcal{E}_p \backslash \{\epsilon^+, \epsilon^-\}$. Then, we can write $\mathcal{E}_p^L = \{\epsilon_1^+, \cdots, \epsilon_{n-1}^+, \epsilon_1^-, \cdots, \epsilon_{n-1}^-\}$ the $(n-1)$ 1-dimensional pairs in $\mathcal{E}_p$ which are different from $\{\epsilon^+, \epsilon^-\}$. Let

$$R := \langle \alpha(\epsilon_1^+), \ \cdots, \ \alpha(\epsilon_{n-1}^+), \ x \rangle.$$

By Definition 2.4 (2), we may assume that $\langle \alpha(\epsilon_1^+), \ldots, \alpha(\epsilon_{n-1}^+) \rangle \subset \mathfrak{t}_{\mathbb{Z}}^* \oplus \mathbb{Z}x$ is a submodule of rank $(n-1)$. Thus, we have

$$R \simeq H^2(BT^{n-1}) \oplus \mathbb{Z}x.$$

Take an element $\epsilon \in \mathcal{E}_p^L$, which becomes an edge in Γ, i.e., $i(\epsilon) = p$, and there exists $t(\epsilon) \in \mathcal{V}$. In other words, $\epsilon \in \mathcal{E}_p^L \cap E_p$ (note that this will be E_p^L of L). By Lemma 2.6, the subset $\nabla_\epsilon(\mathcal{E}_p^L)$ in $\mathcal{E}_{t(\epsilon)}$ consists of exactly $(n-1)$ 1-dimensional pairs. Moreover, because α satisfies the congruence relation on the edge $\epsilon \in \mathcal{E}_p^L \cap E_p$, $\alpha(\nabla_\epsilon(\mathcal{E}_p^L))$ and x span the same subspace R as above. This property holds for all edges $\epsilon \in \mathcal{E}_p^L \cap E_p$. Hence, we can define the following $(2n-2)$-valent subgraph in $\Gamma = (\mathcal{V}, \mathcal{E})$:

$$L_1 = (\mathcal{V}^{L_1}, \mathcal{E}^{L_1})$$

such that

$$\mathcal{V}^{L_1} := \{p, \ t(\epsilon) \mid \epsilon \in \mathcal{E}_p^L \cap E_p\};$$

$$\mathcal{E}^{L_1} := \bigcup_{\epsilon \in \mathcal{E}_p^L \cap E_p} \nabla_\epsilon(\mathcal{E}_p^L) \cup \mathcal{E}_p^L.$$

If we restrict α and ∇ onto L_1, then this becomes a $(2n-2)$-valent $T^*\mathbb{C}^{n-1}$-modeled GKM subgraph, say $\mathcal{L}_1$, in $\mathcal{G}$. If L_1 is maximal, i.e., if there is a $(2n-2)$-valent graph L' such that $L_1 \subset L'$ then $L_1 = L'$, then $\mathcal{L}_1$ is a hyperplane. Assume that L_1 is not maximal. In this case, for every vertex $q \in \mathcal{V}^{L_1}$ and every edge $\epsilon \in \mathcal{E}_q^{L_1} \cap E_q$, we can apply the similar method stated as above. Then, we can construct the $(2n-2)$-valent $T^*\mathbb{C}^{n-1}$-modeled GKM subgraph $\mathcal{L}_2$ which contains $\mathcal{L}_1$. If $\mathcal{L}_2$ is maximal, then this is a hyperplane, which we want to have. Otherwise, by repeating similar arguments, we get the hyperplane $\mathcal{L}$, which contains $\mathcal{E}_p^L$.

Suppose that there are two hyperplanes $\mathcal{L} = (L, \alpha^L, \nabla^L)$ and $\mathcal{L}' = (L', \alpha^{L'}, \nabla^{L'})$ such that $\mathcal{E}_p^L = \mathcal{E}_p^{L'}$. Because $E_p^L = E_p^{L'}$ and two connections are restricted from the connection ∇ of $\mathcal{G}$, we see that the following two subgraphs are the same graph:

$$\bigcup_{\epsilon \in E_p^L} \nabla_\epsilon(\mathcal{E}_p^L) \cup \mathcal{E}_p^L = \bigcup_{\epsilon \in E_p^{L'}} \nabla_\epsilon(\mathcal{E}_p^{L'}) \cup \mathcal{E}_p^{L'}.$$

By iterating this construction along all edges in L and L', finally, we know that $L = L'$. Therefore, such a hyperplane is unique. □

3.2 Pre-halfspace and Its Thom Class

In this section, we introduce a *pre-halfspace* and its *Thom class*.

Take a subgraph $H = (\mathcal{V}^H, \mathcal{E}^H)$ of Γ such that $|\mathcal{E}_p^H| = 2n-1$ or $2n$ for all $p \in \mathcal{V}^H$. We assume that there always exists a vertex $p \in \mathcal{V}^H$ with $|\mathcal{E}_p^H| = 2n-1$. Moreover, we assume that H is *closed* under the connection ∇ of $\mathcal{G} = (\Gamma, \alpha, \nabla)$, that is,

(C1) $\nabla_\epsilon^H := \nabla_\epsilon|_{\mathcal{E}_{i(\epsilon)}^H} : \mathcal{E}_{i(\epsilon)}^H \to \mathcal{E}_{t(\epsilon)}^H$ is bijective, if $|\mathcal{E}_{i(\epsilon)}^H| = |\mathcal{E}_{t(\epsilon)}^H| = 2n-1$ or $2n$.

(C2) $\nabla_\epsilon^H : \mathcal{E}_{i(\epsilon)}^H \to \mathcal{E}_{t(\epsilon)}^H$ is injective, if $|\mathcal{E}_{i(\epsilon)}^H| = 2n-1 < |\mathcal{E}_{t(\epsilon)}^H| = 2n$.

In addition, we also assume that if $|\mathcal{E}_{i(\epsilon)}^H|(= 2n-1) < |\mathcal{E}_{t(\epsilon)}^H|(= 2n)$, then ∇_ϵ^H satisfies the following congruence relation for $\{n^H(i(\epsilon))\} = \mathcal{E}_{i(\epsilon)}^\Gamma - \mathcal{E}_{i(\epsilon)}^H$ (we call such an $n^H(p)$ a *normal edge* or a *normal leg* of H at p):

$$\alpha(n^H(p)) - x \equiv 0 \ (\mathrm{mod}\ \alpha(\epsilon)). \tag{3}$$

Now we may define the pre-halfspace.

Definition 3.4 (pre-halfspace) Let $\mathcal{H} := (H, \alpha^H, \nabla^H)$ be the triple as above, i.e.,

- $H = (\mathcal{V}^H, \mathcal{E}^H)$ is a subgraph Γ such that $|\mathcal{E}_p^H| = 2n - 1$ or $2n$ for all $p \in \mathcal{V}^H$, where there exists a vertex $p \in \mathcal{V}^H$ with $|\mathcal{E}_p^H| = 2n - 1$.
- $\alpha^H := \alpha|_{\mathcal{E}^H} : \mathcal{E}^H \to H^2(BT^n) \oplus \mathbb{Z}x$ is the axial function restricted onto H.
- $\nabla^H := \{\nabla_\epsilon^H \mid \epsilon \in E^H\}$ is the restrected connection of ∇ onto H which satisfies the conditions (C1), (C2), (3) as above.

Then, we call $\mathcal{H}$ a *pre-halfspace* of $\mathcal{G} = (\Gamma, \alpha, \nabla)$.

Example 3.5 The followingleft figure (Fig. 4) shows an example of pre-halfspace of the left graph in Fig. 2. On the other hand, the following right figure (Fig. 5) is an abstract subgraph of the left graph in Fig. 2. However, this is not closed under the connection ∇, because the congruence relation does not hold on the diagonal edge. Therefore, this is not a pre-halfspace.

Let $\mathcal{H} = (H, \alpha^H, \nabla^H)$ be a pre-halfspace of a $T^*\mathbb{C}^n$-modeled GKM graph $\mathcal{G} = (\Gamma, \alpha, \nabla)$. We can define the notion of a *Thom class* for the pre-halfspace (also see [20]).

Definition 3.6 (Thom class) A *Thom class* of $\mathcal{H}$ is defined by the map $\tau_H : \mathcal{V} \to H^2(BT^n) \oplus \mathbb{Z}x$ such that

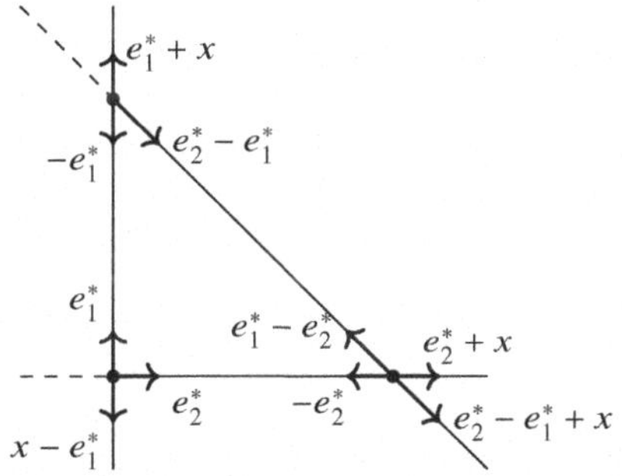

Fig. 4 An example of pre-halfspace of the left graph in Fig. 2

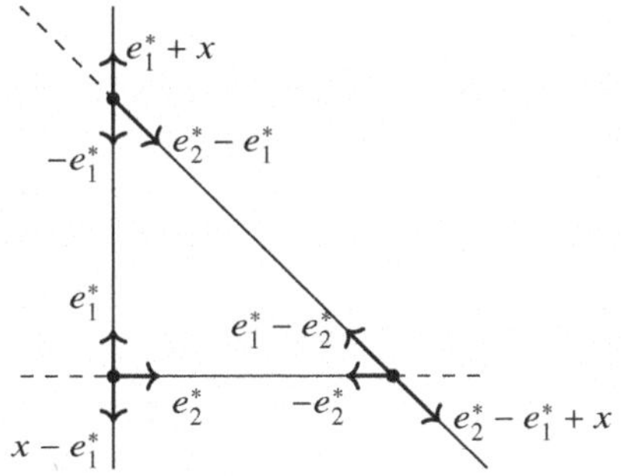

Fig. 5 An abstract subgraph of the left graph in Fig. 2 which is not a pre-halfspace

$$\tau_H(p) = \begin{cases} 0 & \text{if } p \notin \mathcal{V}^H \\ x & \text{if } |\mathcal{E}_p^H| = 2n \\ \alpha(n^H(p)) & \text{if } |\mathcal{E}_p^H| = 2n-1 \end{cases}$$

We also call a vertex $p \in \mathcal{V}$ with $\tau_H(p) = 0$ (resp. $\tau_H(p) = x$, $\tau_H(p) = \alpha(n^H(p))$) an *exterior* (resp. *interior*, *boundary*) *vertex* of H.

For the Thom class τ_H of a pre-halfspace H, we have the following lemma.

Lemma 3.7 *The Thom class τ_H of a pre-halfspace $\mathcal{H}$ is an element of $H^*(\mathcal{G})$.*

Proof Take an edge $\epsilon \in E$. We claim that τ_H satisfies the congruence relation on ϵ, that is, $\tau_H(i(\epsilon)) - \tau_H(t(\epsilon)) \equiv 0 \pmod{\alpha(\epsilon)}$.

We first assume that $i(\epsilon),\ t(\epsilon) \notin \mathcal{V}^H$ or $|\mathcal{E}_{i(\epsilon)}^H| = 2n = |\mathcal{E}_{t(\epsilon)}^H|$. Namely, both of $i(\epsilon)$ and $t(\epsilon)$ are exterior vertices or interior vertices of H. Then, by definition of the Thom class,

$$\tau_H(i(\epsilon)) - \tau_H(t(\epsilon)) = 0 \equiv 0 \bmod \alpha(\epsilon).$$

So the congruence relation holds for these cases.

We next consider the other cases, i.e., the case when both of $i(\epsilon)$ and $t(\epsilon)$ are boundary vertices or the case when $i(\epsilon)$ is a boundary vertex but $t(\epsilon)$ is an exterior or an interior vertex. Put $p = i(\epsilon)$. Assume that $t(\epsilon)(=: q)$ is also a boundary vertex. Because the pre-halfspace is closed by the connection ∇, we have that $\nabla_\epsilon(n^H(p)) = n^H(q)$. By the definition of Thom classes, we have that

$$\begin{aligned} \tau_H(i(\epsilon)) - \tau_H(t(\epsilon)) &= \alpha(n^H(p)) - \alpha(n^H(q)) \\ &= \alpha(n^H(p)) - \alpha(\nabla_\epsilon(n^H(p))) \equiv 0 \bmod \alpha(\epsilon). \end{aligned}$$

Assume that $t(\epsilon)$ is an exterior vertex. Then, $\epsilon = n^H(p)$ and

$$\tau_H(i(\epsilon)) - \tau_H(t(\epsilon)) = \alpha(n^H(p)) - 0 \equiv 0 \bmod \alpha(\epsilon) = \alpha(n^H(p)).$$

Assume that $t(\epsilon)(=: q)$ is an interior vertex. Namely, $\tau_H(q) = x$. In this case, we have

$$\tau_H(p) - \tau_H(q) = \alpha(n^H(p)) - x.$$

It follows from (3) that

$$\alpha(n^H(p)) - x \equiv 0 \bmod \alpha(\epsilon).$$

So the congruence relation also holds for these cases. Consequently, we have $\tau_H \in H^*(\mathcal{G})$. □

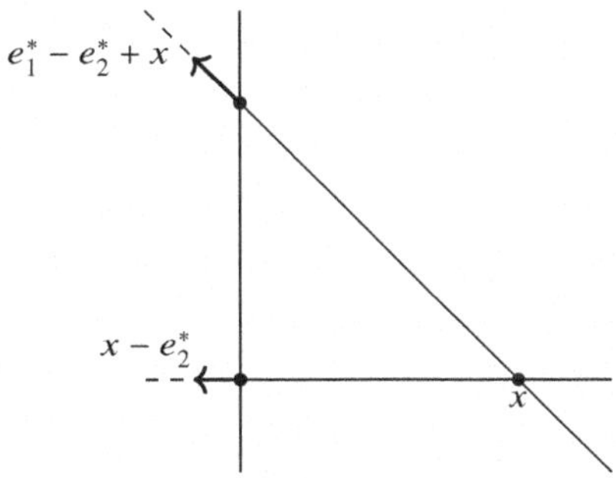

Fig. 6 An example of the Thom class of the pre-halfspace in Fig. 4

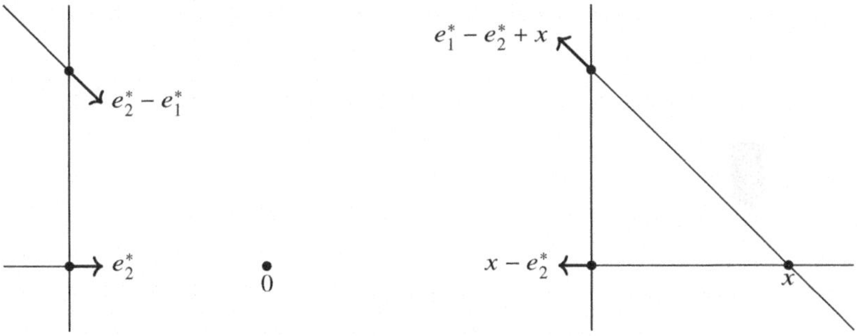

Fig. 7 The above figures are a halfspace and its opposite side of the left GKM graph in Fig. 2. The labels on vertices mean the values of their Thom classes on vertices, where 0 means the value of the Thom class τ_H on the exterior vertex of a halfspace H. We call such a vertex a *fake vertex* when we consider the Thom class τ_H of H. Note that the boundary $\partial H = H \cap \overline{H}$ is connected

Example 3.8 The following figure (Fig. 6) shows an example of the Thom class of the pre-halfspace in Fig. 4.

3.3 *Opposite Side of a Pre-halfspace*

Next, we define the *opposite side* of the pre-halfspace. In order to define it, we need to prove Lemma 3.9 (also see Fig. 7).

In order to prove it, we define a *boundary* of a pre-halfspace. Let H be a pre-halfspace. By the definition of a pre-halfspace, there is a vertex $p \in \mathcal{V}^H$ such that $|\mathcal{E}_p^H| = 2n - 1$. Then, $\mathcal{E}_p^H$ has $(n-1)$ 1-dimensional pairs, say $\{\epsilon_1^+, \cdots, \epsilon_{n-1}^+, \epsilon_1^-, \cdots, \epsilon_{n-1}^-\}$. Because of Lemma 3.3, there exists a unique hyperplane $\mathcal{L} = (L, \alpha^L, \nabla^L)$ in $\mathcal{G}$ such that

$$\alpha^L := \alpha|_{\mathcal{E}^L}, \quad \nabla^L := \{\nabla_\epsilon|_{\mathcal{E}^L_{i(\epsilon)}} \mid \epsilon \in E^L\}.$$

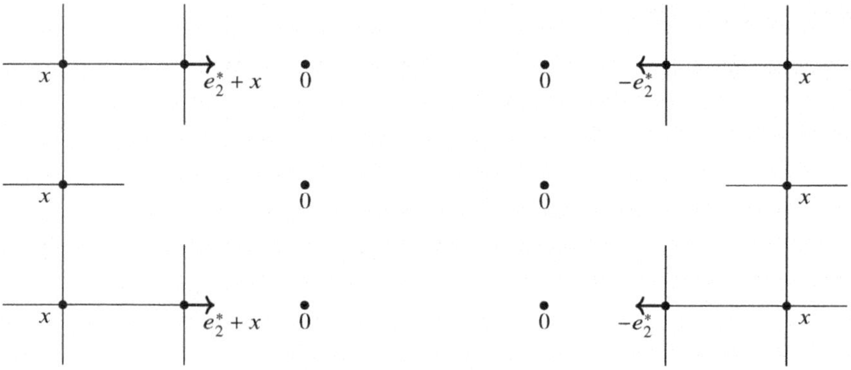

Fig. 8 The above figures are a halfspace and its opposite side of the right GKM graph in Fig. 2, where 0's are the fake vertices. Note that the boundary $\partial H = H \cap \overline{H}$ is not connected

Moreover, $\mathcal{L}$ satisfies that

$$p \in \mathcal{V}^L$$

and

$$\mathcal{E}_p^L = \{\epsilon_1^+, \cdots, \epsilon_{n-1}^+, \epsilon_1^-, \cdots, \epsilon_{n-1}^-\}.$$

We call the union of all such hyperplanes $\mathcal{L}$ a *boundary* of the pre-halfspace $\mathcal{H}$, and we denote it by $\partial\mathcal{H} = (\partial H, \alpha^{\partial H}, \nabla^{\partial H})$. Note that a boundary of $\mathcal{H}$ may not be connected (see Fig. 8). To define the opposite side of $\mathcal{H}$, we need the following lemma:

Lemma 3.9 *Let $\mathcal{H} = (H, \alpha^H, \nabla^H)$ be a pre-halfspace in $\mathcal{G} = (\Gamma, \alpha, \nabla)$ and x be a residual basis. Then, there is a unique pre-halfspace $\mathcal{I} = (I, \alpha^I, \nabla^I)$ such that*

- $H \cup I = \Gamma$
- $\tau_H + \tau_I = \chi \in H^*(\mathcal{G})$

where χ is an element of $H^*(\mathcal{G})$ defined by $\chi(p) = x$ for all $p \in \mathcal{V}$.

Proof Set $H = (\mathcal{V}^H, \mathcal{E}^H)$ and $\hat{I} = (\mathcal{V}^\Gamma - \mathcal{V}^H, \mathcal{E}^\Gamma - \mathcal{E}^H)$. Define

$$I = \hat{I} \cup \partial H.$$

Namely,

$$I = (\mathcal{V}^I, \mathcal{E}^I) = ((\mathcal{V}^\Gamma - \mathcal{V}^H) \cup \mathcal{V}^{\partial H}, (\mathcal{E}^\Gamma - \mathcal{E}^H) \cup \mathcal{E}^{\partial H}).$$

Then, we can easily see that $H \cup I = \Gamma$ and $H \cap I = \partial H$.

We next prove I is a pre-halfspace. Take $p \in \mathcal{V}^I$. If $p \in \mathcal{V}^{\hat{I}} = \mathcal{V}^\Gamma - \mathcal{V}^H$, then $\mathcal{E}_p^H = \emptyset$; therefore, $\mathcal{E}_p^I = \mathcal{E}_p^\Gamma - \emptyset = \mathcal{E}_p^\Gamma$ and $|\mathcal{E}_p^I| = |\mathcal{E}_p^\Gamma| = 2n$. If $p \in \mathcal{V}^{\partial H}$, then $\mathcal{E}_p^I = \mathcal{E}_p^{\partial H} \cup \{n_H(p)\}$, that is, $|\mathcal{E}_p^I| = 2n - 1$. Here $n_H(p)$ is a normal edge (leg) of H on p. Therefore, for an edge $\epsilon \in E^I$ and the restricted connection $\nabla_\epsilon^I := \nabla_\epsilon|_{\mathcal{E}_{i(\epsilon)}^I}$, it follows from the definition of ∇ on Γ that we have

- $\nabla_\epsilon^I : \mathcal{E}_{i(\epsilon)}^I \to \mathcal{E}_{t(\epsilon)}^I$ is bijective, if $|\mathcal{E}_{i(\epsilon)}^I| = |\mathcal{E}_{t(\epsilon)}^I|$.
- $\nabla_\epsilon^I : \mathcal{E}_{i(\epsilon)}^I \to \mathcal{E}_{t(\epsilon)}^I$ is injective, if $|\mathcal{E}_{i(\epsilon)}^I| = 2n - 1 < |\mathcal{E}_{t(\epsilon)}^I| = 2n$.

Take an edge $\epsilon \in E^I$ in I such that $|\mathcal{E}_{i(\epsilon)}^I| = 2n - 1 < |\mathcal{E}_{t(\epsilon)}^I| = 2n$. Put $i(\epsilon) = p \in \mathcal{V}^{\partial H}$ and $n_H(p) = \epsilon^+ (= \epsilon)$. Then, the normal edge (leg) of I on p can be taken as $\epsilon^- = n_I(p)$, where $\{\epsilon^+, \epsilon^-\} = \mathcal{E}_p - \mathcal{E}_p^{\partial H}$ is a 1-dimensional pair in $\mathcal{E}_p$. So we have the following equation:

$$\alpha(n_I(p)) + \alpha(n_H(p)) = \alpha(\epsilon^-) + \alpha(\epsilon^+) = (-\alpha(\epsilon^+) + x) + \alpha(\epsilon^+) = x. \quad (4)$$

Therefore, we have that

$$\begin{aligned}\alpha(n_I(p)) - x &= -\alpha(n_H(p)) \\ &= -\alpha(\epsilon) \equiv 0 \ (\mathrm{mod}\ \alpha(n_H(p)) = \alpha(\epsilon)).\end{aligned}$$

Consequently, we have $\mathcal{I} = (I, \alpha^I, \nabla^I)$ is a pre-halfspace such that $H \cup I = \Gamma$, where $\alpha^I := \alpha|_{\mathcal{E}^I}$ and $\nabla^I := \{\nabla_\epsilon|_{\mathcal{E}^I} \mid \epsilon \in E^I\}$. Moreover, we have that $\tau_H + \tau_I = \chi$, because of the above equation and the definition of the Thom class of the pre-halfspace.

We finally claim the uniqueness of I. By two conditions $H \cup I = \Gamma$, $\tau_H + \tau_I = \chi$, and the definition of the Thom class, we see $\partial I = \partial H$ and $I = (\mathcal{V}^\Gamma - \mathcal{V}^H, \mathcal{E}^\Gamma - \mathcal{E}^H) \cup \partial I$. From Lemma 3.3, the boundary $\partial H = \partial I$ is uniquely determined (though it may not be connected). So we know the uniqueness of I. □

We call $\mathcal{I}$ in Lemma 3.9 an *opposite side* of $\mathcal{H}$ and denote it by $\overline{\mathcal{H}} = (\overline{H}, \alpha^{\overline{H}}, \nabla^{\overline{H}})$. Note that

$$H \cap \overline{H} = \partial H$$

by the proof of Lemma 3.9.

3.4 *Halfspace and Ring* $\mathbb{Z}[\mathcal{G}]$

Under the above preparations, we may define the halfspace.

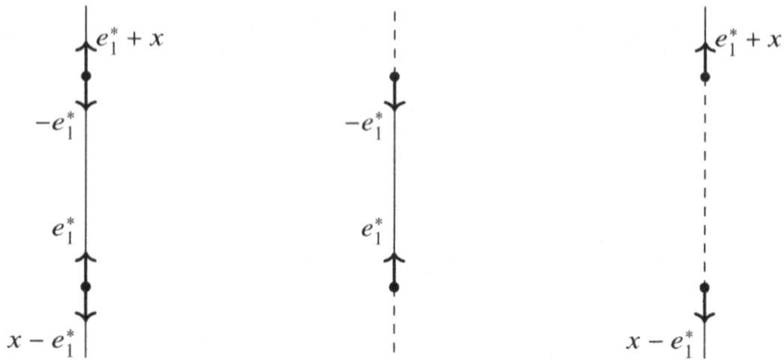

Fig. 9 The middle graph (edge) $\mathcal{H}$ is a pre-halfspace of the left GKM graph. However, its opposite side is the right graph (two legs). This is not connected; therefore, $\mathcal{H}$ is not a halfspace

Definition 3.10 (halfspace) A pre-halfspace $\mathcal{H}$ is said to be a *halfspace*, if H is a connected subgraph and its opposite side is also connected.

Let $\mathbf{H}$ be the set of all halfspaces in $\mathcal{G}$ (Fig. 9). Because the graph Γ is finite and the opposite side of the halfspace is also a halfspace, we may write the set of all halfspaces by

$$\mathbf{H} = \{H_1, \cdots, H_m, \overline{H_1}, \cdots, \overline{H_m}\}.$$

Put

$$\mathbb{Z}[X, \mathbf{H}] = \mathbb{Z}[X, H_1, \cdots, H_m, \overline{H_1}, \cdots, \overline{H_m}]$$

where $\mathbb{Z}[X, H_1, \cdots, H_m, \overline{H_1}, \cdots, \overline{H_m}]$ is a polynomial ring, which is generated by X and all elements in $\mathbf{H}$, and put

$$\mathcal{I} = \left\langle H_i + \overline{H_i} - X, \prod_{H \in \mathbf{H}'} H \,\middle|\, i = 1, \cdots, m, \mathbf{H}' \in \mathbf{I}(\mathbf{H}) \right\rangle$$

which is the ideal in $\mathbb{Z}[X, \mathbf{H}]$ generated by $H_i + \overline{H_i} - X$ ($i = 1, \cdots, m$) and the product

$$\prod_{H \in \mathbf{H}' \in \mathbf{I}(\mathbf{H})} H,$$

where

$$\mathbf{I}(\mathbf{H}) = \{\mathbf{H}' \subset \mathbf{H} \mid \bigcap_{H \in \mathbf{H}'} H = \emptyset\}.$$

We define the following ring $\mathbb{Z}[\mathcal{G}]$:

$$\mathbb{Z}[\mathcal{G}] := \mathbb{Z}[X, \mathbf{H}]\Big/\mathcal{I}.$$

From the next section, we shall prove this ring $\mathbb{Z}[\mathcal{G}]$ is isomorphic to the graph equivariant cohomology ring $H^*(\mathcal{G})$ under some conditions.

4 Ring Structure of the Graph Equivariant Cohomology of a $T^*\mathbb{C}^n$-Modeled GKM Graph

The first goal of this paper is to prove the following theorem.

Theorem 4.1 *Let $\mathcal{G}$ be a 2n-valent $T^*\mathbb{C}^n$-modeled GKM graph, and let $\mathbf{L} = \{L_1, \cdots, L_m\}$ be the set of all hyperplanes in $\mathcal{G}$. Assume that $\mathcal{G}$ satisfies the following two assumptions:*

1. *For each $L \in \mathbf{L}$, there exist the unique pair of the halfspace H and its opposite side $\overline{H}$ such that $H \cap \overline{H} = L$.*
2. *For every subset $\mathbf{L}' \subset \mathbf{L}$, its intersection $\bigcap_{L \in \mathbf{L}'} L$ is empty or connected.*

Then, the following ring isomorphism holds:

$$H^*(\mathcal{G}) \simeq \mathbb{Z}[\mathcal{G}].$$

Henceforth, in this section, the $T^*\mathbb{C}^n$-modeled GKM graph $\mathcal{G} = (\Gamma, \alpha, \nabla)$ satisfies assumptions (1), (2) of Theorem 4.1. For example, the left GKM graph in Fig. 2 satisfies these assumptions; however, the right GKM graph does not satisfy the assumption (1) (also see Fig. 8). We also note that the following example satisfies the assumptions in Theorem 4.1.

Example 4.2 The GKM graph in Fig. 10 can be obtained from the cotangent bundle of a 4-dimensional toric manifold with five fixed points. Note that this can not be realized as a hyperplane arrangement in $\mathbb{R}^2$, because there must be 8 intersection points (i.e., 8 vertices) if all straight lines (five lines) extend to infinity but there are only 5 vertices in Fig. 10. Therefore, there is no corresponding toric hyperKähler manifold because of the fundamental theorem of toric hyperKähler manifolds in [1].

Let $\chi : \mathcal{V} \to H^2(BT) \oplus \mathbb{Z}x$ be the function such that $\chi(p) = x$ for all $p \in \mathcal{V}$ and τ_H be the Thom class of the halfspace H. In order to prove Theorem 4.1, we will prove that the following map is an isomorphism:

$$\Psi : \mathbb{Z}[\mathcal{G}] \to H^*(\mathcal{G})$$

where this map is the induced homomorphism from $\Psi(H) := \tau_H$ and $\Psi(X) := \chi$.

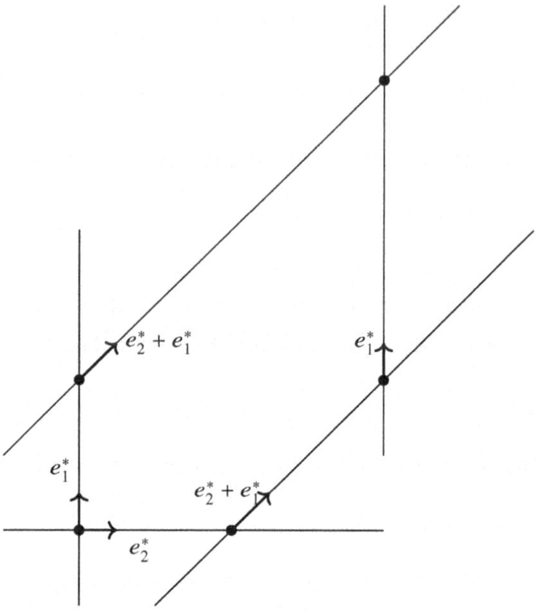

Fig. 10 The $T^*\mathbb{C}^2$-modeled GKM graph defined from the cotangent bundle of a toric manifold. Note that we assume that the axial functions satisfy $\alpha(e) = -\alpha(\overline{e})$ for all edges. We omit the axial functions on legs because it is automatically determined by the definition

We first claim that the map Ψ is well-defined (also see the definition of $\mathbb{Z}[\mathcal{G}]$). By Lemma 3.9, we have

$$\tau_H + \tau_{\overline{H}} = \chi.$$

Let $\mathbf{H}$ be the set of all halfspaces in $\mathcal{G}$. If a subset $\mathbf{H}' \subset \mathbf{H}$ satisfies that $\bigcap_{H \in \mathbf{H}'} H = \emptyset$, then it follows from the definition of the Thom class that

$$\prod_{H \in \mathbf{H}'} \tau_H = 0.$$

Therefore, the map Ψ is a well-defined homomorphism.

From the next section, we start to prove the bijectivity of Ψ. The proof will be divided into two steps:

(I) To study an equivariant graph cohomology of an *x-forgetful graph* $\widetilde{\mathcal{G}}$ and to prove $H^*(\widetilde{\mathcal{G}}) \simeq \mathbb{Z}[\widetilde{\mathcal{G}}]$
(II) To prove Ψ is surjective and injective

In the first step, we will use the technique of [20] (or [21]), which was used to show the ring structure of the graph equivariant cohomology of a certain GKM graph called a *torus graph*. In the second step, we will use the technique of [13], which was applied to show the ring structure of the equivariant cohomology of a toric hyperKähler variety (also referred to as hypertoric variety).

5 An x-Forgetful Graph $\widetilde{\mathcal{G}}$

Let $\mathcal{G} = (\Gamma, \alpha, \nabla)$ be a $T^*\mathbb{C}^n$-modeled graph. We assume that $\mathcal{G}$ satisfies the conditions in Theorem 4.1. In this section, as a preparation to prove Theorem 4.1, we introduce an x-forgetful graph $\widetilde{\mathcal{G}}$ and its graph equivariant cohomology $H^*(\widetilde{\mathcal{G}})$ and prove the ring structure of $H^*(\widetilde{\mathcal{G}})$.

5.1 *x-Forgetful Graph $\widetilde{\mathcal{G}}$ and Its Graph Equivariant Cohomology*

For every $\mathcal{G}$, we may define an *x-forgetful graph* $\widetilde{\mathcal{G}} = (\Gamma, \widetilde{\alpha}, \nabla)$ as follows: Γ and ∇ is the same graph and connection with $\mathcal{G}$, but the function $\widetilde{\alpha}$ is defined as

$$\widetilde{\alpha} = F \circ \alpha : \mathcal{E} \to H^2(BT^n)$$

where $F : H^2(BT^n) \oplus \mathbb{Z}x \to H^2(BT^n)(\simeq (\mathfrak{t}^n)^*_{\mathbb{Z}})$ is the x-forgetful map. We call $\widetilde{\alpha}$ an *x-forgetful axial function* (Fig. 11).

Moreover we define a graph equivariant cohomology of $\widetilde{\mathcal{G}}$ as follows:

$$H^*(\widetilde{\mathcal{G}}) = \{f : \mathcal{V} \to H^*(BT^n) \mid f(i(\epsilon)) - f(t(\epsilon)) \equiv 0 \ (\text{mod}\, \widetilde{\alpha}(\epsilon))\}.$$

Let $L \in \mathbf{L}$ be a hyperplane in $\mathcal{G}$. Fix the halfspace H such that $\partial H = L$. Define the *Thom class* of L by

$$\tau_L = F \circ \tau_H : \mathcal{V} \to H^2(BT^n),$$

where F is the x-forgetful map. Note that for the opposite side $\overline{H}$ of H, the following relation is as follows:

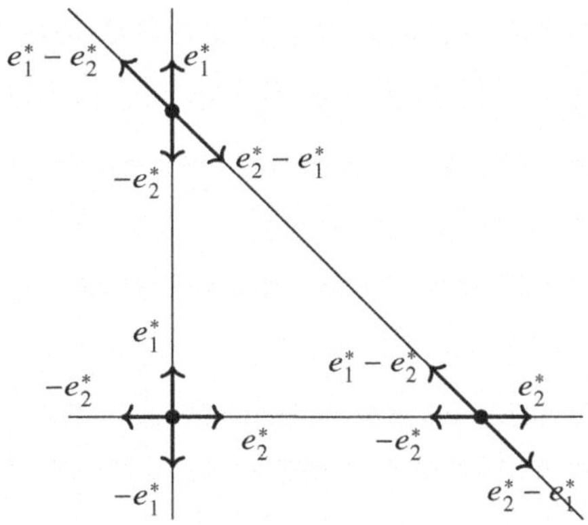

Fig. 11 An example of the x-forgetful graph for the left GKM graph in Fig. 2

$$F \circ \tau_{\overline{H}} = -\tau_L;$$

therefore, the Thom class of L depends on the choice of a halfspace H with $\partial H = L$. So we fix $\{H_1, \cdots, H_m\}$ in the set of all halfspaces $\mathbf{H} = \{H_1, \cdots, H_m, \overline{H}_1, \cdots, \overline{H}_m\}$. By the assumption (1) of Theorem 4.1, there is a one to one corresponding between H and $L = H \cap \overline{H}$. Therefore, we may put the set of all hyperplanes by $\mathbf{L} = \{L_1, \cdots, L_m\}$, where $L_i = H_i \cap \overline{H}_i$ for all $i = 1, \cdots, m$. Let $\mathcal{V}^L$ be the set of all vertices on L. Then, we have

$$\tau_L(p) = \begin{cases} 0 & p \notin \mathcal{V}^L \\ \widetilde{\alpha}(n_H(p)) & p \in \mathcal{V}^L \end{cases}$$

by the definitions of τ_H and the x-forgetful map F, where $n_H(p)$ is a normal edge (or leg) of H on p. Since $\tau_H \in H^*(\mathcal{G})$ (see Lemma 3.7), it is easy to check that

$$\tau_L \in H^*(\widetilde{\mathcal{G}}).$$

5.2 The Ring Structure of $H^*(\widetilde{\mathcal{G}})$

Next, we define the following ring:

$$\mathbb{Z}[\widetilde{\mathcal{G}}] = \mathbb{Z}[L_1, \cdots, L_m] \Big/ \Big\langle \prod_{L \in \mathbf{L}'} L \;\Big|\; \mathbf{L}' \in \mathbf{I}(\mathbf{L}) \Big\rangle,$$

where $\mathbf{I}(\mathbf{L}) = \{\mathbf{L}' \subset \mathbf{L} \mid \bigcap_{L \in \mathbf{L}'} L = \emptyset\}$ and $\langle \prod_{L \in \mathbf{L}'} L \mid \mathbf{L}' \in \mathbf{I}(\mathbf{L})\rangle$ is an ideal, which is generated by the product $\prod_{L \in \mathbf{L}'} L$ for all $\mathbf{L}' \in \mathbf{I}(\mathbf{L})$.

The goal of this section (the first step (I) of the proof of Theorem 4.1) is to prove the following theorem:

Theorem 5.1 *Let $\mathcal{G}$ be a $2n$-valent $T^*\mathbb{C}^n$-modeled GKM graph and $\mathbf{L} = \{L_1, \cdots, L_m\}$ be the set of all hyperplanes in $\mathcal{G}$. Assume that $\mathcal{G}$ satisfies the two assumptions in Theorem* 4.1. *If $\widetilde{\mathcal{G}}$ is the x-forgetful graph, then the following ring isomorphism holds:*

$$H^*(\widetilde{\mathcal{G}}) \simeq \mathbb{Z}[\widetilde{\mathcal{G}}].$$

Define the induced homomorphism

$$\Psi' : \mathbb{Z}[\widetilde{\mathcal{G}}] \to H^*(\widetilde{\mathcal{G}})$$

by $\Psi'(L) = \tau_L$. Obviously, Ψ' is a well-defined homomorphism. In order to show Theorem 5.1, it is enough to prove that this homomorphism is bijective.

5.3 The Localization Map and the Injectivity of Ψ'

We first prove the injectivity of Ψ'. In order to prove it, we introduce the map ρ, which is the analogue of the localization of the equivariant cohomology of a T-manifold to its fixed points.

Let us define the following ring:

$$\mathbb{Z}[\widetilde{\mathcal{G}}]_p = \mathbb{Z}[L_1, \cdots, L_m]/\langle L \mid p \notin \mathcal{V}^L \rangle,$$

where $\langle L \mid p \notin \mathcal{V}^L \rangle$ is an ideal which is generated by L such that $p \notin \mathcal{V}^L$. As a beginning, we prove the following lemma.

Lemma 5.2 *For the x-forgetful graph $\widetilde{\mathcal{G}} = (\Gamma, \widetilde{\alpha}, \nabla)$, we have*

$$I_p : \mathbb{Z}[\widetilde{\mathcal{G}}]_p \simeq \mathbb{Z}[L \mid p \in \mathcal{V}^L] = \mathbb{Z}[L_{p,1}, \cdots, L_{p,n}] \overset{\iota_p}{\simeq} H^*(BT^n),$$

where the last isomorphism ι_p is defined by $\iota_p : L \mapsto \tau_L(p)$.

Proof By the definition of $\mathbb{Z}[\widetilde{\mathcal{G}}]_p$, the first equivalence $\mathbb{Z}[\widetilde{\mathcal{G}}]_p \simeq \mathbb{Z}[L \mid p \in \mathcal{V}^L]$ is obvious. We claim $\mathbb{Z}[L \mid p \in \mathcal{V}^L] = \mathbb{Z}[L_{p,1}, \cdots, L_{p,n}] \overset{\iota_p}{\simeq} H^*_{T^n}(pt)$.

Because Γ is a $2n$-valent graph, we may put

$$\mathcal{E}_p = \{\epsilon_1^+(p), \cdots, \epsilon_n^+(p), \epsilon_1^-(p), \cdots, \epsilon_n^-(p)\}$$

for all $p \in \mathcal{V}$. There is a unique L_i such that

$$\tau_{L_i}(p) = \widetilde{\alpha}(\epsilon_i^+(p)) = -\widetilde{\alpha}(\epsilon_i^-(p))$$

for all $i = 1, \cdots, n$ by Lemma 3.3. Hence, we have

$$\mathbb{Z}[L \mid p \in \mathcal{V}^L] = \mathbb{Z}[L_{p,1}, \cdots, L_{p,n}].$$

Next, by the definition of the axial function of a $T^*\mathbb{C}^n$-modeled GKM graph,

$$\mathbb{Z}\alpha(\epsilon_1^+(p)) \oplus \cdots \oplus \mathbb{Z}\alpha(\epsilon_n^+(p)) \oplus \mathbb{Z}x \simeq H^2(BT^n) \oplus \mathbb{Z}x.$$

Hence, because $\widetilde{\alpha} := F \circ \alpha$ is defined by the x-forgetful map $F : H^2(BT^n) \oplus \mathbb{Z}x \to H^2(BT^n)$, we have that

$$\mathbb{Z}[\widetilde{\alpha}(\epsilon_1^+(p)), \cdots, \widetilde{\alpha}(\epsilon_n^+(p))] \simeq H^*(BT^n).$$

Therefore, ι_p is an isomorphism. □

Next, we shall define a *localization map* $\rho : \mathbb{Z}[\widetilde{\mathcal{G}}] \to \bigoplus_{p \in \mathcal{V}} \mathbb{Z}[\widetilde{\mathcal{G}}]_p$ and prove that it is injective in Lemma 5.3. Since the set $\mathbf{L}' \in \mathbf{I}(\mathbf{L})$ satisfies that $\cap_{L \in \mathbf{L}'} L = \emptyset$, for every $p \in \mathcal{V}$ there is an $L \in \mathbf{L}'$ such that $p \notin \mathcal{V}^L$. Therefore, there exists the following relation for two ideals in $\mathbb{Z}[L_1, \ldots, L_m]$:

$$\langle L \mid p \notin \mathcal{V}^L \rangle \supset \langle \prod_{L \in \mathbf{L}'} L \mid \mathbf{L}' \in \mathbf{I}(\mathbf{L}) \rangle.$$

Hence, the following natural homomorphism is well-defined:

$$\rho_p : \mathbb{Z}[\widetilde{\mathcal{G}}] := \mathbb{Z}[L_1, \cdots, L_m] \Big/ \Big\langle \prod_{L \in \mathbf{L}'} L \Big| \mathbf{L}' \in \mathbf{I}(\mathbf{L}) \Big\rangle \longrightarrow$$
$$\mathbb{Z}[\widetilde{\mathcal{G}}]_p := \mathbb{Z}[L_1, \cdots, L_m] / \langle L \mid p \notin \mathcal{V}^L \rangle.$$

For this projection ρ_p, we can easily show that its kernel is as follows:

$$\operatorname{Ker} \rho_p = \langle L \mid p \notin \mathcal{V}^L \rangle / \langle \prod_{L \in \mathbf{L}'} L \mid \mathbf{L}' \in \mathbf{I}(\mathbf{L}) \rangle.$$

Now we may define the homomorphism ρ as follows:

$$\rho = \bigoplus_{p \in \mathcal{V}} \rho_p : \mathbb{Z}[\widetilde{\mathcal{G}}] \longrightarrow \bigoplus_{p \in \mathcal{V}} \mathbb{Z}[\widetilde{\mathcal{G}}]_p,$$

such that

$$\rho(Y) = \bigoplus_{p \in \mathcal{V}} \rho_p(Y)$$

for $Y \in \mathbb{Z}[\widetilde{\mathcal{G}}]$. We call ρ a *localization map*. The following lemma holds.

Lemma 5.3 *ρ is injective.*

Proof Obviously, we have

$$\operatorname{Ker} \rho = \bigcap_{p \in \mathcal{V}} \operatorname{Ker} \rho_p = \left(\bigcap_{p \in \mathcal{V}} \langle L \mid p \notin \mathcal{V}^L \rangle \right) \Big/ \left\langle \prod_{L \in \mathbf{L}'} L \,\Big|\, \mathbf{L}' \in \mathbf{I}(\mathbf{L}) \right\rangle.$$

Hence, to prove ρ is injective, it is enough to show that $\operatorname{Ker} \rho = \{0\}$, i.e., we shall prove the following relation:

$$\bigcap_{p \in \mathcal{V}} \langle L \mid p \notin \mathcal{V}^L \rangle \subset \left\langle \prod_{L \in \mathbf{L}'} L \mid \mathbf{L}' \in \mathbf{I}(\mathbf{L}) \right\rangle (\subset \mathbb{Z}[L_1, \cdots, L_m]). \tag{5}$$

Take a nonzero polynomial

$$A = \sum_{a_1,\cdots,a_m \in \mathbb{N}\cup\{0\}} k(a_1, \cdots, a_m) L_1^{a_1} \cdots L_m^{a_m}$$
$$\in \bigcap_{p \in \mathcal{V}} \langle L \mid p \notin \mathcal{V}^L \rangle \subset \mathbb{Z}[L_1, \cdots, L_m],$$

where we only consider the case when $k(a_1, \cdots, a_m) \in \mathbb{Z} - \{0\}$. Because A is an element of the monomial ideal $\langle L \mid p \notin \mathcal{V}^L \rangle$ for all $p \in \mathcal{V}$, we have that for each term

$$k(a_1, \cdots, a_m) L_1^{a_1} \cdots L_m^{a_m} \in \langle L \mid p \notin \mathcal{V}^L \rangle.$$

This shows that for each term $k(a_1, \cdots, a_m) L_1^{a_1} \cdots L_m^{a_m}$ of a nonzero element A, there exists $r(= r(p)) \in \{1, \cdots, m\}$ such that $p \notin \mathcal{V}^{L_r}$ and $a_r \neq 0$. Because this satisfies for all $p \in \mathcal{V}$, we have that each term can be written by

$$k(a_1, \cdots, a_m) L_1^{a_1} \cdots L_m^{a_m} = B \prod_{p \in \mathcal{V}} L_{r(p)}^{a_{r(p)}},$$

where B is some monomial in $\mathbb{Z}[L_1, \cdots, L_m]$ and $a_{r(p)} \neq 0$. Since $p \notin \mathcal{V}^{L_{r(p)}}$, we have that

$$\bigcap_{p \in \mathcal{V}} L_{r(p)} = \emptyset.$$

This shows that for each term of A

$$k(a_1, \cdots, a_m) L_1^{a_1} \cdots L_m^{a_m} = B \prod_{p \in \mathcal{V}} L_{r(p)}^{a_{r(p)}} \in \langle \prod_{L \in \mathbf{L}'} L \mid \mathbf{L}' \in \mathbf{I}(\mathbf{L}) \rangle.$$

Therefore, $A \in \langle \prod_{L \in \mathbf{L}'} L \mid \mathbf{L}' \in \mathbf{I}(\mathbf{L}) \rangle$. This establishes the relation (5). □

By using Lemmas 5.2 and 5.3, we can prove the following lemma for the homomorphism $\Psi' : \mathbb{Z}[\widetilde{\mathcal{G}}] \to H^*(\widetilde{\mathcal{G}})$, which is defined from $\Psi'(L) := \tau_L$.

Lemma 5.4 *Ψ' is injective.*

Proof We first define

$$\rho' : H^*(\widetilde{\mathcal{G}}) \to \bigoplus_{p \in \mathcal{V}} H^*(BT^n)$$

by the homomorphism

$$\rho'(f) = \bigoplus_{p \in \mathcal{V}} f(p).$$

Then it is easy to check that the following diagram is commutative:

$$\begin{array}{ccc} \mathbb{Z}[\widetilde{\mathcal{G}}] & \xrightarrow{\ \rho\ } & \bigoplus_{p\in\mathcal{V}} \mathbb{Z}[\widetilde{\mathcal{G}}]_p \\ \Big\downarrow{\scriptstyle \Psi'} & & \Big\downarrow{\scriptstyle \oplus_p I_p} \\ H^*(\widetilde{\mathcal{G}}) & \xrightarrow{\ \rho'\ } & \bigoplus_{p\in\mathcal{V}} H^*(BT^n) \end{array}$$

where $I_p : \mathbb{Z}[\widetilde{\mathcal{G}}]_p \to H^*(BT^n)$ is the isomorphism defined by $I_p(L) := \tau_L(p)$ in Lemma 5.2. Because of Lemma 5.3, ρ is injective. Therefore, the composition map $\oplus_p I_p \circ \rho$ is injective. Because of the commutativity of the diagram, $\rho' \circ \Psi' = \oplus_p I_p \circ \rho$ is also injective. Consequently, Ψ' is injective. □

5.4 The Surjectivity of Ψ'

We next prove the surjectivity of Ψ'. In order to prove it, we will define an ideal $I(K)$ of $H^*(BT^n)$, where K is the non-empty intersection of some hyperplanes, say $K = L_1 \cap \cdots \cap L_k (\neq \emptyset)$. Note that the graph K is connected because of the assumption (2) of Theorem 4.1. Because $L_1, \ldots, L_k$ defines hyperplanes $\mathcal{L}_1, \ldots, \mathcal{L}_k$ (respectively) of $\mathcal{G} = (\Gamma, \alpha, \nabla)$, the subgraph K is also defines a $(2n-2k)$-valent ($T^*\mathbb{C}^{n-k}$-modeled) GKM subgraph of $\mathcal{G}$, say $\mathcal{K} := (K, \alpha^K, \nabla^K)$. Now we may define its x-forgetful graph, i.e., for $\widetilde{\alpha}^K := F \circ \alpha^K$, the labeled graph

$$\widetilde{\mathcal{K}} := (K, \widetilde{\alpha}^K, \nabla^K).$$

We define an ideal $I(K)$ (in $H^*(BT^n)$) on K as follows:

$$I(K) = \langle \widetilde{\alpha}^K(\epsilon)(= \widetilde{\alpha}(\epsilon)) \mid \epsilon \in \mathcal{E}^K \rangle,$$

that is, this ideal is generated by all x-forgetful axial functions of edges and legs in K. The following lemma, which will be used to prove the surjectivity of Ψ', holds for $I(K)$.

Lemma 5.5 *Let f be an element in $H^*(\widetilde{\mathcal{G}})$. If $f(p) \notin I(K)$ for some $p \in \mathcal{V}^K$, then $f(q) \notin I(K)$ for all $q \in \mathcal{V}^K$.*

Proof Let $K := (\mathcal{V}^K, \mathcal{E}^K)$. For $f \in H^*(\widetilde{\mathcal{G}})$, we assume that $f(p) \notin I(K)$ for some $p \in \mathcal{V}^K$. We also assume that there exists a vertex $q \in \mathcal{V}^K$ such that $f(q) \in I(K)$. Since K is connected, there is a path in K from q to p, which consists of edges

$$qr_1,\ r_1r_2,\ \cdots,\ r_{s-1}r_s,\ r_sp \in E^K \subset \mathcal{E}^K.$$

Because of the congruence relations in $H^*(\widetilde{\mathcal{G}})$, there are $A_1, \dots, A_{s+1} \in H^*(BT^n)$ such that

$$\begin{aligned}
&f(q) - f(p)\\
=&(f(q) - f(r_1)) + (f(r_1) - f(r_2)) + \cdots + (f(r_{s-1}) - f(r_s)) + (f(r_s) - f(p))\\
=&A_1\widetilde{\alpha}(qr_1) + A_2\widetilde{\alpha}(r_1r_2)\cdots + A_s\widetilde{\alpha}(r_{s-1}r_s) + A_{s+1}\widetilde{\alpha}(r_s p).
\end{aligned}$$

Therefore, by the definition of $I(K)$, we have

$$f(q) - f(p) \in I(K).$$

However, since $f(q)$, $A_1\widetilde{\alpha}(qr_1), \cdots, A_{s+1}\widetilde{\alpha}(r_s p) \in I(K)$, we have $f(p) \in I(K)$. This gives a contradiction. This established that if $f(p) \notin I(K)$, then $f(q) \notin I(K)$ for all $q \in \mathcal{V}^K$. □

By using this lemma, we can prove the surjectivity of $\Psi' : \mathbb{Z}[\widetilde{\mathcal{G}}] \to H^*(\widetilde{\mathcal{G}})$.

Lemma 5.6 *Ψ' is surjective.*

Proof Let $f \in H^*(\widetilde{\mathcal{G}})$. For some $p \in \mathcal{V}$, we assume that $f(p) \in H^*(BT^n)$ has a nonzero constant term $k \in \mathbb{Z} - \{0\}$, i.e.,

$$f(p) = k + g(p)$$

where $g(p) \in H^{>0}(BT^n) \cup \{0\}$. Note that $H^*(BT^n) \simeq \mathbb{Z}[x_1, \dots, x_n]$, where $|x|_i = 2$ for all $i = 1, \dots n$. Because $f \in H^*(\widetilde{\mathcal{G}})$ satisfies the congruence relation, there exists $g \in H^{>0}(\widetilde{\mathcal{G}}) \cup \{0\}$ such that for all $q \in \mathcal{V}$ we may write

$$f(q) = k + g(q),$$

where $H^{>0}(\widetilde{\mathcal{G}}) \cup \{0\}$ is the set of $g \in H^*(\widetilde{\mathcal{G}})$ whose constant term is 0, i.e., for all $p \in \mathcal{V}$, the constant term of the polynomial $g(p) \in \mathbb{Z}[x_1, \dots, x_n]$ is 0. This shows that for all $f \in H^*(\widetilde{\mathcal{G}})$, there exists the constant term k and $g \in H^{>0}(\widetilde{\mathcal{G}}) \cup \{0\}$ such that

$$f = k + g.$$

Therefore, we can take $k \in \mathbb{Z} \subset \mathbb{Z}[\widetilde{\mathcal{G}}]$ such that

$$f = \Psi'(k) + g.$$

Take $g = f - \Psi'(k)$. Then, $g(p) \in H^{>0}(BT^n) \cup \{0\}$ for all $p \in \mathcal{V}$. Now we may put

$$Z(g) = \{p \in \mathcal{V} \mid g(p) = 0\}.$$

We first assume that $Z(g) = \emptyset$. Then $g(p) \neq 0$ for all $p \in \mathcal{V}$. Note that by Lemma 5.2 we have

$$g(p)(\neq 0) \in H^*(BT^n) = \mathbb{Z}[\tau_{L_{p,1}}(p), \cdots, \tau_{L_{p,n}}(p)],$$

where $L_{p,i}$, $i = 1, \ldots, n$ are the hyperplanes such that $p \in \mathcal{V}^{L_{p,i}}$. This also shows that for the fixed vertex $p \in \mathcal{V}$, we may take an element

$$A \in \mathbb{Z}[\widetilde{\mathcal{G}}]$$

such that

$$\Psi'(A)(p) = g(p).$$

Because $g - \Psi'(A) \in H^*(\widetilde{\mathcal{G}})$ and $g(p) - \Psi'(A)(p) = 0$, we have that

$$p \in Z(g - \Psi'(A)).$$

Next, by taking $h = g - \Psi'(A) = f - \Psi'(k + A)$, we may assume that $Z(h) \neq \emptyset$. Take $p \in \mathcal{V} \backslash Z(h)$, i.e., $h(p) \neq 0$. Let $a\tau_{L_1}^{a_{p,1}} \cdots \tau_{L_n}^{a_{p,n}}(p)$ be a monomial appearing in $h(p)$, where a is a nonzero integer, $p \in \mathcal{V}^{L_{p,i}}$ and $a_i \geq 0$ $(i = 1, \cdots, n)$. Since $h(p) \in H^{>0}(BT^n)$, we may assume that

$$a_1, \cdots, a_b \neq 0, \quad a_{b+1} = \cdots = a_n = 0.$$

Put $K = \cap_{i=1}^{b} L_{p,i}$. Then, we have

$$h(p) \notin I(K) = \langle \widetilde{\alpha}^K(\epsilon) \mid \epsilon \in \mathcal{E}^K \rangle \subset H^*(BT^n)$$

because $h(p)$ contains the nonzero monomial $a\tau_{L_{p,1}}^{a_1} \cdots \tau_{L_{p,b}}^{a_b}(p)$ such that $\tau_{L_{p,i}}(p)$ $(i = 1, \cdots, b)$ is defined by the axial function of the normal edge or leg of K on p (which are not the edges or legs in $\mathcal{E}^K$). Therefore, by Lemma 5.5, we have that for all $q \in \mathcal{V}^K$,

$$h(q) \notin I(K).$$

In particular, $h(q) \neq 0$ for all $q \in \mathcal{V}^K$. Let $r \notin \mathcal{V}^K$. Because $K = L_{p,1} \cap \cdots \cap L_{p,b}$, we see that

$$a\tau_{L_{p,1}}^{a_1} \cdots \tau_{L_{p,b}}^{a_b}(r) = 0.$$

Therefore, if we put

$$h' = h - a\tau_{L_{p,1}}^{a_1} \cdots \tau_{L_{p,b}}^{a_b} = h - \Psi'(aL_{p,1}^{a_1} \cdots L_{p,b}^{a_b})$$
$$= f - \Psi'(k + A + aL_{p,1}^{a_1} \cdots L_{p,b}^{a_b}),$$

then $h'(r) = h(r)$ for all $r \notin \mathcal{V}^K$. Namely, $h(q) \neq 0$ for all $q \in \mathcal{V}^K$ and $h'(r) = h(r)$ for all $r \notin \mathcal{V}^K$. This shows that

$$Z(h') \supset Z(h).$$

Note that by the definition of h', the number of monomials in $h'(p)$ is strictly smaller than that in $h(p)$. If $h'(p) = 0$, and then, we have $Z(h') \supsetneq Z(h)$. If $h'(p) \neq 0$, then we may apply the same argument as above for $h' \in H^*(\widetilde{\mathcal{G}})$ and the vertex $p \in \mathcal{V}$ again because $Z(h') \neq \emptyset$. Then, we have that there exists hyperplanes $L_{p,i_1}, \dots, L_{p,i_c}$ in $\{L_{p,1}, \dots, L_{p,n}\}$ and a nonzero integer a' such that

$$h'' = h' - \Psi'(a'L_{p,i_1}^{a'_1} \cdots L_{p,i_c}^{a'_c})$$

which satisfies that

$$Z(h'') \supset Z(h')$$

and the number of monomials in $h''(p)$ is strictly smaller than that in $h'(p)$, where $a'_1, \dots, a'_c$ are positive integers. If $h''(p) \neq 0$, then we repeat the same argument again. Because the number of monomials in $h(p)$ is strictly smaller than smaller in each step, finally we have an element

$$B \in \mathbb{Z}[\widetilde{\mathcal{G}}]$$

such that

$$Z(h - \Psi'(B)) \supsetneq Z(h).$$

Moreover, repeating this procedure, we can find an element $C \in \mathbb{Z}[\widetilde{\mathcal{G}}]$ such that $Z(h - \Psi'(C)) = \mathcal{V}$. This shows that

$$h - \Psi'(C) = f - \Psi'(k + A + C) = 0.$$

Therefore, for all $f \in H^*(\widetilde{\mathcal{G}})$, there exists an element $k + A + C \in \mathbb{Z}[\widetilde{\mathcal{G}}]$ such that $f = \Psi'(k + A + C)$. This establishes that Ψ' is surjective. □

Consequently, Ψ' is an isomorphic map by Lemmas 5.4 and 5.6, and we have

Fig. 12 An example of the $T^*\mathbb{C}^2$-modeled graph and its x-forgetful graph which does not satisfy the assumption (2) in Theorem 4.1. Geometrically, this graph can be defined from T^*S^4 with the $T^2 \times S^1$-action

$$H^*(\widetilde{\mathcal{G}}) \simeq \mathbb{Z}[\widetilde{\mathcal{G}}].$$

This establishes Theorem 5.1.

Remark 5.7 From the above argument, we know that the assumption (2) of Theorem 4.1 is not needed to prove the "injectivity" of Ψ'; however, it is needed to prove the "surjectivity" of Ψ'. Hence, the assumption (2) of Theorem 4.1 means that $H^*(\widetilde{\mathcal{G}})$ (resp. $H^*(\mathcal{G})$) is generated by elements of $H^2(\widetilde{\mathcal{G}})$ (resp. $H^2(\mathcal{G})$), that is, $\tau_L \in H^2(\widetilde{\mathcal{G}})$ (resp. $\tau_H,\ \chi \in H^2(\mathcal{G})$). For example, the Fig. 12 shows the $T^*\mathbb{C}^2$-modeled GKM graph, which does not satisfy the assumption (2) of Theorem 4.1 and its x-forgetful graph. In this case, we need a generator which is not in $H^2(\widetilde{\mathcal{G}})$.

6 Proof of Theorem 4.1

In this section, we prove Theorem 4.1. We first recall the statement of Theorem 4.1. Let $\mathcal{G}$ be a $2n$-valent $T^*\mathbb{C}^n$-modeled GKM graph and $\mathbf{L} = \{L_1, \cdots, L_m\}$ be the set of all hyperplanes in $\mathcal{G}$. Assume the following two assumptions for $\mathcal{G}$:

1. For each $L \in \mathbf{L}$, there exist the unique pair of the halfspace H and its opposite side $\overline{H}$ such that $H \cap \overline{H} = L$.
2. For every subset $\mathbf{L}' \subset \mathbf{L}$, its intersection $\bigcap_{L \in \mathbf{L}'} L$ is empty or connected.

Then, we will prove the following isomorphism:

$$\mathbb{Z}[\mathcal{G}] \simeq H^*(\mathcal{G}).$$

Recall the ring homomorphism in Sect. 4

$$\Psi : \mathbb{Z}[\mathcal{G}] \to H^*(\mathcal{G})$$

is defined by

$$\Psi(X) = \chi, \quad \Psi(H) = \tau_H. \tag{6}$$

To prove Theorem 4.1, we claim that Ψ is an isomorphism.

By the assumption (1) of Theorem 4.1, we can put the set of all halfspaces in $\mathcal{G}$ by

$$\mathbf{H} = \{H_1, \cdots, H_m, \overline{H_1}, \cdots, \overline{H_m}\},$$

where $L_i = H_i \cap \overline{H_i}$. We prepare the following diagram:

$$\begin{array}{ccc}
\mathbb{Z}[X, H_1, \cdots, H_m, \overline{H}_1, \cdots, \overline{H_m}] & \xrightarrow{\hat{\pi}} & \mathbb{Z}[\mathcal{G}] \\
\downarrow{\scriptstyle \phi'} & \nearrow{\scriptstyle \widetilde{\pi}} & \downarrow{\scriptstyle \Psi} \\
\mathbb{Z}[X, H_1, \cdots, H_m] & \xrightarrow{\pi} & H^*(\mathcal{G}) \\
\downarrow{\scriptstyle \phi} & & \downarrow{\scriptstyle \widetilde{F}} \\
\mathbb{Z}[L_1, \cdots, L_m] & \xrightarrow{\pi'} & H^*(\widetilde{\mathcal{G}})
\end{array} \tag{7}$$

where the map in the diagram is defined as follows:

- $\hat{\pi}$ is the natural projection.
- Ψ is defined by (6) as before.
- ϕ' is the surjective homomorphism induced from

$$\phi'(X) = X, \quad \phi'(H_i) = H_i, \quad \phi'(\overline{H_i}) = X - H_i, \ i = 1, \ldots, m.$$

- π is the homomorphism induced from

$$\pi(X) = \chi, \quad \pi(H_i) = \tau_{H_i}, \ i = 1, \ldots, m.$$

- $\widetilde{F}$ is the homomorphism defined by

$$\widetilde{F}(f)(p) := F \circ f(p)$$

  for $f \in H^*(\mathcal{G})$ and $p \in \mathcal{V}$, where $F : H^*(BT^n \times BS^1) \to H^*(BT^n)$ is the x-forgetful map for the fixed generator x of $H^2(BS^1) \simeq \mathbb{Z}x$.
- ϕ is the surjective homomorphism induced from

$$\phi(X) = 0, \quad \phi(H_i) = L_i, \ i = 1, \ldots, m.$$

- π' is the homomorphism induced from

$$\pi'(L_i) = \tau_{L_i},\ i = 1, \ldots, m.$$

It easily follows from the definitions of homomorphisms as above and Lemma 3.9 that the top diagram is commutative. By Sect. 5.1, we may choose $H_1, \ldots, H_m$ as $\tau_{L_i} = F \circ \tau_{H_i}$ for $i = 1, \ldots, m$. Therefore, we may assume that the bottom diagram is also commutative. Therefore, this diagram is commutative.

By the proof of Theorem 5.1, i.e., $\mathbb{Z}[\widetilde{\mathcal{G}}] \simeq H^*(\widetilde{\mathcal{G}})$, we have that π' is surjective. This shows that $\pi' \circ \phi = \widetilde{F} \circ \pi$ is surjective; therefore, $\widetilde{F}$ is also surjective.

6.1 Surjectivity of Ψ

We first prove the surjectivity of Ψ. By the commutativity of the top diagram, it is enough to prove that the homomorphism

$$\pi : \mathbb{Z}[X, H_1, \ldots, H_m] \to H^*(\mathcal{G})$$

is surjective (see Lemma 6.4). To do that, we will prove the following three lemmas.

The following first lemma is about the kernel of $\widetilde{F} : H^*(\mathcal{G}) \to H^*(\widetilde{\mathcal{G}})$.

Lemma 6.1 *Let χ be the element in $H^*(\mathcal{G})$ such that $\chi(p) = x$ for all $p \in \mathcal{V}$, where x is the residual basis. Then, we have* Ker $\widetilde{F} = \langle \chi \rangle$, *i.e., the ideal generated by χ.*

Proof Let $f \in$ Ker $\widetilde{F}$. By the definition of $\widetilde{F}$, we have $\widetilde{F}(f)(p) = F \circ f(p) = 0$ for all $p \in \mathcal{V}$. Since $F : H^*(BT^n \times BS^1) = \mathbb{Z}[\alpha_1, \cdots, \alpha_n, x] \to \mathbb{Z}[\beta_1, \cdots, \beta_n] = H^*(BT^n)$ is defined by $F(x) = 0$ and $F(\alpha_i) = \beta_i$ for all $i = 1, \cdots, n$, we have

$$f(p) \in \text{Ker } F = \langle x \rangle \subset H^*(BT^n \times BS^1).$$

Therefore, for every $p \in \mathcal{V}$, there exists a polynomial $g(p) \in H^*(BT^n \times BS^1)$ such that

$$f(p) = g(p)x.$$

Because $f \in H^*(\mathcal{G})$, it satisfies the congruence relation

$$f(i(\epsilon)) - f(t(\epsilon)) = g(i(\epsilon))x - g(t(\epsilon))x = (g(i(\epsilon)) - g(t(\epsilon)))x \equiv 0 \pmod{\alpha(\epsilon)}$$

for every edge ϵ. Because x is the residual basis, by definition of $T^*\mathbb{C}^n$-modeled GKM graph (see Definition 2.4), we see that $\alpha(\epsilon) \neq x$ for every edge $\epsilon \in E$. Hence, because the polynomial ring is an integral domain, we have

$$g(i(\epsilon)) - g(t(\epsilon)) \equiv 0 \pmod{\alpha(\epsilon)}$$

for every edge ϵ. This implies that $g \in H^*(\mathcal{G})$. Therefore, for all $f \in \mathrm{Ker}\,\widetilde{F}$, there exists an element $g \in H^*(\mathcal{G})$ such that $f = g\chi$. Hence, $\mathrm{Ker}\,\widetilde{F} \subset \langle \chi \rangle$. On the other hand, we can easily check that $\mathrm{Ker}\,\widetilde{F} \supset \langle \chi \rangle$. This establishes that $\mathrm{Ker}\,\widetilde{F} = \langle \chi \rangle$. □

The following second lemma is about the degree-wise decomposition of an element in $H^*(\mathcal{G})$.

Lemma 6.2 *For every $f \in H^*(\mathcal{G})$, there exists a nonnegative integer l and an element $f_{2i} \in H^{2i}(\mathcal{G})$ for each $0 \leq i \leq l$ which satisfy*

$$f = f_0 + f_2 + \cdots + f_{2l},$$

where $H^{2i}(\mathcal{G})$ consists of the element, say h_{2i}, which satisfies $h_{2i}(p) \in H^{2i}(BT^n \times BS^1)$ for all $p \in \mathcal{V}$.

Proof Since $f(p) \in H^*(BT^n \times BS^1)$, for every $p \in \mathcal{V}$ there exists a nonnegative integer $l(p)$ and an element $f_{2i}(p) \in H^{2i}(BT^n \times BS^1)$ such that

$$f(p) = f_0(p) + \cdots + f_{2l(p)}(p).$$

If we take the maximal integer $l = \max\{l(p) \mid p \in \mathcal{V}\}$, then we may write

$$f(p) = f_0(p) + \cdots + f_{2l}(p).$$

for all $p \in \mathcal{V}$. Therefore, we can define the map $f_{2i} : \mathcal{V} \to H^{2i}(BT^n \times BS^1)$ by $p \mapsto f_{2i}(p)$ for all $0 \leq i \leq l$. We claim that $f_{2i} \in H^*(\mathcal{G})$. Because f satisfies the congruence relation for all edges ϵ, we see that

$$f(i(\epsilon)) - f(t(\epsilon)) = (f_0(i(\epsilon)) - f_0(t(\epsilon))) + \cdots + (f_{2l}(i(\epsilon)) - f_{2l}(t(\epsilon))) = A\alpha(\epsilon) \tag{8}$$

for some $A \in H^*(BT^n \times BS^1)$. Moreover, there is a monomial $A_{2i} \in H^{2i}(BT^n \times BS^1)$ for each $0 \leq i \leq l-1$ such that

$$A = A_0 + \cdots + A_{2l-2}.$$

Comparing the same degree monomials of both sides in (8), we have

$$f_{2i}(i(\epsilon)) - f_{2i}(t(\epsilon)) = A_{2i-2}\alpha(\epsilon) \equiv 0 \mod \alpha(\epsilon).$$

Because this relation satisfies for all $\epsilon \in E$, we have $f_{2i} \in H^*(\mathcal{G})$ for all $i = 0, \cdots, l$. This establishes the statement. □

We call each f_{2i} in Lemma 6.2 a $2i$ *degree homogeneous term* of f for $i = 0, \cdots, l$. We denote $\deg f_{2i} = 2i$. Of course, $f_{2i} \in H^{2i}(\mathcal{G})$.

The following third lemma is about the map $\pi : \mathbb{Z}[X, H_1, \ldots, H_m] \to H^*(\mathcal{G})$. This will be a technical part to show that π is surjective (Lemma 6.4).

Lemma 6.3 *Assume that there exists an element $f \in H^*(\mathcal{G})$ such that $f \notin \mathrm{Im}\,\pi$. Then, there are $A \in \mathbb{Z}[X, H_1, \cdots, H_m]$ and some integer j_k such that*

$$\pi(A) - f = \chi \sum_k g_{2j_k},$$

where $g_{2j_k} \in H^{2j_k}(\mathcal{G})$ but $g_{2j_k} \notin \mathrm{Im}\,\pi$ with $j_0 < j_1 < \cdots < j_k < \cdots$.

Proof Assume $f \notin \mathrm{Im}\,\pi$. Recall that the following two homomorphisms in (7) are surjective by the assumption (1) of Theorem 4.1 and Theorem 5.1:

$$\phi : \mathbb{Z}[X, H_1, \ldots, H_m] \longrightarrow \mathbb{Z}[L_1, \ldots, L_m];$$
$$\pi' : \mathbb{Z}[L_1, \ldots, L_m] \longrightarrow H^*(\widetilde{\mathcal{G}}).$$

Therefore, there exists a nonzero polynomial

$$B \in \mathbb{Z}[X, H_1, \cdots, H_m]$$

such that for $\widetilde{F} : H^*(\mathcal{G}) \to H^*(\widetilde{\mathcal{G}})$,

$$\widetilde{F}(f) = \pi' \circ \phi(B).$$

Because $\pi' \circ \phi = \widetilde{F} \circ \pi$ in the diagram (7), we have

$$\pi' \circ \phi(B) = \widetilde{F} \circ \pi(B) = \widetilde{F}(f).$$

Hence, $\pi(B) - f \in \mathrm{Ker}\,\widetilde{F}$. Because of Lemma 6.1, i.e., $\mathrm{Ker}\widetilde{F} = \langle \chi \rangle$, there is a $g' \in H^*(\mathcal{G})$ such that

$$\pi(B) - f = g'\chi. \tag{9}$$

Since $f \notin \mathrm{Im}\,\pi$ and $\pi(X) = \chi$, we have

$$g' \notin \mathrm{Im}\,\pi.$$

Because of Lemma 6.2, this element g' can be divided into

$$g' = g_0 + \cdots + g_{2l},$$

where g_{2i} is a $2i$ degree homogeneous term, for $0 \leq i \leq l$. If $g_{2i} \in \mathrm{Im}\,\pi$, then $g' - g_{2i} \notin \mathrm{Im}\,\pi$. Therefore, g' can be divided into two terms $(0 \neq) g = \sum_k g_{2j_k}$ for all $g_{2j_k} \notin \mathrm{Im}\,\pi$ and $h = \sum_{k'} g_{2i_{k'}}$ for all $g_{2i_{k'}} \in \mathrm{Im}\,\pi$ such that

$$g' = g + h.$$

Since

$$g'\chi = g\chi + h\chi = g\chi + \pi(CX)$$

for some $C \in \mathbb{Z}[X, H_1, \cdots, H_m]$, together with (9), we see that there is an element $A = B - CX \in \mathbb{Z}[X, H_1, \cdots, H_m]$ such that $\pi(A) - f = g\chi$. □

Now we may prove Lemma 6.4.

Lemma 6.4 *The homomorphism* $\pi : \mathbb{Z}[X, H_1, \cdots, H_m] \to H^*(\mathcal{G})$ *is surjective.*

Proof By Lemma 6.3, it is enough to show that every homogeneous term of $f \in H^*(\mathcal{G})$ is an element of Im π.

Assume that $H^*(\mathcal{G})\backslash \text{Im } \pi \neq \emptyset$. Let f be a minimal degree homogeneous element in $H^*(\mathcal{G})\backslash \text{Im } \pi$. Because of Lemma 6.3, there exists a polynomial $A \in \mathbb{Z}[X, H_1, \cdots, H_m]$ and an element $g \in H^*(\mathcal{G})\backslash \text{Im } \pi$ such that

$$f = \pi(A) - g\chi.$$

By using Lemma 6.3 again, we also have that g is a sum of homogeneous elements in $H^*(\mathcal{G})\backslash \text{Im } \pi$.

We claim that $\pi(A)(\in \text{Im } \pi)$ and $g\chi(\in H^*(\mathcal{G})\backslash \text{Im } \pi)$ are also homogeneous elements in $H^*(\mathcal{G})$, whose degrees are the same with the degree of f. Assume that $\pi(A) = \sum_k h_{2i_k}$ and $g\chi = \sum_k g_{2j_k}\chi$, where $h_{2i_k} \in H^{2i_k}(\mathcal{G}) \cap \text{Im } \pi$ for all $i_0 < i_1 < \cdots$ and $g_{2j_k} \in H^{2j_k}(\mathcal{G}) \setminus \text{Im } \pi$ for all $j_0 < j_1 < \cdots$. Because f is a minimal homogeneous element in $H^*(\mathcal{G})\backslash \text{Im } \pi$, we see that $|g|_{2j_0}\chi = |f|$ and $|h|_{2i_0} \geq |f|$; moreover, the higher terms of $\pi(A) \in \text{Im } \pi$ and $g\chi \notin \text{Im } \pi$ are the same, i.e., they must be 0. Hence, both of $\pi(A)$ and g are also homogeneous elements.

However, in this case, we have

$$|g| = |g|\chi - |\chi| = |f| - 2 < |f|.$$

This gives a contradiction to that f is a minimal homogeneous element in $H^*(\mathcal{G})\backslash \text{Im } \pi$. Hence, there does not exist any homogeneous elements in $H^*(\mathcal{G})\backslash \text{Im } \pi$. Consequently, by Lemma 6.3, we have that $H^*(\mathcal{G})\backslash \text{Im } \pi = \emptyset$, i.e., π is surjective. □

Therefore, by the commutativity of the top diagram in (7), the following lemma holds:

Lemma 6.5 Ψ *is surjective.*

6.2 Injectivity of Ψ

Finally, in this section, we will prove the injectivity of Ψ. In this section, we use the following notation:

$$I_j = \{1, \cdots, l\} - \{j\}$$

for $j = 1, \cdots, l$. We first prove Lemma 6.7. In order to prove Lemma 6.7, we prepare the following lemma.

Lemma 6.6 *Assume that* $\cap_{k=1}^{l} L_k = \emptyset$ *and* $L_k = H_k \cap \overline{H_k}$ $(k = 1, \cdots, l)$*. Then, for all* $j = 1, \ldots, l$*, one of the following holds:*

- $H_j \cap (\cap_{k \in I_j} L_k) = \emptyset$.
- $\overline{H_j} \cap (\cap_{k \in I_j} L_k) = \emptyset$.

Proof Assume $\cap_{k=1}^{l} L_k = \emptyset$. For $j \in \{1, \cdots, l\}$, if the following relation holds:

$$L_j \cap (\cap_{k \in I_j} L_k) = \cap_{k \in I_j} L_k = \emptyset,$$

then it follows from $\cap_{k \in I_j} L_k = \emptyset$ that for each H_j and $\overline{H_j}$ we have

$$H_j \cap (\cap_{k \in I_j} L_k) = \overline{H_j} \cap (\cap_{k \in I_j} L_k) = \emptyset.$$

So we may take $j \in \{1, \cdots, l\}$ such that

$$\bigcap_{k \in I_j} L_k \neq \emptyset.$$

In this case, there exists a vertex $p \in \mathcal{V}^{\cap_{k \in I_j} L_k}$. Since $L_j \cap (\cap_{k \in I_j} L_k) = \cap_{k=1}^{l} L_k = \emptyset$, we have that $p \notin \mathcal{V}^{L_i}$. Therefore, for all vertices $p \in \mathcal{V}^{\cap_{k \in I_j} L_k}$, the following equation holds:

$$\tau_{H_j}(p) = \begin{cases} 0 \text{ (if } p \notin \mathcal{V}^{H_j}) \\ x \text{ (if } p \in \mathcal{V}^{H_j}) \end{cases}$$

where τ_{H_j} is the Thom class of H_j.

If there are two vertices $p, q \in \mathcal{V}^{\cap_{k \in I_j} L_k}$ such that

$$\tau_{H_j}(p) = 0; \quad \tau_{H_j}(q) = x.$$

By the assumption (2) in Theorem 4.1, there exists a path from p to q in $\cap_{k \in I_j} L_k$, i.e., we may take the following sequence in $E^{\cap_{k \in I_j} L_k}$:

$$\epsilon_1, \cdots, \epsilon_s \in E^{\cap_{k\in I_j} L_k}$$

such that $i(\epsilon_1) = p$ and $t(\epsilon_s) = q$. By the definition of the $T^*\mathbb{C}^n$-modeled GKM graph, the axial function satisfies that $\alpha(\epsilon) \neq x$ for all $\epsilon \in E$. Moreover, τ_{H_j} satisfies the congruence relation. Therefore, there exists an edge $\epsilon \in \{\epsilon_1, \ldots, \epsilon_s\}$ such that $r = i(\epsilon_t)$ satisfies that

$$\tau_{H_j}(r) \neq 0,\ x.$$

By the definition of the Thom class of the halfspace H_j, the vertex $r \in \partial H_j = L_j$. However, this gives that $r \in L_j \cap (\cap_{k\in I_j} L_k) = \cap_{k=1}^{l} L_k$. This gives a contradiction to that $\cap_{k=1}^{l} L_k = \emptyset$.

Therefore, we may assume $\tau_{H_j}(p) = 0$ (resp. x) for all $p \in \mathcal{V}^{\cap_{k\in I_j} L_k}$. Then, by definition of the halfspace, we have $H_j \cap \mathcal{V}^{\cap_{k\in I_j} L_k} = \emptyset$ (resp. $\overline{H_j} \cap \mathcal{V}^{\cap_{k\in I_j} L_k} = \emptyset$). This establishes the statement of this lemma. □

From Lemma 6.6, we have the following key fact.

Lemma 6.7 *Assume the $T^*\mathbb{C}^n$-modeled GKM graph $\mathcal{G}$ satisfies two assumptions (1), (2) of Theorem* 4.1. *If $\cap_{k=1}^{l} L_k = \emptyset$ and $L_k = H_k \cap \overline{H_k}$ $(k = 1, \cdots, l)$, then we can take a halfspace H_k such that $\cap_{k=1}^{l} H_k = \emptyset$.*

Proof If $\cap_{k=1}^{l} L_k = \emptyset$ and $L_k = H_k \cap \overline{H_k}$ $(k = 1, \cdots, l)$, we can take H_j as $H_j \cap (\cap_{k\in I_j} L_k) = \emptyset$ for all $j = 1, \cdots, l$ from Lemma 6.6. Now we may set

$$\mathbf{H}' = \left\{H_1, \cdots, H_l \mid H_j \cap (\cap_{k\in I_j} L_k) = \emptyset,\ j = 1, \ldots, l\right\}.$$

We claim that $\cap_{H\in\mathbf{H}'} H = \cap_{j=1}^{l} H_j = \emptyset$. If there exists a vertex $p \in \mathcal{V}^{\cap_{j=1}^{l} H_j}$, it follows from the assumption $\cap_{k=1}^{l} L_k = \emptyset$ that we have $\tau_{H_j}(p) = x$ for all $j = 1, \ldots, l$; therefore,

$$\prod_{j=1}^{l} \tau_{H_j}(p) = x^l.$$

Because $\prod_{j=1}^{l} \tau_{H_j} \in H^*(\mathcal{G})$, $\prod_{j=1}^{l} \tau_{H_j}$ satisfies the congruence relations for all edges $\epsilon \in E$. By definition of $T^*\mathbb{C}^n$-modeled GKM graph, the axial function satisfies $\alpha(\epsilon) \neq x$ for all edges $\epsilon \in E$. This shows that for all edge $\epsilon \in E_p$ the following equation holds:

$$\prod_{j=1}^{l} \tau_{H_j}(t(\epsilon)) = x^l.$$

Because the graph Γ is connected, we can apply the same argument for all vertices; therefore, we have

$$\prod_{j=1}^{l} \tau_{H_j}(q) = x^l$$

for all $q \in \mathcal{V}$. This shows that $\mathcal{V} = \mathcal{V}^{\cap_{j=1}^{l} H_j}$. However, by definition of the halfspace, it is obvious that $\mathcal{V} \neq \mathcal{V}^{\cap_{j=1}^{l} H_j}$ and this gives a contradiction. Hence, we have $\cap_{j=1}^{l} H_j = \cap_{H \in \mathbf{H}'} H = \emptyset$. □

We next will prove Lemma 6.10. In order to prove it, we prepare some notations and two lemmas: Lemmas 6.8 and 6.9.

Let $\widetilde{\pi} : \mathbb{Z}[X, H_1, \cdots, H_m] \to \mathbb{Z}[\mathcal{G}]$ be the natural homomorphism such that $\widetilde{\pi}(X) = X, \widetilde{\pi}(H_i) = H_i$ for $i = 1, \cdots, m$. Because $\overline{H_i} = X - H_i$ in $\mathbb{Z}[\mathcal{G}]$, we have

$$\widetilde{\pi} \circ \phi' = \hat{\pi} : \mathbb{Z}[X, H_1, \cdots, H_m, \overline{H}_1, \cdots, \overline{H_m}] \to \mathbb{Z}[\mathcal{G}].$$

Since $\hat{\pi}$ is surjective, $\widetilde{\pi}$ is also surjective. Moreover, we have

$$\Psi \circ \widetilde{\pi} = \pi : \mathbb{Z}[X, H_1, \cdots, H_m] \to H^*(\mathcal{G})$$

by definitions of Ψ and π. Hence, we have the following commutative diagram:

$$\begin{array}{ccc}
\mathbb{Z}[X, H_1, \cdots, H_m, \overline{H}_1, \cdots, \overline{H_m}] & \xrightarrow{\hat{\pi}} & \mathbb{Z}[\mathcal{G}] \\
\downarrow \phi' & \nearrow \widetilde{\pi} & \downarrow \Psi \\
\mathbb{Z}[X, H_1, \cdots, H_m] & \xrightarrow{\pi} & H^*(\mathcal{G}) \\
\downarrow \phi & & \downarrow \widetilde{F} \\
\mathbb{Z}[L_1, \cdots, L_m] & \xrightarrow{\pi'} & H^*(\widetilde{\mathcal{G}})
\end{array}$$

Define the following ideal in $\mathbb{Z}[X, H_1, \cdots, H_m, \overline{H}_1, \cdots, \overline{H_m}]$:

$$\mathcal{I} = \left\langle H_i + \overline{H_i} - X, \prod_{H \in \mathbf{H}'} H \;\middle|\; i = 1, \cdots, m, \mathbf{H}' \in \mathbf{I}(\mathbf{H}) \right\rangle,$$

where $\mathbf{I}(\mathbf{H}) = \{\mathbf{H}' \subset \mathbf{H} \mid \cap_{H \in \mathbf{H}'} H = \emptyset\}$. For this ideal, the following property holds.

Lemma 6.8 *For the ideal $\mathcal{I} \subset \mathbb{Z}[X, H_1, \cdots, H_m, \overline{H}_1, \cdots, \overline{H_m}]$, the following two properties hold:*

(i) $\mathrm{Ker}\,\widetilde{\pi} = \phi'(\mathcal{I})$.
(ii) $\mathrm{Ker}\,\pi' = \phi \circ \phi'(\mathcal{I})$.

Proof Since $\hat{\pi}$ is the natural projection, it follows from the definition of $\mathbb{Z}[\mathcal{G}]$ that

$$\mathcal{I} = \mathrm{Ker}\,\hat{\pi}.$$

So, by the commutativity of the diagram, we have that

$$\widetilde{\pi}(\phi'(\mathcal{I})) = \hat{\pi}(\mathcal{I}) = \hat{\pi}(\mathrm{Ker}\,\hat{\pi}) = \{0\}.$$

Hence, $\phi'(\mathcal{I}) \subset \mathrm{Ker}\,\widetilde{\pi}$. Let A be an element in $\mathrm{Ker}\,\widetilde{\pi}$. Because ϕ' is surjective, there is an element $B \in \mathbb{Z}[X, H_1, \cdots, H_m, \overline{H}_1, \cdots, \overline{H_m}]$ such that $\phi'(B) = A$. By the commutativity of the diagram, we also have

$$\hat{\pi}(B) = \widetilde{\pi} \circ \phi'(B) = \widetilde{\pi}(A) = 0.$$

So $B \in \mathrm{Ker}\,\hat{\pi} = \mathcal{I}$. Hence, $A = \phi'(B) \in \phi'(\mathcal{I})$, that is, $\mathrm{Ker}\,\widetilde{\pi} \subset \phi'(\mathcal{I})$. Therefore, we establish the first property: $\mathrm{Ker}\,\widetilde{\pi} = \phi'(\mathcal{I})$.

By Theorem 5.1, we know

$$\mathrm{Ker}\,\pi' = \langle \prod_{L \in \mathbf{L}'} L \mid \mathbf{L}' \in \mathbf{I}(\mathbf{L}) \rangle,$$

where $\mathbf{I}(\mathbf{L}) = \{\mathbf{L}' \subset \mathbf{L} \mid \cap_{L \in \mathbf{L}'} L = \emptyset\}$. Take a generator $\prod_{L \in \mathbf{L}'} L \in \mathrm{Ker}\,\pi'$. From Lemma 6.7, for $\mathbf{L}' = \{L_1, \cdots, L_l\} \in \mathbf{I}(\mathbf{L})$, there exists a set of halfspaces $\mathbf{H}' = \{H_1, \cdots, H_l\} \in \mathbf{I}(\mathbf{H})$ such that $H_k \cap \overline{H_k} = L_k$. By the definition of the ideal $\mathcal{I}$, a product $\prod_{k=1}^{l} H_k$ is one of the generators of $\mathcal{I}$. Moreover, by the definitions of ϕ' and ϕ, we see that

$$\phi \circ \phi'(\mathcal{I}) \ni \phi \circ \phi'(\prod_{k=1}^{l} H_k) = \pm \prod_{k=1}^{l} L_k.$$

Because this satisfies for all generators $\prod_{L \in \mathbf{L}'} L$ in $\mathrm{Ker}\,\pi'$, we have that

$$\mathrm{Ker}\,\pi' \subset \phi \circ \phi'(\mathcal{I}).$$

On the other hand, because $\phi'(H + \overline{H} - X) = 0$ and $\phi \circ \phi'(\prod_{H \in \mathbf{H}'} H) = \pm \prod_{L \in \mathbf{L}'} L \in \mathrm{Ker}\,\pi'$, for all $A \in \mathcal{I}$ we have

$$\pi' \circ \phi \circ \phi'(A) = \{0\}.$$

So we have $\mathrm{Ker}\,\pi' \supset \phi \circ \phi'(\mathcal{I})$. Therefore, we conclude the second property: $\mathrm{Ker}\,\pi' = \phi \circ \phi'(\mathcal{I})$. □

In order to prove Lemma 6.10, we also prepare the following technical lemma for general polynomial rings.

Lemma 6.9 *Let $\mathcal{I} \subset \mathbb{Z}[x_1, \cdots, x_l]$ be an ideal generated by homogeneous polynomials, that is, $\mathcal{I} = \langle p_1, \cdots, p_m \rangle$, where p_i is a homogeneous polynomial of $\mathbb{Z}[x_1, \cdots, x_l]$ such that* $\deg p_i \leq \deg p_j$ *for $i < j$. For every element $A \in \mathcal{I}$, if we denote $A = A_1 + \cdots + A_n$, where A_i is a homogeneous term ($i = 1, \cdots, n$) and* $\deg A_i < \deg A_j$ *for $i < j$, then $A_i \in \mathcal{I}$ for all $i = 1, \cdots, n$.*

Proof Because $A \in \mathcal{I} = \langle p_1, \cdots, p_m \rangle$, there exists $X_k \in \mathbb{Z}[x_1, \cdots, x_l]$, $k = 1, \ldots, m$, such that

$$A = X_1 p_1 + \cdots + X_m p_m.$$

Then, we can put $X_k = X_{k1} + X_{k2} + \cdots + X_{ks_k}$ where X_{ki} is a homogeneous term ($i = 1, \cdots, s_k$) and $\deg X_{ki} < \deg X_{kj}$ for $i < j$. Hence, by changing the order of monomials, we may rewrite

$$\begin{aligned} A &= (X_{11} + \cdots + X_{1s_1})p_1 + \cdots + (X_{m1} + \cdots + X_{ms_m})p_m \\ &= X_{11}p_1 + \cdots X_{ms_m}p_m \end{aligned}$$

as

$$A = A_1 + \cdots + A_n,$$

where $|A|_i < |A|_j$ if $i < j$. Because A_i is a homogeneous term, we have

$$A_i = \sum_{j \in \mathbf{D}_i} X_{jh_j} p_j$$

where $\mathbf{D}_i = \{j \mid |X|_{jh_j} + |p|_j = |A|_i\}$. Therefore, $A_i \in \mathcal{I}$ for all $i = 1, \cdots, n$. □

Using two lemmas as above, we have the following lemma.

Lemma 6.10 $\mathrm{Ker}\, \widetilde{\pi} = \mathrm{Ker}\, \pi$.

Proof By Lemma 6.8 (i), $\mathrm{Ker}\, \widetilde{\pi} = \phi'(\mathcal{I})$. Therefore, by using the commutativity of the diagram and $\mathrm{Ker}\, \hat{\pi} = \mathcal{I}$, we have

$$\pi(\mathrm{Ker}\, \widetilde{\pi}) = \pi \circ \phi'(\mathcal{I}) = \Psi \circ \hat{\pi}(\mathcal{I}) = 0.$$

Hence,

$$\mathrm{Ker}\, \widetilde{\pi} = \phi'(\mathcal{I}) \subset \mathrm{Ker}\, \pi.$$

We claim that $\mathrm{Ker}\, \pi \subset \mathrm{Ker}\, \widetilde{\pi} (\subset \mathbb{Z}[X, H_1, \cdots, H_m])$. Assume that $\mathrm{Ker}\, \pi \backslash \phi'(\mathcal{I}) \neq \emptyset$. Let $A \in \mathrm{Ker}\, \pi \backslash \phi'(\mathcal{I}) \subset \mathbb{Z}[X, H_1, \cdots, H_m]$ be a minimal degree homogeneous polynomial. By the previous diagram,

$$\pi' \circ \phi(A) = \widetilde{F} \circ \pi(A) = 0.$$

Hence, by Lemma 6.8 (ii),

$$\phi(A) \in \operatorname{Ker} \pi' = \phi \circ \phi'(\mathcal{I}).$$

Therefore, we can take $B \in \phi'(\mathcal{I})(\subset \operatorname{Ker} \pi)$ such that $\phi(A) = \phi(B)$. Because $\phi'(\mathcal{I})$ is an ideal in $\mathbb{Z}[X, H_1, \ldots, H_m]$ and A is a homogeneous polynomial, it follows from Lemma 6.9 that we may also take B as the homogeneous polynomial in $\phi'(\mathcal{I})$ such that

$$|A| = |B|.$$

By the definition of ϕ, it is easy to check that $\operatorname{Ker} \phi = \langle X \rangle \subset \mathbb{Z}[X, H_1, \ldots, H_m]$; therefore, we have

$$A - B \in \operatorname{Ker} \phi = \langle X \rangle.$$

This means that there exists a polynomial $C \in \mathbb{Z}[X,\ H_1,\ \cdots,\ H_m]$ such that

$$A - B = CX.$$

Because $A = B + CX \notin \phi'(\mathcal{I})$ and $B \in \phi'(\mathcal{I})$, we have $CX \notin \phi'(\mathcal{I})$. Moreover, because we take $|A| = |B|$, CX is also a homogeneous polynomial with $|A| = |B| = |C|X$. Therefore, because $A,\ B \in \operatorname{Ker} \pi$, we have CX is a homogeneous polynomial in $\operatorname{Ker} \pi \setminus \phi'(\mathcal{I})$. Then, we have $|A| = |C|X = |C| + |X| = |C| + 2$. Moreover, because A is a minimal homogeneous polynomial in $\operatorname{Ker} \pi \setminus \phi'(\mathcal{I})$, we have that

$$C \in \phi'(\mathcal{I})(\subset \operatorname{Ker} \pi).$$

However, because $\phi'(\mathcal{I})$ is an ideal in $\mathbb{Z}[X, H_1, \ldots, H_m]$, we see that

$$CX \in \phi'(\mathcal{I}).$$

This shows that $A = B + CX \in \phi'(\mathcal{I})$, and this gives the contradiction to that there is an element in $\operatorname{Ker} \pi \backslash \phi'(\mathcal{I})$. Hence, we have $\operatorname{Ker} \pi \backslash \phi'(\mathcal{I}) = \emptyset$, that is, $\operatorname{Ker} \pi = \phi'(\mathcal{I}) = \operatorname{Ker} \widetilde{\pi}$ by Lemma 6.8 (i). □

So we can prove the injectivity of Ψ.

Lemma 6.11 Ψ *is injective.*

Proof Let A be in Ker Ψ. Since $\widetilde{\pi}$ is surjective, there is an element $B \in \mathbb{Z}[X, H_1, \cdots, H_m]$ such that $\widetilde{\pi}(B) = A$. So we have $\pi(B) = \Psi \circ \widetilde{\pi}(B) = \Psi(A) = 0$. Hence $B \in$ Ker π = Ker $\widetilde{\pi}$ by Lemma 6.10. Therefore, we have $A = \widetilde{\pi}(B) = 0$. This concludes that Ψ is injective. □

Because of Lemmas 6.5 and 6.11, we have that Ψ is the isomorphism. Consequently, the proof of Theorem 4.1 is complete, that is, we get

$$H^*(\mathcal{G}) \simeq \mathbb{Z}[\mathcal{G}].$$

7 Generators of $\mathbb{Z}[\widetilde{\mathcal{G}}]$ as $H^*(BT^n)$-Module

Let $\mathcal{G} = (\Gamma, \alpha, \nabla)$ be a $2n$-valent $T^*\mathbb{C}^n$-modeled GKM graph and $\mathbf{L} = \{L_1, \ldots, L_m\}$ be the set of all hyperplanes in $\mathcal{G}$. Assume that $\mathcal{G}$ satisfies the two assumptions of Theorem 4.1 so that $H^*(\mathcal{G}) \simeq \mathbb{Z}[\mathcal{G}]$.

7.1 Simplicial Complex Associated with L

Let $\mathbf{L} = \{L_1, \ldots, L_m\}$. Let $\Delta_{\mathbf{L}}$ denote the simplicial complex associated with $\mathbf{L}$ defined as follows. There is a vertex v_i in $\Delta_{\mathbf{L}}$ corresponding to the hyperplane L_i such that whenever $L_{i_1} \cap \cdots \cap L_{i_k} \neq \emptyset$ in $\mathcal{G}$, the vertices $\{v_{i_1}, \ldots, v_{i_k}\}$ span a simplex in $\Delta_{\mathbf{L}}$. In particular, for $1 \leq i \leq d$, let $\sigma_i = \langle v_{i_1}, \ldots, v_{i_n} \rangle$ be the $(n-1)$-dimensional simplex of $\Delta_{\mathbf{L}}$ corresponding to a vertex $\mathbf{p}_{\sigma_i} := L_{i_1} \cap \cdots \cap L_{i_n}$ of $\mathcal{G}$.

Note that $\Delta_{\mathbf{L}}$ is pure, i.e., all maximal faces (also called facets) are of the same dimension $n-1$. Let $\Delta_{\mathbf{L}}(n-1)$ denote the set of facets of $\Delta_{\mathbf{L}}$. Then, $d = |\Delta_{\mathbf{L}}(n-1)|$ which is also equal to $|\mathcal{V}^{\Gamma}|$ the number of vertices of Γ. For simplices τ and σ in $\Delta_{\mathbf{L}}$, by $\tau \preceq \sigma$ we mean that τ is a face of σ.

We say that $\Delta_{\mathbf{L}}$ is a *shellable simplicial complex* if the following holds: There is an ordering $\sigma_1, \sigma_2, \ldots, \sigma_d$ of $\Delta_{\mathbf{L}}(n-1)$ such that if Δ_j denotes the subcomplex generated by $\sigma_1, \ldots, \sigma_j$ for each $1 \leq j \leq d$, then $\Delta_i \setminus \Delta_{i-1}$ has a unique minimal face μ_i for each $2 \leq i \leq d$. We further let $\mu_1 := \emptyset$ to be the unique minimal face of $\Delta_1 \setminus \Delta_0$ where $\Delta_0 = \emptyset$ (see [23, Section 2.1 p.79], [24, p.387]). (Also see [2, Definition 5.1.11] for other equivalent definitions of shellability of a pure simplicial complex.)

Example 7.1 In the case when $n = 1$, $\mathcal{G}$ is a 2-valent $T^*\mathbb{C}$-modeled GKM graph and the set of hyperplanes $\mathbf{L} = \{L_1, \ldots, L_m\}$ of $\mathcal{G}$ coincides with the set of vertices of $\mathcal{G}$ (see Fig. 13). Here $\mathcal{G}$ corresponds to the hyperplane arrangement of a 4-dimensional toric hyperKähler manifold. Furthermore, since $L_i \cap L_j = \emptyset$ for every $i \neq j$, $1 \leq i, j \leq m$, the associated simplicial complex $\Delta_{\mathbf{L}}$ is a pure 0-dimensional simplicial complex consisting of a vertex (or 0-dimensional simplex)

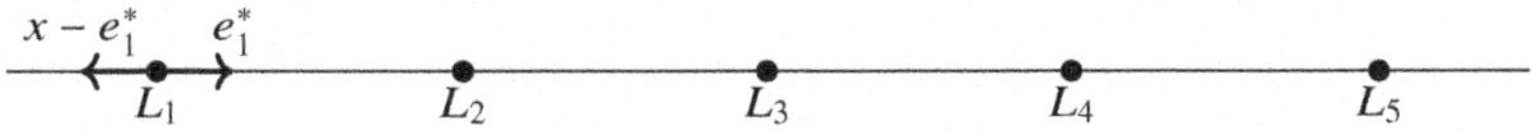

Fig. 13 The $T^*\mathbb{C}$-modeled GKM graph $\mathcal{G}$ defined from a five vertices (hyperplanes) arrangement on the line. In this case, the hyperplane L_i in $\mathcal{G}$ corresponds to the 0-dimensional simplex v_i in $\Delta_{\mathbf{L}}$ for $i = 1, \ldots, 5$. This is defined from a 4-dimensional toric hyperKähler manifold. Note that we omit the axial functions which are automatically determined

Fig. 14 The GKM graph $\mathcal{G}$ in Fig. 10. In this case, the hyperplane L_i (resp. vertex p_i) in $\mathcal{G}$ corresponds to the 0 (resp. 1)-dimensional simplex v_i (resp. σ_i) in $\Delta_{\mathbf{L}}$ for $i = 1, \ldots, 5$

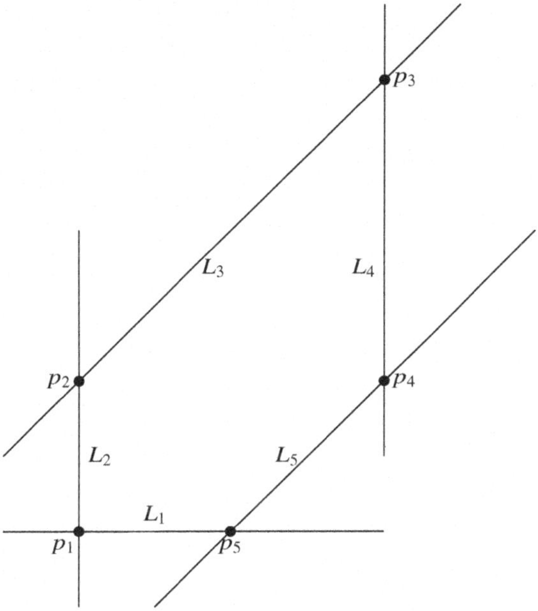

v_i corresponding to L_i for every $1 \le i \le m$. Then, $\Delta_{\mathbf{L}}$ is seen to be trivially shellable for any ordering of the set of vertices $\{v_1, \ldots, v_m\}$, which are also the facets of $\Delta_{\mathbf{L}}$ in this case.

Example 7.2 Consider the simplicial complex $\Delta_{\mathbf{L}}$ corresponding to the $T^*\mathbb{C}^2$-modeled GKM graph given in Fig. 10. We label the set of 5 hyperplanes as $L_1, \ldots, L_5$ where $L_i \cap L_{i+1} \neq \emptyset$ for $1 \le i \le 4$ and $L_5 \cap L_1 \neq \emptyset$ (see Fig. 14). Then, $\Delta_{\mathbf{L}}$ consists, respectively, of the corresponding vertices $v_1, \ldots, v_5$ and the 1-simplices $[v_1, v_2], [v_2, v_3], [v_3, v_4], [v_4, v_5], [v_5, v_1]$, which are its facets. For the ordering

$$\sigma_1 = [v_1, v_2], \sigma_2 = [v_2, v_3], \sigma_3 = [v_3, v_4], \sigma_4 = [v_4, v_5], \sigma_5 = [v_5, v_1]$$

of $\Delta_{\mathbf{L}}(1)$, we see that $\Delta_1 \setminus \Delta_0 = \{\emptyset, \{v_1\}, \{v_2\}, [v_1, v_2]\}$ has the minimal element $\mu_1 = \emptyset$, $\Delta_2 \setminus \Delta_1 = \{\{v_3\}, [v_2, v_3]\}$ has the minimal element $\mu_2 = \{v_3\}$, $\Delta_3 \setminus \Delta_2 = \{\{v_4\}, [v_3, v_4]\}$ has the minimal element $\mu_3 = \{v_4\}$, $\Delta_4 \setminus \Delta_3 = \{\{v_5\}, [v_4, v_5]\}$ has the minimal element $\mu_4 = \{v_5\}$, and $\Delta_5 \setminus \Delta_4 = \{[v_5, v_1]\}$ has the minimal element $\mu_5 = [v_5, v_1]$. Thus, $\Delta_{\mathbf{L}}$ is a shellable simplicial complex.

In Sect. 7.5, we shall show the shellability of $\Delta_{\mathbf{L}}$ for a $T^*\mathbb{C}^2$-modeled GKM graph $\mathcal{G}$ that is induced from the 8-dimensional toric hyperKähler manifold.

For $\gamma \in \Delta_{\mathbf{L}}$, let j be the smallest $1 \leq j \leq d$ such that $\gamma \preceq \sigma_j$. Then, $\gamma \in \Delta_j \setminus \Delta_{j-1}$. Thus, it follows that $\mu_j \preceq \gamma \preceq \sigma_j$. Hence, there exists a unique j, $1 \leq j \leq d$ such that $\gamma \in [\mu_j, \sigma_j]$ where $[\mu_j, \sigma_j] := \{\gamma \mid \mu_j \preceq \gamma \preceq \sigma_j\}$. In other words, we can write

$$\Delta_{\mathbf{L}} = [\mu_1, \sigma_1] \sqcup \cdots \sqcup [\mu_d, \sigma_d]. \tag{10}$$

If a simplicial complex $\Delta_{\mathbf{L}}$ satisfies (10), then it is called partitionable (see [23, p.80, Section 2.1]). In particular, shellable simplicial complexes are partitionable.

Moreover,

$$\mu_i \preceq \sigma_j \Rightarrow j \geq i. \tag{11}$$

Let

$$\mathcal{E}_{\mathbf{p}_{\sigma_i}} = \{\epsilon_{i_1}^+, \ldots, \epsilon_{i_n}^+, \epsilon_{i_1}^-, \ldots, \epsilon_{i_n}^-\}$$

for $1 \leq i \leq d$. Recall from Definition 2.4 that the set $\{\alpha(\epsilon_{i_j}^+), x \mid j = 1, \ldots, n\}$ spans $\mathfrak{t}_{\mathbb{Z}}^* \oplus \mathbb{Z}x$, i.e.,

$$\langle \alpha(\epsilon_{i_1}^+), \ldots, \alpha(\epsilon_{i_n}^+), x \rangle = \mathfrak{t}_{\mathbb{Z}}^* \oplus \mathbb{Z}x. \tag{12}$$

Let $\widetilde{\mathcal{G}}$ denote the x-forgetful graph associated with $\mathcal{G}$ where $\widetilde{\mathcal{G}} = (\Gamma, \widetilde{\alpha}, \nabla)$ having Γ and ∇ same as $\mathcal{G}$ and $\widetilde{\alpha}$ is the x-forgetful axial function defined as $\widetilde{\alpha} = F \circ \alpha : \mathcal{E} \longrightarrow H^2(BT^n)$ where $F : H^2(BT^n) \bigoplus \mathbb{Z}x \longrightarrow H^2(BT^n)$ is the x-forgetful map (see Sect. 5.2). Recall that

$$\mathbb{Z}[\widetilde{\mathcal{G}}] := \frac{\mathbb{Z}[L_1, \ldots, L_m]}{\langle \prod_{L \in \mathbf{L}'} L \mid \mathbf{L}' \in \mathbf{I}(\mathbf{L}) \rangle}$$

where $\mathbf{I}(\mathbf{L}) = \{\mathbf{L}' \subseteq \mathbf{L} \mid \bigcap_{L \in \mathbf{L}'} L = \emptyset\}$.

Let x_γ denote the monomial

$$x_\gamma := \prod_{j=1}^{p} L_{i_j}$$

where $\gamma = \langle v_{i_1}, \ldots, v_{i_p} \rangle \in \Delta_{\mathbf{L}}$.

7.2 The Characteristic Function Associated with the Hyperplane L

Definition 7.3 Let L be a connected $(2n-2)$-valent hyperplane in $\mathcal{G}$. Let H and $\overline{H}$ be the unique halfspace such that $L = H \cap \overline{H}$.

For $p \in \mathcal{V}^L$,

$$\mathcal{E}_p^L = \{\epsilon_1^+, \ldots, \epsilon_{n-1}^+, \epsilon_1^-, \ldots, \epsilon_{n-1}^-\}$$

is the $(n-1)$-pairs and

$$\mathcal{E}_p^\Gamma = \{\epsilon_1^+, \ldots, \epsilon_n^+, \epsilon_1^-, \ldots, \epsilon_n^-\}$$

is the n-pairs so that $n_H(p) = \epsilon_n^+$ and $n_{\overline{H}}(p) = \epsilon_n^-$.

By (12), the axial functions $\widetilde{\alpha}(\epsilon_1^+), \widetilde{\alpha}(\epsilon_2^+), \ldots, \widetilde{\alpha}(\epsilon_n^+)$ form a basis for $(\mathfrak{t}_{\mathbb{Z}}^n)^*$. The *characteristic function* associated with L is defined as the unique element $\lambda(L) \in \mathfrak{t}_{\mathbb{Z}}^n$ such that

$$\langle \widetilde{\alpha}(\epsilon_i^+), \lambda(L) \rangle = \delta_{i,n}.$$

Lemma 7.4 *The definition of $\lambda(L)$ is independent of the choice of a vertex $p \in \mathcal{V}^L$.*

Proof Let $\epsilon = pq \in \mathcal{E}^L$, in particular let $\epsilon = \epsilon_j^+$ for some $1 \leq j \leq n-1$. Here $i(\epsilon) = p$ and $t(\epsilon) = q$. Then under the connection ∇_ϵ the set of edges $\mathcal{E}_p^\Gamma$ around p maps, maps bijectively onto the set of edges $\mathcal{E}_q^\Gamma$ around q. Since the hyperplane L is closed under the connection ∇, $\mathcal{E}_p^L$ maps bijectively onto $\mathcal{E}_q^L$. Moreover, since a halfspace H is closed under ∇, it follows that $\nabla_\epsilon(n_H(p)) = n_H(q)$ so that $\widetilde{\alpha}(\nabla_\epsilon(n_H(p))) \equiv \widetilde{\alpha}(n_H(q)) \mod \widetilde{\alpha}(\epsilon)$. Moreover, since $\epsilon \in \mathcal{E}^L$ and $\nabla_\epsilon(\epsilon) = \overline{\epsilon}$ by definition of ∇_ϵ it follows that the elements $\widetilde{\alpha}(\epsilon_1^+), \ldots, \widetilde{\alpha}(\epsilon_{n-1}^+)$ and $\widetilde{\alpha}(\nabla_\epsilon(\epsilon_1^+)), \ldots, \widetilde{\alpha}(\nabla_\epsilon(\epsilon_{n-1}^+))$ span the same subspace of $(\mathfrak{t}_{\mathbb{Z}}^n)^*$. Further, since $\epsilon \in \mathcal{E}^L$ and $\langle \widetilde{\alpha}(\epsilon), \lambda(L) \rangle = 0$, by the congruence relation we have

$$\langle \widetilde{\alpha}(\nabla_\epsilon(\epsilon_n^+)), \lambda(L) \rangle = \langle \widetilde{\alpha}(\epsilon_n^+), \lambda(L) \rangle.$$

Thus, for the n-pairs $\mathcal{E}_q^\Gamma = \{\nabla_\epsilon(\epsilon_1^+), \ldots, \nabla_\epsilon(\epsilon_n^+), \nabla_\epsilon(\epsilon_1^-), \ldots, \nabla_\epsilon(\epsilon_n^-)\}$, we have

$$\langle \lambda(L), \widetilde{\alpha}(\nabla_\epsilon(\epsilon_i^+)) \rangle = \delta_{i,n}.$$

Hence, without loss of generality, we could have started with the vertex $q \in \mathcal{V}^L$ to define $\lambda(L)$. Moreover, since L is connected, by repeating the above procedure for an edge ϵ' such that $i(\epsilon') = q$, it follows that the definition of $\lambda(L)$ is independent of the choice of $p \in \mathcal{V}^L$. □

7.3 The $H^*(BT^n)$-Algebra Structure of $H^*(\widetilde{\mathcal{G}})$

Since $H^*(\widetilde{\mathcal{G}}) \subset \bigoplus_{p\in\mathcal{V}} H_T^*(p) \simeq \bigoplus_{p\in\mathcal{V}} H^*(BT^n)$, the ring $H^*(\widetilde{\mathcal{G}})$ may be regarded as the $H^*(BT^n)$-submodule of $\bigoplus_{p\in\mathcal{V}} H^*(BT^n)$. In Theorem 7.6 of this section, which is the second main theorem of this paper, we determine module generators of $H^*(\widetilde{\mathcal{G}})$ as a $H^*(BT^n)$-module. For this purpose, we begin with the following lemma (also see [20] for the corresponding statement on torus graphs).

Lemma 7.5

(i) The $H^*(BT^n)$-module structure on $H^*(\widetilde{\mathcal{G}})$ is obtained from the following map from $H^2(BT^n)$ to $H^*(\widetilde{\mathcal{G}})$:

$$H^2(BT^n) \ni u \mapsto \sum_{i=1}^{m} \langle u, \lambda(L_i)\rangle \cdot \tau_{L_i} \in H^*(\widetilde{\mathcal{G}}).$$

Moreover, $\Psi' : H^*(\widetilde{\mathcal{G}}) \to \mathbb{Z}[\widetilde{\mathcal{G}}]$ is an isomorphism of $H^*(BT^n)$-algebras, where the algebra structure on $\mathbb{Z}[\widetilde{\mathcal{G}}]$ is obtained by sending $u \in H^2(BT^n)$ to the element $\sum_{i=1}^{m} \langle u, \lambda(L_i)\rangle \cdot L_i \in \mathbb{Z}[\widetilde{\mathcal{G}}]$.

(ii) We therefore have the following presentation for $H^*(\widetilde{\mathcal{G}})$ as an $H^*(BT^n)$-algebra:

$$H^*(\widetilde{\mathcal{G}}) \simeq \frac{H^*(BT^n)[L_1, \ldots, L_m]}{\langle \prod_{L\in\mathbf{L}'} L \mid \mathbf{L}' \in \mathbf{I}(\mathbf{L})\ ;\ \sum_{i=1}^{m} \langle u, \lambda(L_i)\rangle \cdot L_i - u,\ \forall\, u \in H^2(BT^n)\rangle}$$

Proof Let $p \in \mathcal{V}^L$ where $p = L_{i_1} \cap \cdots \cap L_{i_n}$. Then, by Sect. 5.1, we have

$$\sum_{i=1}^{m} \langle u, \lambda(L_i)\rangle \cdot \tau_{L_i}(p) = \sum_{j=1}^{n} \langle u, \lambda(L_{i_j})\rangle \cdot \widetilde{\alpha}(n_{H_{i_j}}(p)). \tag{13}$$

Note that $n_{H_{i_j}}(p) = \epsilon_j^+(p)$ for $1 \le j \le n$ so that $\widetilde{\alpha}(n_{H_{i_j}}(p))$ for $1 \le j \le n$ form a basis of $\mathfrak{t}_{\mathbb{Z}}^*$. Since $\lambda(L_{i_j}) \in \mathfrak{t}_{\mathbb{Z}}$ for $1 \le j \le n$ is the corresponding dual basis, the right-hand side of (13) is nothing but u. Thus,

$$\left(\sum_{i=1}^{m} \langle u, \lambda(L_i)\rangle \cdot \tau_{L_i}\right)(p) = u \quad \text{for every} \quad p \in \mathcal{V}. \tag{14}$$

Since $p \in \mathcal{V}$ was arbitrary from Sect. 5.3, it follows that the $H^*(BT^n)$-algebra structure defined above is canonical corresponding to the diagonal inclusion of $H^*(BT^n)$ in $\bigoplus_{i=1}^d H^*(BT^n) = (H^*(BT^n))^d$.

Finally, by definition of Ψ' in Sect. 5.2, we also have that the $H^*(BT^n)$-algebra structure on $\mathbb{Z}[\widetilde{\mathcal{G}}]$ is obtained as in the statement. □

The following theorem is the second main theorem in this paper.

Theorem 7.6 *Let $\Delta_{\mathbf{L}}$ be the simplicial complex defined by the hyperplanes $\{L_1, \ldots, L_m\}$ of $\widetilde{\mathcal{G}}$. Suppose that $\Delta_{\mathbf{L}}$ is a shellable simplicial complex with respect to ordering $\sigma_1, \ldots, \sigma_d$ of $\Delta_{\mathbf{L}}(n)$. In particular, $\Delta_{\mathbf{L}}$ is partitionable with partition* (10) *and* (11) *holds. Then, the following statements hold:*

(i) For $\gamma \in \Delta_{\mathbf{L}}$, let $x_\gamma = L_{j_1} \cdots L_{j_p} \in \mathbb{Z}[\widetilde{\mathcal{G}}]$, where $\gamma = \langle v_{j_1}, \ldots, v_{j_p} \rangle$. Then, there exists an element $u \in H^2(BT^n)$ such that

$$L_{j_1} \cdot x_\gamma = -\sum_k \langle u, \lambda(L_{j_k}) \rangle \cdot x_{\gamma_k} + u \cdot x_\gamma$$

where k runs through $1 \leq k \leq m$ such that $k \notin \{j_1, \ldots, j_n\}$ and $\gamma_k = \langle v_k, v_{j_1}, \ldots, v_{j_p} \rangle$.

(ii) Let $\eta \preceq \gamma \preceq \theta$ be simplices in $\Delta_{\mathbf{L}}$. Then, we can write

$$x_\gamma = \sum_k c_k \cdot x_{\eta_k} + c \cdot x_\eta$$

for $c_k, c \in H^*(BT^n)$ and $\eta_k \npreceq \theta$.

(iii) The monomials x_{μ_i} for $1 \leq i \leq d$ form a basis of $\mathbb{Z}[\widetilde{\mathcal{G}}]$ as $H^*(BT^n)$-module.

(iv) Let $f \in \mathbb{Z}[\widetilde{\mathcal{G}}]$ and

$$f = \sum_{j=1}^d a_j \cdot x_{\mu_j} \tag{15}$$

for unique $a_j \in H^*(BT^n)$. Let $i = i(f)$ be the smallest $1 \leq i \leq d$ such that $a_i \neq 0$. Then, we can determine the coefficients a_j, $j \geq i$ iteratively as follows: We have $a_i = \frac{\rho_{\mathbf{p}_{\sigma_i}}(f)}{\rho_{\mathbf{p}_{\sigma_i}}(x_{\mu_i})}$. Suppose $a_i, a_{i+1}, \ldots, a_{j-1}$ are determined by induction, then

$$a_j = \frac{\rho_{\mathbf{p}_{\sigma_j}}(f - \sum_{k=i}^{j-1} a_k \cdot x_{\mu_k})}{\rho_{\mathbf{p}_{\sigma_j}}(x_{\mu_j})}.$$

Proof

(i) Let $\sigma = \langle v_{j_1}, \ldots, v_{j_n}\rangle$ be an $(n-1)$-simplex containing γ, and let $\mathbf{p}_\sigma = L_{j_1} \cap \cdots \cap L_{j_n}$ be the corresponding vertex in $\widetilde{\mathcal{G}}$. Consider $\lambda(L_{j_1}) = e_{j_1} \in \mathfrak{t}^n_{\mathbb{Z}}$ which is dual to $u := \widetilde{\alpha}(\epsilon^+_{j_1})$. In this case, by Lemma 7.5, we have that

$$u = \sum_{i=1}^{m} \langle u, \lambda(L_i)\rangle \cdot L_i.$$

Since $\langle u, e_{j_i}\rangle = \delta_{i,1}$ for $i = 1, \ldots, n$, we have the relation

$$L_{j_1} = -\sum_k \langle u, \lambda(L_k)\rangle \cdot L_k + u \tag{16}$$

as an $H^*(BT^n)$-module, where the sum on the right-hand side is over those $1 \le k \le m$ such that $k \notin \{j_1, \ldots, j_n\}$. Multiplying (16) by x_γ, we get

$$L_{j_1} \cdot x_\gamma = -\sum \langle u, \lambda(L_k)\rangle \cdot x_{\gamma_k} + u \cdot x_\gamma \tag{17}$$

where $\gamma_k \in \Delta_{\mathbf{L}}$ is spanned by the vertices $v_k, v_{j_1}, \ldots, v_{j_p}$ and $x_{\gamma_k} = L_k \cdot L_{j_1} \cdots L_{j_p}$. This proves (i)

(ii) Let $\eta \prec \gamma \preceq \theta$ be simplices in $\Delta_{\mathbf{L}}$. Let $\gamma = \langle v_{j_1}, \ldots, v_{j_p}\rangle$, $\eta = \langle v_{j_1}, \ldots, v_{j_r}\rangle \in \Delta_{\mathbf{L}}$ and $\theta = \langle v_{j_1}, \ldots, v_{j_l}\rangle$ and $r < p \le l$. Because the following argument can also apply for any such l, we can assume that $l = n$, i.e., θ is $(n-1)$-dimensional. Thus, $\lambda(L_{j_1}), \ldots, \lambda(L_{j_n})$ is the basis of $\mathfrak{t}_{\mathbb{Z}}$ dual to the basis $\widetilde{\alpha}(\epsilon_{j_1}), \ldots, \widetilde{\alpha}(\epsilon_{j_n})$ of $\mathfrak{t}^*_{\mathbb{Z}}$. Let $u := \widetilde{\alpha}(\epsilon_{j_p}) \in H^*(BT^n)$. Thus, with the similar reason to obtain (16) as in the proof of (i), we have the following relation in $\mathbb{Z}[\widetilde{\mathcal{G}}]$:

$$L_{j_p} + \sum_k \langle u, \lambda(L_k)\rangle \cdot L_k - u = 0 \tag{18}$$

where k runs through $\{1, \ldots, m\} \setminus \{j_1, \ldots, j_n\}$ in the equation. Multiplying (18) by $L_{j_1} \cdots L_{j_{p-1}}$, we get

$$x_\gamma + \sum_k \langle u, \lambda(L_k)\rangle \cdot x_{\eta_k} - u \cdot x_{\gamma'} = 0 \tag{19}$$

where $\eta_k = \langle v_k, v_{j_1}, \ldots, v_{j_{p-1}}\rangle$, $\gamma' = \langle v_{j_1}, \ldots, v_{j_{p-1}}\rangle \in \Delta_{\mathbf{L}}$. Note that $\eta_k \not\preceq \theta$ since $k \notin \{j_1, \ldots, j_n\}$. Also $\eta \preceq \eta_k$ and $\eta \preceq \gamma'$, since $r \le p-1$. Therefore, $\eta \preceq \gamma' \prec \gamma \preceq \theta$. Now, proceeding by downward induction on p and repeating the above arguments for γ', we arrive at (ii).

(iii) By the ring structure of $\mathbb{Z}[\widetilde{\mathcal{G}}]$ defined in Sect. 5.2, for every element in $\mathbb{Z}[\widetilde{\mathcal{G}}]$ can be written by the sum of x_γ's for $\gamma \in \Delta_{\mathbf{L}}$ with $H^*(BT^n)$-coefficients. Therefore, for every $\gamma \in \Delta_{\mathbf{L}}$, it suffices to show that x_γ lies in the $H^*(BT^n)$-submodule of $\mathbb{Z}[\widetilde{\mathcal{G}}]$ spanned by x_{μ_i} for $1 \le i \le d$, where μ_i is the minimal face, which appears in $\Delta_{\mathbf{L}} = [\mu_1, \sigma_1] \sqcup \cdots \sqcup [\mu_d, \sigma_d]$. Since $\Delta_{\mathbf{L}}$ is a shellable simplicial complex, for every $\gamma \in \Delta_{\mathbf{L}}$ there exists the unique $1 \le i \le d$ such that $\mu_i \preceq \gamma \preceq \sigma_i$ (see (11)). We prove (iii) by downward induction on i.

If $\gamma \in [\mu_d, \sigma_d]$, we are done since $\mu_d = \sigma_d = \gamma$, i.e., $x_\gamma = x_{\mu d}$ and hence lies in the $H^*(BT^n)$-span of x_{μ_d}. Assume that for every $\gamma \in [\mu_{i+1}, \sigma_{i+1}] \sqcup \cdots \sqcup [\mu_d, \sigma_d]$, $x_\gamma \in H^*(BT^n)x_{\mu_{i+1}} \oplus \cdots \oplus H^*(BT^n)x_{\mu_d}$. If $\gamma \in [\mu_i, \sigma_i]$, then by (ii) we can write

$$x_\gamma = \sum_{\mu_i \prec \gamma_j \not\preceq \sigma_i} c_j \cdot x_{\gamma_j} + c \cdot x_{\mu_i} \tag{20}$$

for $c_j, c \in H^*(BT^n)$. Now there is the unique r such that $\mu_r \preceq \gamma_j \preceq \sigma_r$. This implies by (11) that $r > i$.

Thus, by induction assumption x_{γ_j} lies in the $H^*(BT^n)$-span of x_{μ_q} for $q \ge r$. This together with (20) implies that x_γ lies in the $H^*(BT^n)$-span of x_{μ_q} for $q \ge i$.

It remains now to show that x_{μ_i} for $1 \le i \le d$ are linearly independent. Suppose that there exist $a_i \in H^*(BT^n)$ for $1 \le i \le d$ such that

$$\sum_{i=1}^{d} a_i \cdot x_{\mu_i} = 0 \tag{21}$$

in $\mathbb{Z}[\widetilde{\mathcal{G}}]$. Let $i \in \{1, \ldots, n\}$ be the smallest integer such that $a_i \neq 0$.

Recall that $\sigma_i = \langle v_{i_1}, \ldots, v_{i_n} \rangle$, where $\mathbf{p}_{\sigma_i} = L_{i_1} \cap \cdots \cap L_{i_n}$ in $\mathcal{G}$. Consider the localization map $\rho = (\rho_{\mathbf{p}_{\sigma_j}})_{j=1}^{d}$ defined in Sect. 5.3. By (11) and the definition of $\rho_{\mathbf{p}_{\sigma_i}}$, it follows that $\rho_{\mathbf{p}_{\sigma_i}}(x_{\mu_j}) = 0$ for $j > i$ (since $\mu_j \not\preceq \sigma_i$, there exists L_k in $\mathbf{L}$ such that the corresponding vertex $v_k \in \mu_j$ and $v_k \notin \sigma_i$ in $\Delta_{\mathbf{L}}$. Thus, $\rho_{\mathbf{p}_{\sigma_i}}(L_k) = 0$ in $\mathbb{Z}[\widetilde{\mathcal{G}}]_{\mathbf{p}_{\sigma_i}}$). Thus, applying $\rho_{\mathbf{p}_{\sigma_i}}$ on (21), we get

$$\rho_{\mathbf{p}_{\sigma_i}}(\sum_{j=1}^{n} a_j \cdot x_{\mu_j}) = \rho_{\mathbf{p}_{\sigma_i}}(\sum_{j \ge i} a_j \cdot x_{\mu_j}) = \rho_{\mathbf{p}_{\sigma_i}}(a_i \cdot x_{\mu_i}) = \rho_{\mathbf{p}_{\sigma_i}}(a_i) \cdot \rho_{\mathbf{p}_{\sigma_i}}(x_{\mu_i}) = 0$$

in the integral domain $\mathbb{Z}[\widetilde{\mathcal{G}}]_{\mathbf{p}_{\sigma_i}} \simeq \mathbb{Z}[L_{i_1}, \ldots, L_{i_n}]$. Since $\rho_{\mathbf{p}_{\sigma_i}}(x_{\mu_i})$ is the monomial $L_{i_{j_1}}, \ldots, L_{i_{j_p}}$, where $\mu_i = \langle v_{i_{j_1}}, \ldots, v_{i_{j_p}} \rangle$, and hence a nonzero element of $\mathbb{Z}[L_{i_1}, \ldots, L_{i_n}]$, we get that $\rho_{\mathbf{p}_{\sigma_i}}(a_i) = 0$. Moreover, ρ can be seen to be the diagonal embedding when restricted to the subalgebra $H^*(BT^n)$ of $\mathbb{Z}[\widetilde{\mathcal{G}}]$ ($u \in H^*(BT^n)$ is equal to $\sum_{j=1}^{m} \langle u, \lambda(L_j) \rangle \cdot L_j \in \mathbb{Z}[\widetilde{\mathcal{G}}]$ maps to

$\sum_{j=1}^{n}\langle u, \lambda(L_{i_j})\rangle \cdot L_{i_j} \in \mathbb{Z}[L_{i_1}, \ldots, L_{i_n}]$, which is identified with $\sum_{j=1}^{n}\langle u, \lambda(L_{i_j})\rangle \cdot \tau_{L_{i_j}}$ in $H^*(BT^n)$ (see Lemma 5.2), which is equal to u see (14)), and ρ is injective by Lemma 5.3, which implies that $a_i = 0$. This contradicts our original assumption that $a_i \neq 0$. Thus, we cannot have a relation of the type (21) in $\mathbb{Z}[\widetilde{\mathcal{G}}]$ unless $a_i = 0$ in $H^*(BT^n)$ for each $1 \leq i \leq d$. Hence, we conclude that x_{μ_i} for $1 \leq i \leq d$ are linearly independent in $\mathbb{Z}[\widetilde{\mathcal{G}}]$. This proves (iii).

(iv) As for the proof of linear independence of $x_{\mu_i} 1 \leq i \leq d$, our idea is to again use the localization map $\rho = (\rho_{\mathbf{p}_{\sigma_i}}) : \mathbb{Z}[\widetilde{\mathcal{G}}] \longrightarrow \bigoplus_{i=1}^{d} \mathbb{Z}[\widetilde{\mathcal{G}}]_{\mathbf{p}_{\sigma_i}} (\simeq \bigoplus_{i=1}^{d} H^*(BT^n))$ defined in Sect. 5.3. We know that ρ is injective and by (11) and the definition of $\rho_{\mathbf{p}_{\sigma_i}}$ we have $\rho_{\mathbf{p}_{\sigma_i}}(x_{\mu_j}) = 0$ for $j > i$. Also ρ is the diagonal map when restricted to the subalgebra $H^*(BT^n)$ of $\mathbb{Z}[\widetilde{\mathcal{G}}]$. In particular, this implies that $\rho_{\mathbf{p}_{\sigma_j}}(a_k) = a_k$ for $1 \leq j, k \leq d$.

Applying $\rho_{\mathbf{p}_{\sigma_i}}$ on (15), we get

$$\rho_{\mathbf{p}_{\sigma_i}}(f) = \rho_{\mathbf{p}_{\sigma_i}}(\sum_{j=1}^{d} a_j \cdot x_{\mu_j}) = \rho_{\mathbf{p}_{\sigma_i}}(\sum_{j \geq i} a_j \cdot x_{\mu_j}) = a_i \cdot \rho_{\mathbf{p}_{\sigma_i}}(x_{\mu_i})$$

in the unique factorization domain $\mathbb{Z}[L_{i_1}, \ldots, L_{i_n}]$, where $\mathbf{p}_{\sigma_i} = L_{i_1} \cap \cdots \cap L_{i_n}$. Thus, $\rho_{\mathbf{p}_{\sigma_i}}(f)$ is divisible by the irreducible elements $L_{i_{j_1}}, \ldots, L_{i_{j_p}}$ and hence by the monomial $\rho_{\mathbf{p}_{\sigma_i}}(x_{\mu_i}) = L_{i_{j_1}} \cdots L_{i_{j_p}}$ in $\mathbb{Z}[L_{i_1}, \ldots, L_{i_n}]$. Thus, $a_i = \frac{\rho_{\mathbf{p}_{\sigma_i}}(f)}{\rho_{\mathbf{p}_{\sigma_i}}(x_{\mu_i})} \in \mathbb{Z}[L_{i_1}, \ldots, L_{i_n}]$. Now, let

$$f_1 := f - a_{i(f)} \cdot x_{\mu_{i(f)}} \in \mathbb{Z}[\widetilde{\mathcal{G}}].$$

Then, $f_1 = \sum_{j > i(f)} a_j \cdot x_{\mu_j}$. Moreover, now putting $i = i(f_1)$ and repeating the above argument given for determining $a_{i(f)}$, we get

$$a_{i(f_1)} = \frac{\rho_{\mathbf{p}_{\sigma_{i(f_1)}}}(f_1)}{\rho_{\mathbf{p}_{\sigma_{i(f_1)}}}(x_{\mu_{i(f_1)}})}$$

in $\mathbb{Z}[\widetilde{\mathcal{G}}]_{\mathbf{p}_{\sigma_{i(f_1)}}} \simeq H^*(BT^n)$. Proceeding similarly after k steps, we get

$$f_k = f - \sum_{i(f_1)\le j<i(f_k)} a_j \cdot x_{\mu_j} = \sum_{j>i(f_{k-1})} a_j \cdot x_{\mu_j}.$$

Putting $i = i(f_k)$ to be the smallest index in $\{i(f_{k-1}), i(f_{k-1})+1, \ldots, d\}$ such that $a_{i(f_k)} \neq 0$ and following similar arguments as above, we get

$$a_{i(f_k)} = \frac{\rho_{\mathbf{p}_{\sigma_{i(f_k)}}}(f_k)}{\rho_{\mathbf{p}_{\sigma_{i(f_k)}}}(x_{\mu_{i(f_k)}})}$$

in $\mathbb{Z}[\widetilde{\mathcal{G}}]_{\mathbf{p}_{\sigma_{i(f_k)}}} \simeq H^*(BT^n)$. This proves (iv).

□

7.4 The Ordinary Cohomology Ring of $\widetilde{\mathcal{G}}$ and an Example

In geometry, if the equivariant cohomology $H_T^*(M;\mathbb{Z})$ has the structure of a free $H^*(BT)$-algebra, then we can compute its ordinary cohomology $H^*(M;\mathbb{Z})$ by $H_T^*(M;\mathbb{Z}) \otimes_{H^*(BT)} \mathbb{Z}$. By Theorem 7.6, we know that $H^*(\widetilde{\mathcal{G}})$ is a free $H^*(BT)$-algebra of rank d. So we may define the "ordinary" cohomology of $\widetilde{\mathcal{G}}$ by $H^*(\widetilde{\mathcal{G}}) \otimes_{H^*(BT)} \mathbb{Z}$; we denote it by $H^*_{ord}(\widetilde{\mathcal{G}})$. The precise computation of $H^*_{ord}(\widetilde{\mathcal{G}})$ is given as the following corollary.

Corollary 7.7

(i) *The following is the presentation:*

$$H^*_{ord}(\widetilde{\mathcal{G}}) \simeq \frac{\mathbb{Z}[L_1, \ldots, L_m]}{\langle \prod_{L\in\mathbf{L}'} L \mid \mathbf{L}' \in \mathbf{I}(\mathbf{L})\ ;\ \sum_{i=1}^m \langle u, \lambda(L_i)\rangle \cdot L_i,\ \ \forall\, u \in H^2(BT^n)\rangle}$$

for the ordinary cohomology ring $H^*_{ord}(\widetilde{\mathcal{G}})$ as a $\mathbb{Z}$-algebra.

(ii) *The monomials x_{μ_i}, $1 \le i \le d$ form a $\mathbb{Z}$-basis for $H^*_{ord}(\widetilde{\mathcal{G}})$.*

Proof

(i) The $H^*(BT^n)$-algebra structure on $H^*(\widetilde{\mathcal{G}})$ is given by Lemma 7.5 and $\mathbb{Z}$ has $H^*(BT)$-algebra structure given by augmentation, which sends each $u \in H^2(BT^n)$ to 0. Since $H^*(\widetilde{\mathcal{G}})$ is free, the corollary now follows from Lemma 7.5 due to the $H^*(BT)$-algebra isomorphism Ψ' of $\mathbb{Z}[\widetilde{\mathcal{G}}]$ with $H^*(\widetilde{\mathcal{G}})$.

(ii) This follows by Theorem 7.6 (iii) and by the isomorphism

$$H^*_{ord}(\widetilde{\mathcal{G}}) \simeq \mathbb{Z}[\widetilde{\mathcal{G}}] \otimes_{H^*(BT^n)} \mathbb{Z}.$$

□

7.5 $H^*(\widetilde{\mathcal{G}})$ for $\mathcal{G}$ Induced From the 8-Dimensional Toric HyperKähler Manifold

By using the fundamental theorem of toric hyperKähler manifolds in [1], the 8-dimensional toric hyperKähler manifold M is completely classified up to equivariant diffeomorphism by the hyperplane arrangement $\mathcal{L}_{k,l,m}$ in $\mathbb{R}^2$ consisting of k horizontal lines $\{Hor_1, \ldots, Hor_k\}$ which is ordered from the bottom, l virtical lines $\{Vir_1, \ldots, Vir_l\}$ which is ordered from the left, and m diagonal lines $\{Dia_1, \ldots, Dia_m\}$ which is ordered from the left in $\mathbb{R}^2$ (also see Fig. 16).

It is easy to check that every set of hyperplanes in $\mathcal{L}_{k,l,m}$ have the non-empty intersections except the following cases:

$$Hor_r \cap Vir_s \cap Dia_t = \emptyset \quad \text{for } 1 \le r \le k,\ 1 \le s \le l \text{ and } 1 \le t \le m$$

and

$$Hor_i \cap Hor_j = \emptyset \quad \text{for } 1 \le i, j \le k;$$
$$Vir_r \cap Vir_s = \emptyset \quad \text{for } 1 \le r, s \le l;$$
$$Dia_p \cap Dia_q = \emptyset \quad \text{for } 1 \le p, q \le m.$$

This hyperplane arrangement $\mathcal{L}_{k,l,m}$ induces the $T^*\mathbb{C}^2$-modeled GKM graph $\mathcal{G}$. We can see that the characteristic functions associated with the hyperplanes are given by $\lambda(Hor_r) = e_1$ for all $1 \le r \le k$, $\lambda(Vir_s) = e_2$ for all $1 \le s \le l$ and $\lambda(Dia_p) = -e_1 - e_2$ for all $1 \le p \le m$, where $H_2(BT^2) = \mathbb{Z} \cdot e_1 \bigoplus \mathbb{Z} \cdot e_2$. Therefore, the x-forgetful graph $\widetilde{\mathcal{G}}$ is given by Fig. 15.

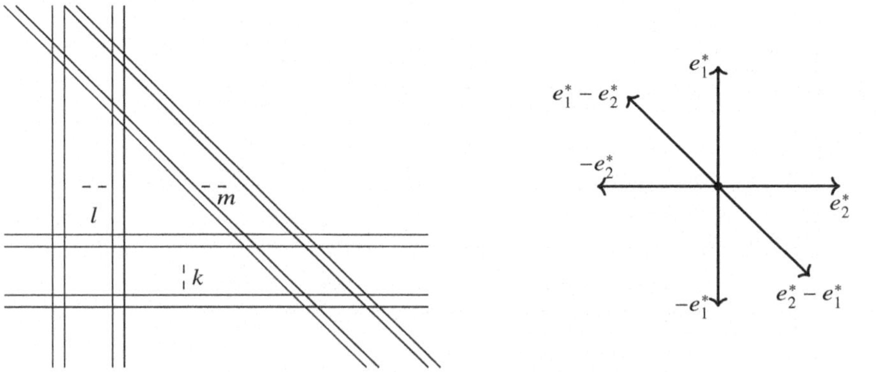

Fig. 15 The x-forgetful graph induced from $\mathcal{L}_{k,l,m}$. The axial functions on four edges around each vertex (each intersection of two lines) are defined by choosing the labels in the right figure for each direction, where $\{e_1^*, e_2^*\} \subset \mathfrak{t}_{\mathbb{Z}}^*$ is the dual basis of $\{e_1, e_2\} \subset \mathfrak{t}_{\mathbb{Z}}$. For example, Fig. 11 is the x-forgetful graph of $\mathcal{L}_{1,1,1}$

Consider the polynomial ring

$$R := \mathbb{Z}[X_1, \ldots, X_k, Y_1, \ldots, Y_l, Z_1, \ldots, Z_m]$$

in $k+l+m$ variables. Let I be the ideal in R generated by the following monomials:

$$X_r Y_s Z_t \quad \text{for } 1 \leq r \leq k,\ 1 \leq s \leq l \text{ and } 1 \leq t \leq m;$$

$$X_i X_j \quad \text{for } 1 \leq i \neq j \leq k;$$

$$Y_r Y_s \quad \text{for } 1 \leq r \neq s \leq l;$$

$$Z_p Z_q \quad \text{for } 1 \leq p \neq q \leq m.$$

It follows from Theorem 5.1 that R/I is isomorphic to $H^*(\widetilde{\mathcal{G}})$ under the map, which is defined by the following correspondences:

$$X_r \mapsto \tau_{Hor_r} \quad \text{for } 1 \leq r \leq k;$$

$$Y_s \mapsto \tau_{Vir_s} \quad \text{for } 1 \leq s \leq l;$$

$$Z_t \mapsto \tau_{Dia_t} \quad \text{for } 1 \leq t \leq m.$$

Now we determine the structure of $H^*(\widetilde{\mathcal{G}})$ as an $H^*(BT^2)$-algebra.

Let $u \in H^2(BT^2) = \mathbb{Z} \cdot e_1^* \bigoplus \mathbb{Z} \cdot e_2^*$ and $u = a \cdot e_1^* + b \cdot e_2^*$. Then, under the $H^*(BT^2)$-algebra structure on $\mathbb{Z}[\widetilde{\mathcal{G}}]$, u corresponds to the element

$$\sum_{r=1}^{k} a \cdot X_r + \sum_{s=1}^{l} b \cdot Y_s - \sum_{t=1}^{m} (a+b) \cdot Z_t$$

$$= a \cdot (X_1 + \cdots + X_k - Z_1 - \cdots - Z_m) + b \cdot (Y_1 + \cdots + Y_l - Z_1 - \cdots - Z_m).$$

Let $\mathcal{R} := H^*(BT^2)[X_1, \ldots, X_k, Y_1, \ldots, Y_l, Z_1, \ldots, Z_m]$ and $\mathcal{I}$ be the ideal in $\mathcal{R}$ generated by the monomials generating the ideal I in R, together with the following two linear polynomials:

$$X_1 + \cdots + X_k - Z_1 - \cdots - Z_m - e_1^*; \quad Y_1 + \cdots + Y_l - Z_1 - \cdots - Z_m - e_2^*.$$

Then, it follows from Lemma 7.5 that the ring $\mathcal{R}/\mathcal{I}$ is isomorphic to $H^*(\widetilde{\mathcal{G}})$ as an $H^*(BT^2)$-algebra.

We now note that the simplicial complex $\Delta_{\mathcal{L}}$ dual to the hyperplane arrangement $\mathcal{L}$ has vertices $u_1, \ldots, u_k$ corresponding to the hyperplanes $Hor_1, \ldots, Hor_k$, $v_1, \ldots, v_l$ corresponding to the hyperplanes $Vir_1, \ldots, Vir_l$, and $w_1, \ldots, w_m$ corresponding to the hyperplanes $Dia_1, \ldots Dia_m$.

Moreover, $\Delta_{\mathcal{L}}$ is a 1-dimensional simplicial complex, where the number of 1-simplices, i.e., the vertices of the x-forgetful graph, in $\Delta_{\mathcal{L}}$ is $kl + km + lm$. We can see that $\Delta_{\mathcal{L}}$ is shellable with the following shelling order of the 1-dimensional simplices:

$$\sigma_1 = [u_1, v_1] < \sigma_2 = [u_1, v_2] < \cdots < \sigma_l = [u_1, v_l]$$
$$< \sigma_{l+1} = [u_1, w_1] < \sigma_{l+2} = [u_1, w_2] < \cdots < \sigma_{l+m} = [u_1, w_m]$$
$$< \sigma_{l+m+1} = [u_2, v_1] < \sigma_{l+m+2} = [u_2, v_2] < \cdots < \sigma_{2l+m} = [u_2, v_l]$$
$$< \sigma_{2l+m+1} = [u_2, w_1] < \sigma_{2l+m+2} = [u_2, w_2] < \cdots < \sigma_{2l+2m} = [u_2, w_m]$$
$$\vdots$$
$$< \sigma_{(k-1)\cdot l+(k-1)\cdot m+1} = [u_k, v_1] < \sigma_{(k-1)\cdot l+(k-1)\cdot m+2} = [u_k, v_2] < \cdots < \sigma_{kl+(k-1)\cdot m} = [u_k, v_l]$$
$$< \sigma_{kl+(k-1)\cdot m+1} = [u_k, w_1] < \sigma_{kl+(k-1)\cdot m+2} = [u_k, w_2] < \cdots < \sigma_{kl+km} = [u_k, w_m]$$
$$< \sigma_{kl+km+1} = [v_1, w_1] < \sigma_{kl+km+2} = [v_1, w_2] < \cdots < \sigma_{kl+km+m} = [v_1, w_m]$$
$$\vdots$$
$$< \sigma_{kl+km+(l-1)\cdot m+1} = [v_l, w_1] < \sigma_{kl+km+(l-1)\cdot m+2} = [v_l, w_2] < \cdots < \sigma_{kl+km+lm} = [v_l, w_m].$$

For example, we give the order on vertices in $\widetilde{\mathcal{G}}$ induced from $\mathcal{L}_{2,1,2}$, i.e., 1-simplicies in $\Delta_{\mathcal{L}}$ as in Fig. 16.

In order to find the module generators of $H^*(\widetilde{\mathcal{G}})$, it is enough to find the minimal vertices of $\Delta_i \setminus \Delta_{i-1}$, where Δ_i is the subcomplex generated by $\sigma_1, \ldots, \sigma_i$ in $\Delta_{\mathcal{L}}$. For example, as a set $\Delta_1 = [u_1, v_1] = \{u_1, v_1, \sigma_1\}$ and $\Delta_2 = [u_1, v_1] \cup [u_1, v_2] = \{u_1, v_1, v_2, \sigma_1, \sigma_2\}$; therefore, $\Delta_2 \backslash \Delta_1 = \{v_2, \sigma_2\}$ such that $v_2 \prec \sigma_2$ and the minimal face is $\mu_2 := v_2$. Similarly, we obtain the following shelling:

$$\mu_1 = \emptyset, \mu_2 = \{v_2\}, \mu_3 = \{v_3\}, \ldots, \mu_l = \{v_l\},$$
$$\mu_{l+1} = \{w_1\}, \mu_{l+2} = \{w_2\}, \ldots, \mu_{l+m} = \{w_m\},$$
$$\mu_{l+m+1} = \{u_2\}, \mu_{l+m+2} = [u_2, v_2] \ldots, \mu_{2l+m} = [u_2, v_l],$$

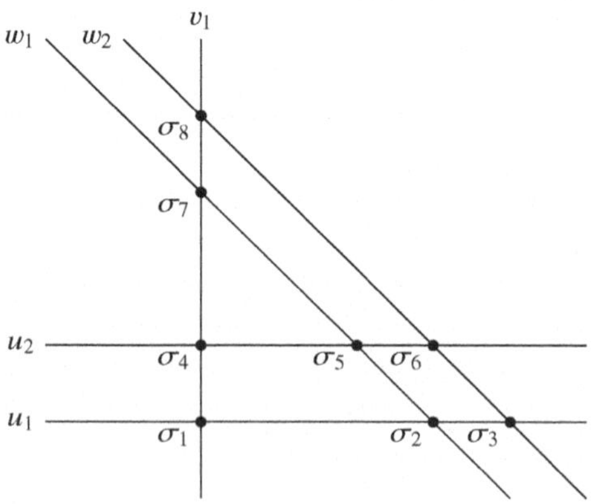

Fig. 16 Ordering the vertices of the x-forgetful graph induced from $\mathcal{L}_{2,1,2}$. This is equivalent to choose shelling of $\Delta_{\mathcal{L}}$

$$\mu_{2l+m+1} = [u_2, w_1], \mu_{2l+m+2} = [u_2, w_2] \ldots, \mu_{2l+2m} = [u_2, w_m],$$

$$\mu_{2l+2m+1} = \{u_3\}, \mu_{2l+2m+2} = [u_3, v_2], \ldots, \mu_{3l+2m} = [u_3, v_l],$$

$$\mu_{3l+2m+1} = [u_3, w_1], \mu_{3l+2m+2} = [u_3, w_2], \ldots, \mu_{3l+3m} = [u_3, w_m],$$

$$\vdots$$

$$\mu_{(k-1)\cdot l+(k-1)\cdot m+1}=\{u_k\}, \mu_{(k-1)\cdot l+(k-1)\cdot m+2}=[u_k, v_2], \ldots, \mu_{kl+(k-1)\cdot m}=[u_k, v_l],$$

$$\mu_{kl+(k-1)\cdot m+1} = [u_k, w_1], \mu_{kl+(k-1)\cdot m+2} = [u_k, w_2] \ldots, \mu_{kl+km} = [u_k, w_m]$$

$$\mu_{kl+km+1} = [v_1, w_1], \mu_{kl+km+2} = [v_1, w_2], \ldots, \mu_{kl+km+m} = [v_1, w_m],$$

$$\vdots$$

$$\mu_{kl+km+(l-1)\cdot m+1}=[v_l, w_1], \mu_{kl+km+(l-1)\cdot m+2}=[v_l, w_2], \ldots, \mu_{kl+km+lm}=[v_l, w_m].$$

By Theorem 7.6, the monomial basis for $\mathcal{R}/\mathcal{I}$ as a $H^*(BT^2)$-module is as follows:

$$1, Y_2, \ldots, Y_l, Z_1, \ldots, Z_m, X_2, \ldots, X_k,$$
$$X_2Y_2, \ldots, X_2Y_l, X_2Z_1, \ldots, X_2Z_m,$$
$$\vdots$$
$$X_kY_2, \ldots, X_kY_l, X_kZ_1, \ldots, X_kZ_m,$$
$$Y_1Z_1, \ldots, Y_1Z_m, \ldots, Y_lZ_1, \ldots, Y_lZ_m$$

For example, in the case when $k = 2$, $l = 1$ and $m = 2$, the equivariant cohomology ring $H^*(\widetilde{\mathcal{G}})$ of $\widetilde{\mathcal{G}}$ in Fig. 16 is isomorphic as $H^*(BT^2)$-algebra to

$$\frac{H^*(BT^2)[X_1, X_2, Y_1, Z_1, Z_2]}{\langle X_1X_2, Z_1Z_2, X_1Y_1Z_1, X_1Y_1Z_2, X_2Y_1Z_1, X_2Y_1Z_2; X_1+X_2-Z_1-Z_2-e_1^*, Y_1-Z_1-Z_2-e_2^*.\rangle}$$

The shelling order of $\Delta_{\mathcal{L}_{2,1,2}}$ is given by

$$\sigma_1 = [u_1, v_1] < \sigma_2 = [u_1, w_1] < \sigma_3 = [u_1, w_2] < \sigma_4 = [u_2, v_1]$$
$$< \sigma_5 = [u_2, w_1] < \sigma_6 = [u_2, w_2] < \sigma_7 = [v_1, w_1] < \sigma_8 = [v_1, w_2].$$

Here we have

$$\mu_1 = \emptyset,\ \mu_2 = \{w_1\},\ \mu_3 = \{w_2\},\ \mu_4 = \{u_2\},$$
$$\mu_5 = [u_2, w_1],\ \mu_6 = [u_2, w_2], \mu_7 = [v_1, w_1],\ \mu_8 = [v_1, w_2].$$

From Theorem 7.6(iii) we have the following basis of $H^*(\widetilde{\mathcal{G}}) \simeq \mathbb{Z}[\widetilde{\mathcal{G}}] \simeq \mathcal{R}/\mathcal{I}$ as a free $H^*(BT^2)$-module:

$$\begin{aligned} &x_{\mu_1} = 1, \quad x_{\mu_2} = Z_1, \quad x_{\mu_3} = Z_2, \quad x_{\mu_4} = X_2, \\ &x_{\mu_5} = X_2 Z_1, \quad x_{\mu_6} = X_2 Z_2, \quad x_{\mu_7} = Y_1 Z_1, \quad x_{\mu_8} = Y_1 Z_2. \end{aligned} \tag{22}$$

We shall now apply Theorem 7.6 (iv) to determine some of the multiplicative structure constants of the basis (22). We first consider Z_1^2. Let $Z_1^2 = \sum_{i=1}^{8} a_i \cdot x_{\mu_i}$. Note first that $\rho_{\mathbf{p}_{\sigma_i}}(Z_1^2) = 0$ for $i = 1, 3, 4, 6, 8$ so that $a_1 = a_3 = a_4 = a_6 = a_8 = 0$. We further see that $\rho_{\mathbf{p}_{\sigma_2}}(Z_1^2) = Z_1^2$ in $H_{T^2}^*(x_{\sigma_2}) \simeq \mathbb{Z}[X_1, Z_1]$. Also $x_{\mu_2} = Z_1$ and $\rho_{\mathbf{p}_{\sigma_2}}(Z_1) = Z_1$. Thus, $a_2 = \frac{\rho_{\mathbf{p}_{\sigma_2}}(Z_1^2)}{\rho_{\mathbf{p}_{\sigma_1}}(Z_1)} = Z_1$, which corresponds to the element $-e_2^*$ under the isomorphism $H_{T^2}^*(x_{\sigma_1}) \simeq H^*(BT^2)$. Thus, $a_2 = -e_2^*$. Proceeding as in Theorem 7.6(iv), we next consider $Z_1^2 + e_2^* \cdot Z_1$. Using the relation $e_2^* = Y_1 - Z_1 - Z_2$ and $Z_1 \cdot Z_2 = 0$ in $\mathcal{R}/\mathcal{I}$, we get $Z_1^2 + e_2^* \cdot Z_1 = Z_1^2 + (Y_1 - Z_1 - Z_2) \cdot Z_1 = Y_1 Z_1 - Z_1 Z_2 = Y_1 Z_1 = x_{\mu_7}$. Thus, $Z_1^2 = -e_2^* \cdot Z_1 + Y_1 Z_1 = -e_2^* \cdot x_{\mu_2} + x_{\mu_7}$.

Next, we consider X_2^2. If $X_2^2 = \sum_{i=1}^{8} a_i \cdot x_{\mu_i}$, then $\rho_{\mathbf{p}_{\sigma_i}}(X_2^2) = 0$ for $i = 1, 2, 3, 7, 8$. Thus, by Theorem 7.6(iv), we get $a_i = 0$ for $i = 1, 2, 3, 7, 8$. To find a_4, a_5, a_6, we first apply $\rho_{\mathbf{p}_{\sigma_4}}(X_2^2) = X_2^2$ in $H_{T^2}^*(x_{\sigma_4}) = \mathbb{Z}[X_2, Y_1]$. Since $x_{\mu_4} = X_2$, we get $a_4 = \frac{\rho_{\mathbf{p}_{\sigma_4}}(X_2^2)}{\rho_{\mathbf{p}_{\sigma_4}}(X_2)} = X_2 = e_1^*$ under the isomorphism $H_{T^2}^*(x_{\sigma_4}) \simeq H^*(BT^2)$. We then consider $X_2^2 - e_1^* \cdot X_2 = X_2^2 - (X_1 + X_2 - Z_1 - Z_2) \cdot X_2 = Z_1 X_2 + Z_2 X_2$ using the relations $e_1^* = X_1 + X_2 - Z_1 - Z_2$ and $X_1 X_2 = 0$ in $\mathcal{R}/\mathcal{I}$. This implies from Theorem 7.6 that $a_5 = a_6 = 1$ so that $X_2^2 = e_1^* \cdot x_{\mu_4} + x_{\mu_5} + x_{\mu_6}$.

Using similar arguments, we have the following in the $H^*(BT^2)$-algebra, $\mathcal{R}/\mathcal{I}$:

$$\begin{aligned} &X_2^2 = e_1^* \cdot X_2 + 1 \cdot X_2 Z_1 + 1 \cdot X_2 Z_2; \\ &X_2 Z_1 = 1 \cdot X_2 Z_1; \\ &X_2 Z_2 = 1 \cdot X_2 Z_2; \\ &Z_1^2 = -e_2^* \cdot Z_1 + 1 \cdot Y_1 Z_1; \\ &Z_1 Z_2 = 0; \\ &Z_2^2 = -e_2^* \cdot Z_2 + 1 \cdot Y_1 Z_2. \end{aligned}$$

By Corollary 7.7 the ordinary cohomology ring $H_{ord}^*(\widetilde{\mathcal{G}})$ is isomorphic to

$$R'/I' \simeq \mathcal{R}/\mathcal{I} \otimes_{H^*(BT^2)} \mathbb{Z}$$

where $\mathbb{Z}$ is viewed as a $H^*(BT^2) = \mathbb{Z}[e_1^*, e_2^*]$-module via the augmentation map which sends e_i^* to 0 for $i = 1, 2$. Hence $R' = \mathbb{Z}[X_1, X_2, Y_1, Z_1, Z_2]$ and

$$I'=\langle X_1X_2, Z_1Z_2, X_1Y_1Z_1, X_1Y_1Z_2, X_2Y_1Z_1, X_2Y_1Z_2, X_1+X_2-Z_1-Z_2, \\ Y_1-Z_1-Z_2\rangle.$$

By Corollary 7.7(ii) and (22), we see that $H^*_{ord}(\widetilde{\mathcal{G}})$ is isomorphic as a graded $\mathbb{Z}$-module to

$$\mathbb{Z} \oplus \mathbb{Z}Z_1 \oplus \mathbb{Z}Z_2 \oplus \mathbb{Z}X_2 \oplus \mathbb{Z}X_2Z_1 \oplus \mathbb{Z}X_2Z_2 \oplus \mathbb{Z}Y_1Z_1 \oplus \mathbb{Z}Y_1Z_2.$$

It therefore follows that the Euler characteristic of M is $kl + km + lm = 8$, which is the number of elements in the monomial basis. Also the rank of $H^2_{ord}(M;\mathbb{Z})$ is $k+l+m-2 = 3$, which are the number of monomials of degree 1 in the basis. By Theorem 7.6, using the fact that every $u \in H^*(BT^2)$ is equated to zero in the graded ring R'/I', it follows that every monomial of degree r is a linear combination of the monomials x_{μ_i} of degree greater than or equal to r (see 19). Since there are no monomials x_{μ_i} of degree greater than or equal to 3, we get that $H^{2n}_{ord}(\mathcal{G}) = 0$ for all $n \geq 3$. This also implies in particular that $H^6_{ord}(M;\mathbb{Z}) = 0$ and $H^8_{ord}(M;\mathbb{Z}) = 0$. Using this fact or directly observing that there are 4 monomial basis elements of degree 2, it follows that $\mathrm{rank}(H^4_{ord}(M;\mathbb{Z})) = kl+km+lm-(k+l+m-2)-1 = 4$.

Further, the multiplicative structure constants for the basis

$$1, Z_1, Z_2, X_2, X_2Z_1, X_2Z_2, Y_1Z_1, Y_1Z_2$$

of R'/I' can be derived to be as follows:

$$\begin{aligned}
X_2^2 &= 1 \cdot X_2Z_1 + 1 \cdot X_2Z_2;\\
X_2Z_1 &= 1 \cdot X_2Z_1;\\
X_2Z_2 &= 1 \cdot X_2Z_2;\\
Z_1^2 &= 1 \cdot Y_1Z_1;\\
Z_1Z_2 &= 0;\\
Z_2^2 &= 1 \cdot Y_1Z_2.
\end{aligned}$$

More generally, we can compute the multiplicative structure constants of the $\mathbb{Z}$-algebra $H^*_{ord}(\widetilde{\mathcal{G}}) \simeq R'/I'$ for the x-forgetful graph $\widetilde{\mathcal{G}}$ induced from $\mathcal{L}_{k,l,m}$ with repect to the basis

$$\begin{gathered}
1, Y_2, \ldots, Y_l, Z_1, \ldots, Z_m, X_2, \ldots, X_k,\\
X_2Y_2, \ldots, X_2Y_l, X_2Z_1, \ldots, X_2Z_m,\\
\vdots\\
X_kY_2, \ldots, X_kY_l, X_kZ_1, \ldots, X_kZ_m,\\
Y_1Z_1, \ldots, Y_1Z_m, \ldots, Y_lZ_1, \ldots, Y_lZ_m.
\end{gathered} \tag{23}$$

Here again we note that the Euler characteristic of M is $kl + km + lm$, which is the number of elements in the monomial basis. Also the rank of $H^2_{ord}(M;\mathbb{Z})$ is $k + l + m - 2$, which are the number of monomials of degree 1 in the basis. By Theorem 7.6, using the fact that every $u \in H^*(BT^2)$ is equated to zero in the graded ring R'/I', it follows that every monomial of degree r is a linear combination of the monomials x_{μ_i} of degree greater than or equal to r (see (19)). Since there are no monomials x_{μ_i} of degree greater than or equal to 3, we get that $H^{2n}_{ord}(\mathcal{G}) = 0$ for all $n \geq 3$. This implies in particular that $H^6_{ord}(M;\mathbb{Z}) = 0$ and $H^8_{ord}(M;\mathbb{Z}) = 0$. Thus, it suffices to compute the structure constants when we multiply two monomial basis elements of degree 1, which gives us a degree 2 monomial. This can be done as follows.

To compute the structure constants of $H^*_{ord}(\mathcal{L}_{k,l,m})$ with respect to the basis (23), firstly we observe by the following steps given in Theorem 7.6(iv) (as explained in detail above for the case when $k = 2$ $l = 1$ and $m = 2$) that in $\mathcal{R}/\mathcal{I}$ we have the following relations for $2 \leq r \leq k$, $2 \leq s \leq l$ and $1 \leq t \leq m$:

$$X_r^2 = e_1^* \cdot X_r + \sum_{t=1}^{m} 1 \cdot X_r Z_t;$$

$$Y_s^2 = e_2^* \cdot Y_s + \sum_{t=1}^{m} 1 \cdot Y_s Z_t;$$

$$Z_t^2 = -e_2^* \cdot Z_t + \sum_{s=1}^{l} 1 \cdot Y_s Z_t.$$

Other products of degree 1 monomials in $\mathcal{R}/\mathcal{I}$ multiply to give square free monomials of degree 2, $X_r Y_s$ or $X_r Z_t$ or $Y_s Z_t$, which are already part of the basis. Note also that $X_r X_{r'} = 0$, $Y_s Y_{s'} = 0$ and $Z_t Z_{t'} = 0$ for $r \neq r'$, $s \neq s'$ and $t \neq t'$.

We therefore arrive at the following relations in $R'/I' \simeq H^*_{ord}(\widetilde{\mathcal{G}})$:

$$X_r^2 = \sum_{t=1}^{m} 1 \cdot X_r Z_t;$$

$$Y_s^2 = \sum_{t=1}^{m} 1 \cdot Y_s Z_t;$$

$$Z_t^2 = \sum_{s=1}^{l} 1 \cdot Y_s Z_t;$$

$$X_r Y_s = 1 \cdot X_r Y_s;$$

$$X_r Z_t = 1 \cdot X_r Z_t;$$

$$Y_s Z_t = 1 \cdot Y_s Z_t,$$

where $2 \leq r \leq k$, $2 \leq s \leq l$ and $1 \leq t \leq m$.

We have the following corollary of Theorem 7.6 for the hyperplane arrangements $\mathcal{L}_{k,l,m}$ classifying the 8-dimensional toric hyperKähler manifolds.

Corollary 7.8 *The ordinary cohomology $H^*_{ord}(\mathcal{L}_{k,l,m})$ is isomorphic to a free $\mathbb{Z}$-module generated by the elements* (23). *Furthermore, all structure constants of their multiplications are* 1, *except for the case when they are equal to* 0.

Acknowledgments The first author is grateful to Professors Megumi Harada and Mikiya Masuda for their invaluable advice and comments for the previous version of this paper [17]. The first author was supported by JSPS KAKENHI Grant Number 17K14196 and 21K03262.

References

1. R. Bielowski, A. Dancer, The geometry and topology of toric hyperKäler. Commun. Anal. Geom. **8**, 727–759 (2000)
2. W. Bruns, J. Herzog, *Cohen-Macaulay Rings*. Cambridge Studies in Advanced Mathematics, vol. 39 (Cambridge University Press, Cambridge, 1993)
3. V.M. Buchstaber, T.E. Panov, *Toric Topology*. Mathematical Surveys and Monographs, vol. 204 (American Mathematical Society, Providence, 2015)
4. A. Darby, S. Kuroki, J. Song, Equivariant cohomology of torus orbifolds. Canad. J. Math. **74**(2), 299–328 (2022)
5. T. Delzant, Hamiltoniens périodiques et images convexes de l'application moment. Bull. Soc. Math. France **116**, 315–339 (1988)
6. M. Franz, H. Yamanaka, Graph equivariant cohomological rigidity for GKM graphs. Proc. Jpn. Acad. Ser. A Math. Sci. **95**, 107–110 (2019)
7. Y. Fukukawa, H. Ishida, M. Masuda, The cohomology ring of the GKM graph of a flag manifold of classical type. Kyoto J. Math. **54**, 653–677 (2014)
8. M. Goresky, R. Kottwitz, R. MacPherson, Equivariant cohomology, Koszul duality, and the localization theorem. Invent. Math. **131**, 25–83 (1998)
9. V. Guillemin, S. Sabatini, C. Zara, Balanced fiber bundles and GKM theory. Int. Math. Res. Not. **17**, 3886–3910 (2013)
10. V. Guillemin, T.S. Holm, C. Zara, A GKM description of the equivariant cohomology ring of a homogeneous space. J. Algebraic Comb. **23**, 21–41 (2006)
11. R. Goto, *On Toric Hyper-Kähler Manifolds Given by the Hyper-Kähler Quotient Method. Infinite analysis, Part A, B (Kyoto, 1991)*. Advanced Series in Mathematical Physics, vol. 16 (World Scientific Publishing, River Edge, 1992), pp. 317–338
12. V. Guillemin, C. Zara, 1-skeleta, Betti numbers, and equivariant cohomology. Duke Math. J. **107**, 283–349 (2001)
13. M. Harada, N. Proudfoot, Properties of the residual circle action on a hypertoric variety. Pac. J. Math. **214**, 263–284 (2004)
14. H. Konno, Equivariant cohomology rings of toric hyperkahler manifolds, in *Quaternionic Structures in Mathematics and Physics*, Rome, 1999 (Univ. Studi Roma "La Sapienza", Rome, 1999), pp. 231–240 (electronic)
15. H. Konno, Cohomology rings of toric hyperKähler manifolds. Int. J. Math. **11**, 1001–1026 (2000)
16. H. Konno, Variation of toric hyperKähler manifolds. Int. J. Math. **14**, 289–311 (2003)
17. S. Kuroki, Hypertorus graphs and graph equivariant cohomologies, preprint (2010)
18. S. Kuroki, Equivariant cohomology distinguishes the geometric structures of toric hyperKähler manifolds. Proc. Steklov Inst. Math. **275**, 251–283 (2011)

19. S. Kuroki, Upper bounds for the dimension of tori acting on GKM manifolds. J. Math. Soc. J. **71**, 483–513 (2019)
20. H. Maeda, M. Masuda, T. Panov, Torus graphs and simplicial posets. Adv. Math. **212**, 458–483 (2007)
21. M. Masuda, T. Panov, On the cohomology of torus manifolds, Osaka J. Math. **43**, 711–746 (2006)
22. N. Proudfoot, A survey of hypertoric geometry and topology. Proc. Toric Topology Contemp. Math. **460**, 323–338 (2008)
23. R.P. Stanley, *Combinatorics and Commutative Algebra*. Progress in Mathematics, vol. 41, 2nd edn. (Birkhauser, Boston, 1996)
24. V. Uma, K-theory of torus manifolds, in *Toric Topology International Conference* 28 May–3 June 2006, Osaka City University, Osaka. Contemporary Mathematics, vol. 460 (AMS, 2006), pp. 385–389

On the Genera of Moment-Angle Manifolds Associated with Dual-Neighborly Polytopes: Combinatorial Formulas and Sequences

Santiago López de Medrano

Introduction

To every simple polytope P, there is associated a manifold $Z(P)$ of the same dimension known in different works as its (*real*) *moment-angle manifold, universal abelian cover* [6], *polyhedral product* [1], or *intersection of quadrics* (*more precisely, of coaxial ellipsoids*) [13, 15].

The topology of $Z(P)$ cannot be described in full generality, but it has been described for some large families of polytopes P. One of them is the family of *dual-neighborly* polytopes P of even dimension $2p$ for which it was conjectured in [3] that they are connected sums of copies of the sphere product $S^p \times S^p$. This has been shown to be true if $p > 2$ [7], for a subfamily of those polytopes if $p = 2$ [8], and is evident for $p = 1$. For P of odd dimension $2p + 1$, it was proven for $p > 2$ (under a certain additional hypothesis, probably unnecessary) that they are connected sums of copies of the sphere product $S^p \times S^{p+1}$ [7].

Together with them, for each such polytope P of dimension of at least 5, there is an infinite lattice of polytopes obtained by applying iteratively in all possible

---

This work was partially supported by grant PAPIIT IN102918, UNAM.

S. L. de Medrano (✉)
Instituto de Matemáticas. Universidad Nacional Autónoma de México, Mexico City, Mexico
e-mail: santiago@im.unam.mx

A. Bahri et al. (eds.), *Toric Topology and Polyhedral Products*, Fields Institute Communications 89, https://doi.org/10.1007/978-3-031-57204-3_12

ways a well-known operation $P \mapsto P'$, known as the *book*[1] construction for which it was shown that the associated manifold is a connected sum of sphere products $S^a \times S^b$ with factors of different dimensions. The number of combinatorially different dual-neighborly polytopes of even dimension grows enormously fast with their dimension and number of facets [17], and all their corresponding infinite lattices are disjoint, so we know that a huge part of the simple polytopes have associated manifolds which are connected sums of sphere products.

However, only when $p = 1$ or when the number of facets of the polytope is at most $2p+3$ did we know the exact number of summands and precisely which sphere products appear. For $p = 1$ and P the n-gon, the number of those summands (i.e., the genus of the surface $Z(P)$) is known to be $2^{n-3}(n-4)+1$. The same sequence of numbers appears in many other geometric and combinatorial questions (see [18] and [9]). It appeared for the first time (to our knowledge) in a 1935 paper by Coxeter as the genera of surfaces obtained by a certain construction of his [5] and was found independently around 1980 by Hirzebruch (unpublished, but see [12]) as the genera of a certain family of real surfaces that are intersections of quadrics. Only much later was Coxeter's construction recognized as a precursor of what is now called a *polyhedral product* and that intersections of quadrics of the type that Hirzebruch had considered are another instance of the same construction. I have seen no evidence that these two great geometers were ever conscious of that coincidence.

We will give now a generalization of this formula for all even dimensions that gives the number of terms in the connected sum, which is natural to call the *genus* of $Z(P)$. The formula is actually valid homologically even in those cases where it can be conjectured (but not yet proven) that Z is a connected sum of sphere products and can be extended to all the polytopes obtained from P through the book construction. It is still not understood how the combinatorics of P determines the precise products that appear in the connected sum after several applications of the book construction and not only their number.

The genus formula follows, in the case of a dual-neighborly polytope of dimension $d = 2p$ and n facets, from known combinatorial formulas for the number of the faces of a neighborly polytope in each dimension. One can obtain from them the Euler characteristic of $Z(P)$ and therefore its genus.

This direct result is useful for computations, but very messy and not too useful. The real work consisted in the search of a better formula. After several failed attempts, two elements opened the road for a solution. First, the appearance of the sequences of genera for a few small values of p in the *Sloane Encyclopedia of Sequences* [18], which included generating functions for them that suggested immediately a nice and simple conjecture for all dimensions. Second, a specific direct formula for the number of faces of a dual-neighborly polytope in the book by Brøndsted [4] that was more suited for our computations. From these facts, a long computation led us to a proof of the conjecture (Theorems 2.1 and 3.1).

[1] This name has been used for many years in the theory of intersection of quadrics. In the literature on polytopes, this construction is called the *wedge* on P.

This gave a collateral proof of various combinatorial identities that we had tried to prove in our first attempts. Other by-products are a new interpretation and some new formulas for the cases appearing in the *Sloane Encyclopedia of Sequences*, as well as an infinite generalization of them. One could search for extensions of the various interpretations of those few cases.

Additionally, it was shown in [7] that other geometric operations on the polytopes, such as truncation of vertices, induce in the associated manifolds operations that preserve connected sums of sphere products. We give the genus also for all P obtained from the ones above by iterated vertex truncations and book constructions in any order.

So this work is a quantitative continuation of [7], giving the explicit number of sphere products in the connected sums. Alas, this time Samuel Gitler was no more among us to participate and enjoy this extension of our work.

1 Background

The construction of $Z(P)$ for a simple d-polytope P with n facets can be described as follows: one can assume that P is embedded in a d-dimensional affine subspace A of $\mathbb{R}^n$ in such a way that $A \cap \mathbb{R}^n_+ = P$ and that A intersects transversely every coordinate subspace of $\mathbb{R}^n_+$. Then $Z(P)$ is the union of all the images of P under all compositions of reflections of $\mathbb{R}^n$ on its coordinate subspaces $\{x_i = 0\}$. $Z(P)$ is a combinatorial d-manifold that can be easily smoothed as a transversal intersection of ellipsoids [10, 13]. $Z(P)$ can also be constructed abstractly as a quotient of $P \times \mathbb{Z}_2^n$ under the identifications in the facets of P corresponding to the fixed points under the reflections on the coordinate subspaces [6, 13].

The book construction consists in taking the product $P \times [0, 1]$ and one of its facets F and, for each given point $x \in F$, identifying all points (x, t) for $t \in [0, 1]$ into a single point. Under this operation, the dimension and the number of facets of the polytope increases by 1, and we denote by P' the polytope so obtained and by Z' the corresponding manifold. A geometric construction of Z' and manipulations with homology exact sequences shows that the total homology (i.e., the direct sum of the homology groups) of Z and Z' is the same.

One can consider compositions of an arbitrary number of book constructions along different facets. Following [2] we denote by P^J, Z^J, where $J = (j_1, j_2, \dots, j_n)$, the result of applying j_i times the book construction on the i-th facet of P for $i = 1, \dots, n$. See [2] for details of a more general construction and a combinatorial description of P^J in the dual context of simplicial complexes.

The simple polytope P is called *dual-neighborly* if every collection of k facets of P has a non-empty intersection, for all $k \leq d/2$ (Cf. [4, p.92] and [3, p.114]). They are dual to the much studied *neighborly* ones. It can be proven that $Z(P)$ is $[d/2 - 1]$-connected if, and only if, P is dual-neighborly. So, for a $2p$-dimensional ($(2p + 1)$-dimensional dual-neighborly P, respectively), $Z(P)$ has homology only in dimension p (in dimensions p and $p+1$, respectively), other than the 0 and top dimensional ones.

It is known that if two dual-neighborly polytopes have the same dimension d and the same number of facets n, then they have the same number of k-dimensional faces for all k from 0 to d (see [4, p.113] and [11, p.124], for the neighborly polytopes). Various explicit formulas are known for this number of $k-$faces as a function of (d, n), one of which will be more suited to our purposes. We will give this formula in the next section.

Now suppose that P is dual-neighborly polytope and of even dimension $d = 2p$ and n facets. Then the homology of $Z(P)$ is free and is nontrivial only in the middle dimension p, so it has the homology[2] of a connected sum of copies of the sphere product $S^p \times S^p$. It was shown in [7] that it is actually diffeomorphic to such a connected sum if $d > 4$ (but the number of those products was not given) and that the book construction preserves connected sums of sphere products. Applied once, this construction gives again a dual-neighborly polytope P' of dimension $2p+1$, and the manifold $Z(P')$ is a connected sum of copies of $S^d \times S^{d+1}$ (even in the case $d = 4$), and after any number of further applications of it, one obtains again manifolds that are connected sums of products $S^a \times S^b$ for various pairs of dimensions (a, b).

2 The Euler Characteristic $\chi(Z(P))$ for Dual-Neighborly Polytopes P of Even Dimension

Let P be a dual-neighborly polytope of even dimension $d = 2p$ with n facets. Since $n \geq d + 1$ (with equality only for the simplex), it is better to use the parameters $p = d/2$ and m defined as

$$m = n - d - 1 = n - 2p - 1$$

that starts with $m = 0$. The number f_k of k-faces of P (for $k = 0, \dots, d$) is determined only by the numbers p, m. Explicit formulas for them can be deduced from the formulas for the number of k-faces of a neighborly d-polytope with n vertices in [11, section 9.2]. A direct explicit formula for f_k is given in [4, p.113], which in our notation becomes

$$f_k = \sum_{j=0}^{p} \binom{j}{k} \binom{m+j}{j} + \sum_{j=0}^{p-1} \binom{2p-j}{k} \binom{m+j}{j}$$

[2] The homology of $Z(P)$ can be computed by splitting it in terms of the combinatorics of P. The proof in [13] for those with $d+3$ facets is equally valid in general (see also [14]). Several proofs of this splitting have been given in the more general context of moment-angle complexes, culminating in the geometric splitting of the their suspension proven in [1], giving a splitting for any generalized homology theory.

Now, the polytope P is reflected on the coordinate hyperplanes of $\mathbb{R}^n$ to give a cell decomposition of $Z(P)$ formed by 2^n cells of dimension d which are all copies of P. A face of dimension k has $d-k$ coordinates equal to zero, so it is reflected only on $n-d+k$ hyperplanes and therefore produces $2^{n-d+k}=2^{m+k+1}$ cells.

Thus, the total number of k-cells of $Z(P)$ is $f_k \times 2^{m+k+1}$, and therefore, the Euler characteristic of $Z(P)$, which we denote by $\chi(p,m)$), is the alternating sum:

$$\chi(p,m)=\sum_{k=0}^{2p}(-1)^k 2^{m+k+1}\left(\sum_{j=0}^{p}\binom{j}{k}\binom{m+j}{j}+\sum_{j=0}^{p-1}\binom{2p-j}{k}\binom{m+j}{j}\right)$$

This formula is useful for computations, even for d, m in the thousands, since it can be easily programmed in the computer. It also shows that, for any fixed p, $\chi(p,m)$ is of the form 2^{m+1} times a polynomial in m of degree p. But otherwise it is quite messy and opaque. For example, it is easy to see directly that $\chi(p,0)=2$ (P is the simplex Δ^{2p} and $Z(P)$ is the sphere S^{2p}), $\chi(p,1)=2(1+(-1)^p)$ (P is $\Delta^p\times\Delta^p$ and $Z(P)=S^p\times S^p$), and it is known that $\chi(p,2)=2+(-1)^p(4p+6)$ [13]. But these facts are not clear from the formula.

In any case, it is convenient to simplify it—factoring 2^{m+1}, our formula can be rearranged as follows:

$$\chi(p,m)=2^{m+1}\left(\sum_{j=0}^{p}\sum_{k=0}^{2p}(-2)^k\binom{j}{k}\binom{m+j}{j}+\sum_{j=0}^{p-1}\sum_{k=0}^{2p}(-2)^k\binom{2p-j}{k}\binom{m+j}{j}\right)$$

Now, since $\binom{j}{k}=0$ if $k>j$, we have

$$\sum_{k=0}^{2p}(-2)^k\binom{j}{k}=\sum_{k=0}^{j}(-2)^k\binom{j}{k}=(-2+1)^j=(-1)^j$$

And, since $\binom{2p-j}{k}=0$ if $k>2p-j$, we have

$$\sum_{k=0}^{2p}(-2)^k\binom{2p-j}{k}=\sum_{k=0}^{2p-j}(-2)^k\binom{2p-j}{k}=(-2+1)^{2p-j}=(-1)^{2p-j}=(-1)^j$$

And therefore we obtain a better formula

$$\chi(p,m)=2^{m+1}\left(\sum_{j=0}^{p}(-1)^j\binom{m+j}{j}+\sum_{j=0}^{p-1}(-1)^j\binom{m+j}{j}\right)$$

An even simpler formula can be obtained by computing the generating function of the above expression parametrized by m for a fixed p:

$$\sum_{m\geq 0} \chi(p,m) z^m$$

Since the formula for $\chi(p,m)$ involves two sums which differ only in their length, we can cover both cases with the general sequence

$$S(r,m) = 2\sum_{j=0}^{r}(-1)^j 2^m \binom{m+j}{j}$$

and its corresponding generating function

$$\sum_{m\geq 0} S(r,m) z^m$$

which is the sum for $j = 0, \ldots r$, of the generating functions

$$2\sum_{m\geq 0}(-1)^j 2^m \binom{m+j}{j} z^m = 2(-1)^j \sum_{m\geq 0}\binom{m+j}{j}(2z)^m .$$

Now, it is well known (and easy to prove) that

$$\sum_{m\geq 0}\binom{m+j}{j} y^m = \frac{1}{(1-y)^{j+1}} \tag{*}$$

which gives

$$\sum_{m\geq 0} S(r,m) z^m = 2\sum_{j=0}^{r}(-1)^j \frac{1}{(1-2z)^{j+1}} = -2\sum_{j=0}^{r}\frac{1}{(2z-1)^{j+1}}$$

This is a geometric progression with sum

$$-2\frac{\frac{1}{(2z-1)^{r+2}} - \frac{1}{2z-1}}{\frac{1}{2z-1} - 1} = -2\frac{\frac{1}{(2z-1)^{r+1}} - 1}{1-(2z-1)} = -2\frac{\frac{1}{(2z-1)^{r+1}} - 1}{2-2z}$$

$$= \frac{1}{(1-z)} + \frac{1}{(z-1)(2z-1)^{r+1}}$$

The generating function for $\chi(p,m)$ is the sum of two instances of the above expression evaluated a $r = p$ and $r = p - 1$, which add up to

$$\frac{2}{1-z} + \frac{1}{(z-1)}\left(\frac{1}{(2z-1)^{p+1}} + \frac{1}{(2z-1)^p}\right)$$
$$= \frac{2}{1-z} + \frac{2z}{(z-1)(2z-1)^{p+1}}$$

We can obtain another expression for the general term of this series, by working each part separately:

The first term $\frac{2}{1-z}$ is $\sum_{m\geq 0} 2z^m$.

And the second term has two factors: $2z/(z-1)$ and $1/(2z-1)^{p+1}$. The first one is simply $\frac{2z}{z-1}$, so in its series the coefficient of z^i is -2 for $i > 1$ and for the second factor (using formula (*) again),

$$\frac{1}{(2z-1)^{p+1}} = \frac{(-1)^{p+1}}{(1-2z)^{p+1}} = (-1)^{p+1} \sum_{j\geq 0} \binom{j+p}{p} (2z)^j$$

so the coefficient of z^j is $(-1)^{p+1} \binom{j+p}{p} 2^j$

So, in the product, the coefficient of z^m is

$$\sum_{j=0}^{m-1} (-1)^p 2^{j+1} \binom{j+p}{p}$$

since for $j \geq m$, the coefficient of the first factor is 0.

To which we still have to add the first term. So, finally, the coefficient of z^m, which is $\chi(p, m)$, is

$$\chi(p, m) = 2 + (-1)^p \sum_{j=0}^{m-1} 2^{j+1} \binom{j+p}{p}.$$

We have proven

Theorem 2.1 *If P is a dual-neighborly polytope of even dimension $d = 2p$ and $n = d + m + 1$ facets, then $\chi(p, m)$, the Euler characteristic of $Z(P)$, can be expressed in any of the following equivalent forms:*

(i) $\chi(p, m) = 2^{m+1} \left(\sum_{j=0}^{p} (-1)^j \binom{m+j}{j} + \sum_{j=0}^{p-1} (-1)^j \binom{m+j}{j} \right)$

(ii) $\chi(p, m) = (-1)^p \sum_{j=0}^{m-1} 2^{j+1} \binom{j+p}{p} + 2$

(iii) *$\chi(p, m)$, as a sequence parametrized by m, has generating function*

$$\frac{2}{1-z} + \frac{2z}{(z-1)(2z-1)^{p+1}}$$

These formulas are valid for any p and m, and in the cases where we know that $Z(P)$ is a connected sum of sphere products $S^p \times S^p$, we can derive the number of products in the connected sum from the formulas.

But these formulas do not extend to the manifolds obtained by the book construction on P: already the first application gives an odd-dimensional manifold with $\chi = 0$. And also, the even-dimensional ones obtained by iteration of the book construction on P may include products of two odd-dimensional spheres that contribute negative terms to χ, so χ will not depend only on the dimension and number of facets of the corresponding polytope.

We will solve these problems by the introduction of the concept of *genus* of such a connected sum.

3 The *genus* $g(Z(P))$ for Dual-Neighborly Polytopes P of Even Dimension and Associated Polytopes

For a connected sum of sphere products M, we can naturally define its *genus* $g(M)$, as the number of products in the sum, as in the case of surfaces.

Let $\beta(M)$ the sum of the Betti numbers $\beta_i(M)$, then $\beta(M) = 2g(M) + 2$, or

$$g(M) = \frac{\beta(M)}{2} - 1\,.$$

This definition can be extended to any manifold or even any topological space with finite β by the above formula. In some cases, this may not be an integer (if X is a point then $g(X) = -1/2$). Some properties of this generalized genus (e.g., that it is additive with respect to the connected sum of manifolds) will be considered elsewhere.

In the case that all the summands are of the form $S^p \times S^p$, there is a simple relation between $g(M)$ and the Euler characteristic of M:

$$\chi(M) = 2 + (-1)^p 2g(M)$$

or

$$g(p, m) = (-1)^p \chi(p, m)/2 - (-1)^p$$

and again, this relation is valid for any manifold with the same homology groups. This includes the manifolds $Z(P)$ where P is a dual-neighborly polytope and for which the formulas for the Euler characteristic are still valid. So with this extension of the concept of genus, we can state

Theorem 3.1 *If P is a dual-neighborly polytope of even dimension $d = 2p$ and $n = d + m + 1$ facets, then $g(p, m)$, the genus of $Z(P)$, can be expressed in any of the following equivalent forms:*

(i) $$g(p,m) = (-1)^p 2^m \left(\sum_{j=0}^{p} (-1)^j \binom{m+j}{j} + \sum_{j=0}^{p-1} (-1)^j \binom{m+j}{j} \right) - (-1)^p$$

(ii) $$g(p,m) = \sum_{j=0}^{m-1} \binom{j+p}{p} 2^j$$

(iii) *$g(p, m)$, as a sequence parametrized by m, has generating function*

$$\frac{z}{(1-z)(1-2z)^{p+1}}$$

The same expressions are valid for any polytope P^J obtained from P by iterated book operations, where J is any n-tuple of nonnegative integers.

The last part of the theorem follows from the fact that the book construction preserves the total homology of the manifold Z. In particular, in the first application, P' is a dual-neighborly polytope of dimension $2p + 1$; the homology of $Z(P')$ is concentrated in dimensions p and $p + 1$ and the above formula also gives its genus. This includes many odd-dimensional dual-neighborly polytopes, including all the cyclic ones.

Probably this is true also for any dual-neighborly polytope of odd dimension greater than 3, but we do not know yet how the genus of $Z(P)$ depends only on the combinatorics of P.

For a 3-dimensional simple polytope P, the genus of $Z(P)$ is not determined by the number of facets (the simplest examples are the cube and the pentagonal book), but this is a particularity of this dimension where every simple polyhedron is dual-neighborly.

It must be mentioned that the experts in the field consider that, viewed from different angles, a large proportion of the simple polytopes are dual-neighborly. See [11, pp.129, 129a, 129b], [19, section 4], and [17, section 1]. And these are only the roots of infinite lattices of polytopes P^J stemming from them, latices that can be shown to be disjoint for two non-combinatorially equivalent roots P.

The formula for the genus of other polytopes obtained from P and its derivates by applying other operations, such as the truncation of vertices or edges (see [7]), can be derived from the same formulas. So the result applies to a large number of simple polytopes. We illustrate this with the case of the operation of truncating a vertex:

If P with associated manifold $Z(P)$ is any simple d-polytope with n facets, and $P^\vee$ the result of truncating one of its vertices, then [7]

$$Z(P^{\vee}) = 2Z(P)\#(2^{n-d} - 1)(S^1 \times S^{d-1})$$

so the genus is duplicated and then increased by $2^{n-d} - 1$.

A simple induction shows that if the number of vertex truncations is t, then the genus of the resulting Z is

$$2^t(g(Z(P)) - 1) + t2^{n+t-d-1} + 1$$

or, in terms of the parameter $m = n - d - 1$ used in Sect. 2,

$$2^t(g(Z(P)) - 1) + t2^{m+t} + 1$$

showing that the result depends only on the parameters m and t.

Observe that P' is a $(d+1)$-polytope with $n+1$ facets, so truncating one of its vertices duplicates the genus and increases it by $2^{n-1-(d-1)} - 1$. So it produces the same effect on the genus of $Z(P')$ as the truncation of a vertex of P on the genus of $Z(P)$. Since the book construction P' does not affect the genus of $Z(P)$, it follows that the effect of compositions of both operations in any order has the same effect on the genus of $Z(P)$ as the application of the truncations only. The combinatorial type of the result of such a chain of operations on P depends on the choice of facets for the book constructions, the choice of vertices to be truncated, and the order of their application. The dimension of the associated manifold depends only on the number of book constructions, while its genus depends only on the number of vertex truncations.

We illustrate this with some examples: when P is a d-simplex, the result of t vertex truncations and any number of book constructions in any order is

$$Z(P') = (2^t(t-1) + 1)(S^1 \times S^{d-1})$$

which gives the same sequence of genera $g(1, m)$ of surfaces (which can all be obtained by vertex truncation of the triangle) as well as that of many 3-dimensional polytopes for which $Z(P)$ is connected sums of copies of $S^1 \times S^2$. The fact that the topology of the manifold obtained as the result of several vertex truncations does not depend on the vertices chosen, while in dimension greater than 2 one can get many combinatorially different polytopes, was first observed (in the context of intersection of real quadrics in complex space) in [3].

When P is the square, $Z(P)$ is the torus, and applying to P the book construction any number of times one gets a product of simplices whose corresponding manifold is a product of spheres $S^a \times S^b$, so in a given dimension we get many combinatorial types of polyhedra and many topological types of manifolds with one truncation of the triangle and the same number of book constructions. If we start with the pentagon, we get many different connected sums of five sphere products, and in this case, it may happen that combinatorially different polytopes give topologically equivalent manifolds (cf. [13]).

Cutting off an edge of the polytope is a more complicated operation, and the resulting manifold may depend on which edge is chosen. A simple example is when P is a triangular prism (a vertex truncation of the tetrahedron and therefore $Z(P) = S^1 \times S^2$): if we slice a vertical edge of P (i.e., one joining the top and bottom triangles), we obtain a polyhedron combinatorially equivalent to the cube, and the corresponding manifold is $S^1 \times S^1 \times S^1$. If we slice one of the sides of a triangle, one sees easily that combinatorially we get the same result as if we had truncated a vertex, so the corresponding manifold is the connected sum of five copies of $S^1 \times S^2$, by our previous formula (cf. remark (2) before Theorem 2.4 in [7]). The first one is not a connected sum, and its *genus* (in the generalized sense defined above) is different from that of the second one. In [7], we showed that if $Z(P)$ is simply connected, then this operation does preserve connected sums, the topology of the result does not depend on the edge chosen, and the genus of the result can be computed from that of $Z(P)$. Additionally, combined carefully with book constructions and vertex truncations, it gives many more examples of connected sums of sphere products, including many whose genus we can now compute explicitly, starting from even-dimensional dual-neighborly polytopes using Theorem 3.1. The same must be true in the non-simply connected case if the edge generates a trivial homology class of $Z(P)$ when reflected on the coordinate hyperplanes containing its boundary, and one can envisage similar results for deeper truncations (with more hypotheses), but all of this has not been studied.

4 On the Sequences of Genera

Formula (i) for the sequence of genera in Theorem 3.1 shows that $g(p, m)$ is of the form 2^m times a polynomial $R_p(m)$ of degree p minus $(-1)^p$. For any value of p, this polynomial can be computed easily. Formulas (ii) and (iii) are more compact, but do not immediately reflect this property.

For $p = 1$, P is a polygon and formula (i) gives $g(1, m) = 2^m(m - 1) + 1$, which in terms of the number of sides n of P ($n = m + 3$) gives the usual formula $2^{n-3}(n - 4) + 1$ for the genus of $Z(P)$ [9]. Formula (ii) gives the sum $g(1, m) = \Sigma_{j=0}^{m-1}(j + 1)2^j$. Both formulas and the generating function appear in the *Sloane Encyclopedia of Sequences* [18, https://oeis.org/A000337] together with a long list of appearances of this sequence in questions of topology, combinatorics, polytope theory, and algebraic geometry.

For $p = 2, 3, 4$ and 5, the corresponding sequences also appear as entries /A055580, /A027608, /A211386, and /A21138, of [18] with their generating sequences, some formulas, and various appearances in combinatorial problems.

Sequences for higher values of p do not seem to appear. It is a curious fact that the sequences of the Euler characteristics that we have obtained above do not seem to appear at all in the *Sloane Encyclopedia of Sequences*.

The generating functions given in the *Sloane Encyclopedia of Sequences* for those few cases gave us the clue to solve our problem. Our debt to the *Encyclopedia* is partially covered by giving new formulas and a new topological interpretation to some of its sequences, as well as an infinite family of sequences generalizing them, and, hopefully, by suggesting generalizations of their interpretations.

As a by-product of our computations, we have also obtained some combinatorial identities. For example, from the formula for the number of faces of a given dimension of a neighborly polytope in [11, p.166], after dualizing and taking their alternating sum, we get the following formula for the Euler characteristic of the manifold associated to the dual-neighborly one:

$$\sum_{k=0}^{2p}(-1)^k 2^{m+k+1} \sum_{j=0}^{p} \frac{m+2p+1}{m+k+1} \binom{m+2p-j}{2p-k-j} \binom{m+k+1}{2j-2p+k}$$

So this expression is equal to any of the three versions of $\chi(p, m)$ given in Theorem 2.1. Perhaps this is easy to see for the experts on combinatorial identities, which is not the case of this author.

The article [16], which solved a topological problem by daring to deal with complicated combinatorial identities, was very stimulating for not giving up in our struggle with them.

References

1. A. Bahri, M. Bendersky, F.R. Cohen, S. Gitler, The polyhedral product functor: a method of computation for moment-angle complexes, arrangements and related spaces. Adv. Math. **22**, 1634–1668 (2010)
2. A. Bahri, M. Bendersky, F.R. Cohen, S. Gitler, Operations on Polyhedral Products and a new topological construction of infinite families of toric manifolds. Homol. Homotopy Appl. **17**, 137–160 (2015)
3. F. Bosio, Meersseman, L., Real quadrics in $\mathbb{C}^n$, complex manifolds and convex polytopes. Acta Math. **197**, 53–127 (2006)
4. A. Brøndsted, *An Introduction to Convex Polytopes* (Springer, Berlin, 1983)
5. H. Coxeter, Regular skew polyhedra in three and four dimension, and their topological analogues. Proc. Lond. Math. Soc. **2**(1), 33–62 (1938)
6. M. Davis, T. Januszkiewicz, Convex polytopes, Coxeter orbifolds and torus actions. Duke Math. J. **62**, 417–451 (1991)
7. S. Gitler, S. López de Medrano, Intersections of quadrics, moment-angle manifolds and connected sums. Geometry and Topology **17**, 1497–1534 (2013)
8. V. Gómez Gutiérrez, Cyclic polytopes and intersections of quadrics. Boletín de la Sociedad Matemática Mexicana **20**(2), 237–255 (2014)
9. V. Gómez Gutiérrez, S. López de Medrano, Surfaces as complete intersections, in *Riemann and Klein Surfaces, Automorphisms, Symmetries and Moduli Spaces*. Contemporary Mathematics, vol. 629 (2014), pp. 171–180
10. V. Gómez Gutiérrez, S. López de Medrano, Topology of the intersections of ellipsoids in $\mathbb{R}^n$. RACSAM **112**, 879–891 (2018). https://doi.org/10.1007/s13398-018-0552-6
11. B. Grünbaum, Convex polytopes, in *Graduate Texts in Mathematics*, vol. 221 (Springer, Berlin, 2003)

12. F. Hirzebruch, private conversation (1986), based on *Arrangements of lines and algebraic surfaces*, in Arithmetic and Geometry, vol. II (= Progr. Math. 36) (Birkhauser, Boston, 1983), pp. 113–140
13. S. López de Medrano, The Topology of the Intersection of Quadrics in $\mathbb{R}^n$, in *Algebraic Topology* (Arcata Ca,1986). Lecture Notes in Mathematics, vol. 1370 (Springer, Berlin, 1989), pp. 280–292
14. S. López de Medrano, Singular Intersections of Quadrics I, in *Singularities in Geometry, Topology, Foliations and Dynamics*, ed. by J. Cisneros-Molina, D. Tráng Lê, M. Oka, J. Snoussi. Trends in Mathematics. Birkhäuser, Trends in Mathematics (2017), pp. 155–170
15. S. López de Medrano, From singularities to the polyhedral product functor, in *Handbook of Geometry and Topology of Singularities II*, ed. by J. José Cisneros-Molina, L. Tráng, v Seade (Springer, Berlin, 2021)
16. A. Ortiz-Rodríguez, F. Sánchez-Bringas, On some combinatorial formulae coming from Hessian Topology. Discret. Math. **342**, 615–622 (2019)
17. A. Padrol, Many neighborly polytopes and oriented matroids. Discrete Comput. Geom. **50**, 865–902 (2013)
18. N.J.A. Sloane, *On-Line Encyclopedia of Integer Sequences (OEIS)*. https://oeis.org
19. G.M. Ziegler, Recent progress on polytopes, in *Advances in Discrete and Computational Geometry*, ed. by B. Chazelle, J.E. Goodman, R. Pollack. Contemporary Mathematics **223**, 395–406 (1999)

Homeomorphic Model for the Polyhedral Smash Product of Disks and Spheres

Arnaud Ngopnang Ngompé

1 Introduction

In all the following, $m \in \mathbb{N}$ is any natural number and $[m] = \{1, \cdots, m\}$. Also, we set K to be an abstract simplicial complex whose vertex set is contained in $[m]$, that is, K is a family of subsets $\sigma \subseteq [m]$, called simplices, such that whenever $\sigma \in K$ and $\tau \subseteq \sigma$, then $\tau \in K$.

Definition 1.1 ([**Polyhedral Smash Product**][3, Construction 8.3.1]) Let $(\underline{X}, \underline{A}) = \{(X_i, A_i)\}_{i \in [m]}$ be a family of pointed CW-pairs, that is, the X_i are CW-complexes, and $A_i \hookrightarrow X_i$ are subcomplexes, for all $i \in [m]$. The **polyhedral smash product** of $(\underline{X}, \underline{A})$ over K, denoted $\widehat{Z}(K; (\underline{X}, \underline{A})) \subseteq \bigwedge_{i=1}^{m} X_i$, is the space given by

$$\widehat{Z}(K; (\underline{X}, \underline{A})) = \bigcup_{\sigma \in K} \widehat{D}(\sigma), \text{ where}$$

$$\widehat{D}(\sigma) = \bigwedge_{i=1}^{m} Y_i, \tag{1}$$

$$\text{with } Y_i = \begin{cases} X_i \text{ if } i \in \sigma, \\ A_i \text{ otherwise} \end{cases}, \quad \forall \sigma \in K. \tag{2}$$

A. Ngopnang Ngompé (✉)
Department of Mathematics and Statistics, University of Regina, Regina, SK, Canada
e-mail: ann037@uregina.ca

A. Bahri et al. (eds.), *Toric Topology and Polyhedral Products*, Fields Institute Communications 89, https://doi.org/10.1007/978-3-031-57204-3_13

Using categorical language, consider $\text{CAT}(K)$ to be the face category of K, that is, objects are simplices and morphisms are inclusions. Define the $\text{CAT}(K)$-diagram given by

$$\begin{aligned} \widehat{D} : \text{CAT}(K) &\to \text{Top} \\ \sigma &\mapsto \widehat{D}(\sigma), \end{aligned} \tag{3}$$

where $\widehat{D}(\sigma)$ is given by (1) and the functor $\widehat{D}$ maps the morphism $\rho \subseteq \sigma$ to the inclusion $\widehat{D}(\rho) \subseteq \widehat{D}(\sigma)$. Then,

$$\widehat{Z}(K; (\underline{X}, \underline{A})) = \operatorname*{colim}_{\sigma \in K} \widehat{D}(\sigma). \tag{4}$$

Below we recall some well-known operations on spaces.

Definition 1.2 ([4, §0]) For $n \in \mathbb{N}$, let (X, x_0) and (Y, y_0) be two pointed topological spaces.

- The **join** $X * Y$ of X and Y is the quotient space defined by $X * Y = X \times Y \times I / \sim$, where $I = [0, 1]$ and $\sim$ is the equivalence relation generated by

$$(x, y, 0) \sim (x, y', 0), \ \forall x \in X \text{ and } \forall y, y' \in Y,$$
$$(x, y, 1) \sim (x', y, 1), \ \forall x, x' \in X \text{ and } \forall y \in Y.$$

- The **wedge sum** $X \vee Y$ of X and Y is the quotient space defined by

$$X \vee Y = X \amalg Y / (x_0 \sim y_0).$$

- The **smash product** $X \wedge Y$ of X and Y is the quotient space defined by

$$X \wedge Y = X \times Y / X \vee Y.$$

- The **(unreduced) suspension** ΣX of X is the space defined by $\Sigma X = S^0 * X$, where S^0 denotes the 0-sphere.
- The **(unreduced) cone** CX of X is the space defined by $CX = c * X$, where c is a single point.

David stone made the following conjecture.

Conjecture 1.3 If F is a compact subspace of $\mathbb{R}^n$, then there is a homeomorphism

$$\widehat{Z}(K; (c * \Sigma F, \Sigma F)) \cong \Sigma \left(\circledast^m F \right) * |K|,$$

where $\circledast^m F$ is defined as the m-fold join of m copies of F.

As it is mentioned in [1, Remark 2.20], David Stone used a geometrical argument to prove a particular case of his conjecture by taking $F = S^0$. Hence, he proved the following [5].

Theorem 1.4 *There is a homeomorphism*

$$\widehat{Z}\left(K; \left(D^2, S^1\right)\right) \cong \Sigma^{m+1}|K|.$$

In this paper, we apply the same technique to a more general case. For $k \in \mathbb{N}$, we consider $F = S^{k-1}$, which is compact (as a closed and bounded subspace of $\mathbb{R}^k$), and we have $\Sigma F \cong S^k$, $c * \Sigma F \cong D^{k+1}$ and $\circledast^m F \cong S^{km-1}$ (since $S^i * S^j \cong S^{i+j+1}$). Hence,

$$\begin{aligned}\left(\circledast^m F\right) * |K| &\cong S^{km-1} * |K| \\ &\cong \left(\circledast^{km} S^0\right) * |K| \\ &\cong \Sigma^{km}|K|.\end{aligned}$$

So we can state a generalization of Stone's result.

Theorem 1.5 *For any* $k \in \mathbb{N} \cup \{0\}$, *there is a homeomorphism*

$$\widehat{Z}\left(K; \left(D^{k+1}, S^k\right)\right) \cong \Sigma^{km+1}|K|.$$

The goal of this paper is firstly to generalize David Stone's technique for the proof of Theorem 1.5 and secondly to provide a further generalization (see Theorem 6.6) of the latter result for a set of disks and spheres of different dimensions.

Theorem 1.6 *For any m-tuple* $J = (j_1, \cdots, j_m)$ *in* $(\mathbb{N} \cup \{0\})^m$, *there is a homeomorphism*

$$\widehat{Z}\left(K; \left(\underline{D^{J+1}, S^J}\right)\right) \cong \Sigma^{j_1+\cdots+j_m+1}|K|,$$

where $\left(\underline{D^{J+1}, S^J}\right) = \left\{\left(D^{j_1+1}, S^{j_1}\right), \cdots, \left(D^{j_m+1}, S^{j_m}\right)\right\}$.

In order to prove Theorem 1.5, we need to put together some topological and combinatorial tools; hence, the rest of the paper is organized as follows. In Sects. 2 and 3, we describe, respectively, the necessary topological and combinatorial tools. Section 4 is devoted to the proof of Theorem 1.5, for $k \geq 1$. The case $k = 0$, namely, Theorem 5.1, is treated in Sect. 5 using a more categorical argument. Finally, in Sect. 6, we prove the main result, Theorem 6.6, using an inductive argument based on the case $k = 0$.

2 Topological Tools

Among the tools we use in the proof of Theorem 1.5, the homeomorphisms $\Psi : C\Delta^{n-1} \to C^n$ and $\overline{\Psi} : \Sigma\Delta^{n-1} \to \widetilde{D}^n$, described below, are both playing an important role. They were defined by David Stone in [5], and we recycle them here to prove this more general case. Before we introduce them, let us first recall the usual homeomorphism $\widetilde{\Theta} : CX/X \to \Sigma X$.

Given a space X, we identify X with the base $c \times X \times \{1\}$ of $CX = c * X$. Set $S^0 = \{s_1, s_2\}$ to be the 0-sphere, and consider the map

$$\begin{aligned} \Theta &: CX \to \Sigma X \\ [c, x, \lambda] &\mapsto \Theta[c, x, \lambda] = \begin{cases} (s_1, x, 2\lambda), & \text{if } 0 \leq \lambda \leq \frac{1}{2} \\ (s_2, x, 2 - 2\lambda), & \text{if } \frac{1}{2} \leq \lambda \leq 1. \end{cases} \end{aligned} \tag{5}$$

Then Θ factors through a map $\widetilde{\Theta} : CX/X \to \Sigma X$; see Fig. 1.

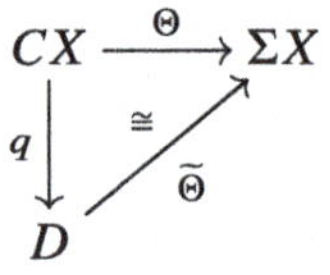

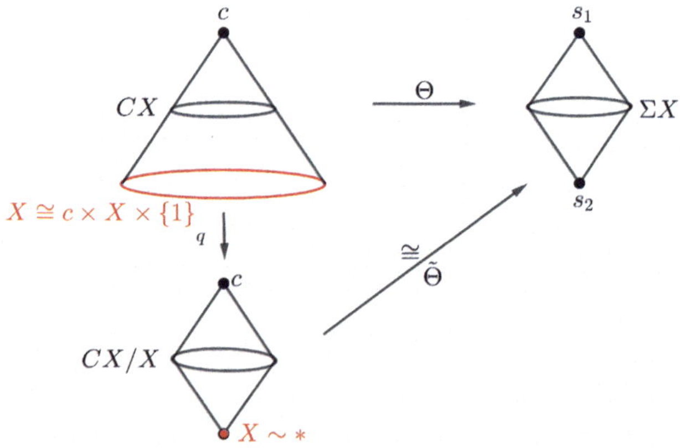

Fig. 1 Factorization of Θ through $\widetilde{\Theta}$

Lemma 2.1 *The map $\widetilde{\Theta}$ is a homeomorphism.*

Notation 2.2 In $\mathbb{R}^n$,

- let e_i be the ith standard basis vector. Let c denote the origin, and let $t_1, \cdots, t_n$ denote the coordinates of a point $x \in \mathbb{R}^n$. We identify $x \sim \overrightarrow{cx}$ and so $x = \sum_{i=1}^n t_i e_i$.
- Set $C = [0, 2]$, with based point 2, and consider
  $C^n = [0,2]^n = \{\sum_{i=1}^n t_i e_i \in \mathbb{R}^n : 0 \leq t_i \leq 2\}$, the n-cube of side 2.
  $\partial_+ C^n = \{\sum_{i=1}^n t_i e_i \in C^n : \max t_i = 2\}$, the outer boundary of C^n.
  $\partial_- C^n = \{\sum_{i=1}^n t_i e_i \in C^n : \min t_i = 0\}$, the inner boundary of C^n.
  $\partial C^n = \partial_+ C^n \cup \partial_- C^n$, the boundary of C^n.
  $\widetilde{D}^n = C^n / \partial_+ C^n$, with the quotient map $\omega : C^n \to \widetilde{D}^n$.

Lemma 2.3 *The quotient space $\widetilde{D}^n$ is a topological disk.*

Proof Considering the CW-pair $(C^n, \partial_+ C^n)$, we have

$$\begin{aligned}(C^n, \partial_+ C^n) &\cong (c * \partial_+ C^n, \partial_+ C^n)\\ &\cong (c * D^{n-1}, D^{n-1}),\end{aligned}$$

where the latter pair is the inclusion of the base of the cone

$$c \times D^{n-1} \times \{1\} \cong D^{n-1} \subseteq (c * D^{n-1}).$$

Hence, collapsing the respective subspaces yields a homeomorphism

$$\begin{aligned}\widetilde{D}^n = C^n / \partial_+ C^n &\cong (c * D^{n-1}) / D^{n-1}\\ &\cong \Sigma D^{n-1}\\ &\cong D^n.\end{aligned}$$

Therefore, $\widetilde{D}^n$ is a topological disk. □

Notation 2.4 Let us consider the following setup.

- For any set $X = \{x_1, \cdots, x_p\} \subseteq \mathbb{R}^n$, let $\mathrm{cx}(X)$ denote the convex hull of X, that is,

$$\mathrm{cx}(X) = \left\{ \sum_{i=1}^{p} t_i x_i \in \mathbb{R}^n : t_i \geq 0, \sum_{i=1}^{p} t_i = 1 \right\}.$$

- Set $\Delta^{n-1} = \mathrm{cx}\{e_1, \cdots, e_n\}$ to be the standard $(n-1)$-simplex.
- For any $J \subseteq [n]$, set $\Delta(J) = \mathrm{cx}(\{e_i : i \in J\}) \cong \Delta^{|J|-1}$, where $|J|$ denotes the cardinality of J.

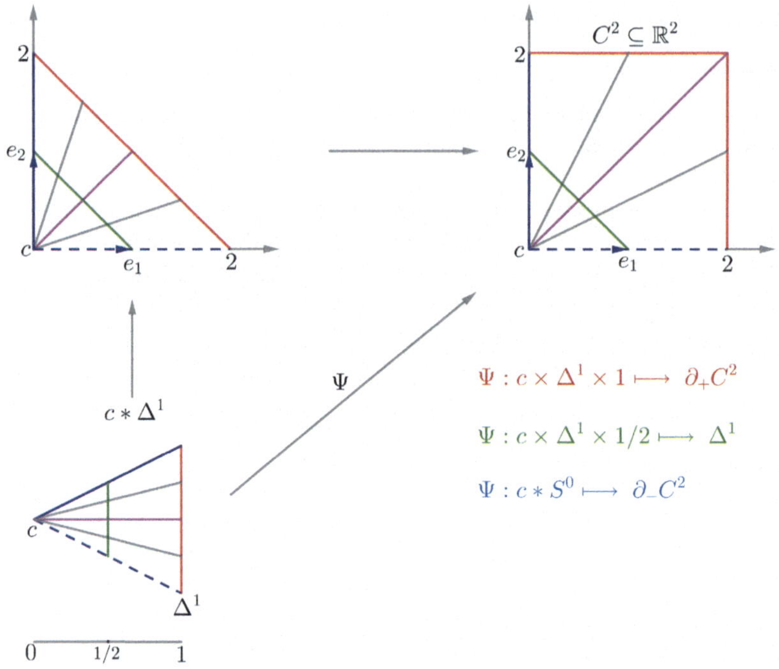

Fig. 2 Illustration of the map $\Psi : C\Delta^1 \to C^2$, that is, for $n = 2$

Remark 2.5 The abstract cone $C\Delta^{n-1}$ can be realized as a subspace of $\mathbb{R}^n$, a subspace which is homeomorphic to the n-cube C^n by reparametrization as we can observe in Fig. 2. This motivates the existence of a bijection $\Psi : C\Delta^{n-1} \to C^n$, defined by Eq. (6).

For $x = \sum_{i=1}^{n} t_i e_i \in \Delta^{n-1}$, set $\bar{t} = \max\{t_i\}$, so $\bar{t} > 0$. Define the map

$$\Psi : C\Delta^{n-1} \to C^n$$

$$[c, x, \lambda] \mapsto \Psi[c, x, \lambda] = \begin{cases} 2\lambda x, \text{ if } 0 \leq \lambda \leq \dfrac{1}{2} \\ \left((2 - 2\lambda) + (2\lambda - 1)\dfrac{2}{\bar{t}}\right) x, \text{ if } \dfrac{1}{2} \leq \lambda \leq 1, \end{cases} \tag{6}$$

where $C^n = [0, 2]^n$ is the n-cube of side 2 set in Notation 2.2.

Remark 2.6 As mentioned in Notation 2.2, the basepoint of $C = [0, 2]$ is 2. The above defined map Ψ does not send the cone point to the basepoint $(2, \cdots, 2)$ of the n-cube C^n, as one might expect, but to the origin c of $\mathbb{R}^n$ for convenience.

By Lemma 2.1, we have $C\Delta^{n-1}/\Delta^{n-1} \cong \Sigma\Delta^{n-1} \cong \Delta^n$. Also $\Psi(c \times \Delta^{n-1} \times \{1\}) = \partial_+ C^n$, and hence, Ψ factors through the map

$$\overline{\Psi} : \Sigma\Delta^{n-1} \to \widetilde{D}^n,$$

where $\widetilde{D}^n = C^n/\partial_+ C^n$ is the topological disk introduced in Notation 2.2.

$$\begin{array}{ccc} C\Delta^{n-1} & \xrightarrow{\Psi} & C^n \\ {\scriptstyle q}\downarrow & & \downarrow{\scriptstyle \omega} \\ \Sigma\Delta^{n-1} & \xrightarrow[\overline{\Psi}]{} & \widetilde{D}^n \end{array}$$

Lemma 2.7 *The maps* Ψ *and* $\overline{\Psi}$ *are both homeomorphisms.*

Proof As a continuous bijection from the compact space $C\Delta^{n-1}$ to the Hausdorff space C^n, Ψ is a homeomorphism. Hence, Ψ gives us the homeomorphism of the pairs $(C\Delta^{n-1}, \Delta^{n-1}) \cong (C^n, \partial_+ C^n)$, so that the induced map $\overline{\Psi} : \Sigma\Delta^{n-1} \to \widetilde{D}^n$ is a homeomorphism. □

Remark 2.8 If we consider $([0, 2], 2)$ to be a pointed space, then collapsing $\partial_+ C^{2(k+1)}$ in $C^{2(k+1)} \cong C^{k+1} \times C^{k+1}$, we get

$$\begin{aligned} \widetilde{D}^{k+1} \bigwedge \widetilde{D}^{k+1} &\cong D^{k+1} \bigwedge D^{k+1} \\ &\cong D^{2(k+1)}. \end{aligned}$$

This can be generalized to the case of $C^{p(k+1)} \cong \underbrace{C^{k+1} \times \cdots \times C^{k+1}}_{p \text{ times}}$, and so collapsing $\partial_+ C^{p(k+1)}$ corresponds to $\bigwedge^p \widetilde{D}^{k+1} \cong \bigwedge^p D^{k+1} \cong D^{p(k+1)}$. Hence,

$$\omega\left(\prod^{p} C^{k+1}\right) \cong \bigwedge^{p} \widetilde{D}^{k+1} \cong \bigwedge^{p} D^{k+1} \text{ and}$$

$$\omega\left(\prod^{p} \partial_- C^{k+1}\right) \cong \bigwedge^{p} \partial\widetilde{D}^{k} \cong \bigwedge^{p} S^{k}, \text{ where } \partial\widetilde{D}^{k} \text{ denotes the boundary of } \widetilde{D}^{k+1}.$$

$$\begin{array}{ccc} \left(\prod^p C^{k+1}\right)/\partial_+ C^{p(k+1)} & \xrightarrow{\cong} & \bigwedge^p D^{k+1} \\ {\scriptstyle \cong}\downarrow & \nearrow{\scriptstyle \cong} & \\ \bigwedge^p \widetilde{D}^{k+1} & & \end{array}$$

Lemma 2.9 *For any compact and Hausdorff spaces* X *and* Y*, there is a homeomorphism*

$$\varphi : C(X * Y) \to CX \times CY.$$

Proof We follow the proof of [2, Lemma 8.1]. We can represent a point in $C(X * Y)$ by $[c, [x, y, t], \lambda]$. We define the homeomorphism φ by

$$\varphi([c, [x, y, t], \lambda]) = ([c, x, 2\lambda \cdot \min\{t, 1/2\}], [c, y, 2\lambda \cdot \min\{1-t, 1/2\}]) \in CX \times CY,$$

where the cone point is at $\lambda = 0$. At $\lambda = 1$, φ reduces to the usual homeomorphism

$$X * Y \cong (CX \times Y) \cup (X \times CY).$$

The map φ is a homeomorphism as a continuous bijection from the compact space $C(X * Y)$ to the Hausdorff space $CX \times CY$. □

3 Combinatorial Tools

One of the main goals of this section is to embed the simplicial complex K in a bigger simplex with vertex set $[(k + 1)m]$. We start by introducing a linearized version of the join of spaces.

Definition 3.1 (Geometrically Joinable)

- Two compact subspaces X and Y of $\mathbb{R}^n$ are said to be **geometrically joinable** if whenever $x, x' \in X$, $y, y' \in Y$, and $\lambda, \lambda' \in I$ are such that $\lambda x + (1 - \lambda)y = \lambda' x' + (1 - \lambda')y'$, then we have one of the three following possibilities
  - $\lambda = \lambda' = 0$, and so $y = y'$;
  - $\lambda = \lambda' = 1$, and so $x = x'$;
  - $0 \neq \lambda = \lambda' \neq 1$, $x = x'$ and $y = y'$.
- More generally, p compact subspaces $X_1, \cdots, X_p \subseteq \mathbb{R}^n$ are **geometrically joinable** if whenever we have an equality between two convex combinations of points of $X_1, \cdots, X_p$, that is, whenever

$$\sum_{i=1}^{p} \lambda_i x_i = \sum_{i=1}^{p} \lambda_i' x_i',$$

  for some $x_i, x_i' \in X_i$ and $\lambda_i, \lambda_i' \in I$ with $\sum_{i=1}^{p} \lambda_i = 1 = \sum_{i=1}^{p} \lambda_i'$, then for all $i = 1, \cdots, p$ such that $\lambda_i \neq 0$ or $\lambda_i' \neq 0$, we have $\lambda_i = \lambda_i'$ and $x_i = x_i'$.
- If p compact subspaces $X_1, \cdots, X_p \subseteq \mathbb{R}^n$ are geometrically joinable, then we define their **geometric join** $\overline{*}_{i=1}^{p} X_i$ to be the set of all convex combinations of elements of X_i, that is,

$$\overline{*}_{i=1}^{p} X_i = \left\{ \sum_{i=1}^{p} \lambda_i x_i \in \mathbb{R}^n : x_i \in X_i \text{ and } \lambda_i \in I \text{ such that } \sum_{i=1}^{p} \lambda_i = 1 \right\}.$$

The notion of geometrically joinable introduced here is also called "in general position". In the following, when we use the notation $X\overline{*}Y$, it is to be understood that X and Y are indeed geometrically joinable.

Example 3.2 Single points $X_1 = \{x_1\}, \cdots, X_p = \{x_p\}$ in $\mathbb{R}^n$ are geometrically joinable if and only if they are affinely independent. In that case, their geometric join $\overline{*}_{i=1}^p X_i$ is their convex hull, which is a $(p-1)$-simplex, that is,

$$\overline{*}_{i=1}^p X_i = \mathrm{cx}\{x_1, \cdots, x_p\} \cong \Delta^{p-1}.$$

Lemma 3.3

1. *Let X and Y be geometrical joinable subspaces of $\mathbb{R}^n$. The map*

$$\begin{aligned} \Phi : X * Y &\to X\overline{*}Y \\ [x, y, \lambda] &\mapsto \lambda x + (1-\lambda)y \end{aligned}$$

   is a homeomorphism.
2. *More generally, for geometrically joinable subspaces $X_1, \cdots, X_p$ of $\mathbb{R}^n$, the map*

$$\begin{aligned} \Phi_p : *_{i=1}^p X_i &\to \overline{*}_{i=1}^p X_i \\ [x_i, \lambda_i]_{i=1}^p &\mapsto \sum_{i=1}^p \lambda_i x_i \end{aligned}$$

   is a homeomorphism.

Remark 3.4 Observe that if U and V are respective subspaces of geometrically joinable spaces X and Y, then U is geometrically joinable to V.

Lemma 3.5 *Let X, Y_1 and Y_2 be three compact subspaces of $\mathbb{R}^n$. If X is geometrically joinable to each Y_i and*

$$X\overline{*}Y_1 \cap X\overline{*}Y_2 = X\overline{*}(Y_1 \cap Y_2), \tag{7}$$

then X is geometrically joinable to $Y_1 \cup Y_2$.

Proof Let $x, x' \in X$, $w, w' \in Y_1 \cup Y_2$ and $\lambda, \lambda' \in I$ be such that

$$\lambda x + (1-\lambda)w = \lambda' x' + (1-\lambda')w'. \tag{8}$$

If we have either $w, w' \in Y_1$ or $w, w' \in Y_2$, then there is nothing to show since X is geometrically joinable to both Y_1 and Y_2. Without loss of generality, suppose $w \in Y_1$ and $w' \in Y_2$, and so (8) gives us

$$X\overline{*}Y_1 \ni \lambda x + (1-\lambda)w = \lambda' x' + (1-\lambda')w' \in X\overline{*}Y_2.$$

Then by (7), there are $x'' \in X$, $w'' \in Y_1 \cap Y_2$ and $\lambda'' \in I$ such that

$$\lambda x + (1-\lambda)w = \lambda'' x'' + (1-\lambda'')w'', \tag{9}$$

$$\lambda' x' + (1-\lambda')w' = \lambda'' x'' + (1-\lambda'')w''. \tag{10}$$

If $\lambda'' = 0$ ($\lambda'' = 1$, respectively), then

- $\lambda = 0$ ($\lambda = 1$, respectively) and $w = w''$ ($x = x''$, respectively) by (9), since X and Y_1 are geometrically joinable.
- $\lambda' = 0$ ($\lambda' = 1$, respectively) and $w' = w''$ ($x' = x''$, respectively) by (10), since X and Y_2 are geometrically joinable.

Thus, $\lambda = 0 = \lambda'$ ($\lambda = 1 = \lambda'$, respectively) and $\omega = \omega'$ ($x = x'$, respectively)

If $0 \neq \lambda'' \neq 1$, then

- $\lambda = \lambda''$ and, $x = x''$ and $w = w''$ by (9), since X and Y_1 are geometrically joinable.
- $\lambda' = \lambda''$ and, $x' = x''$ and $w' = w''$ by (10), since X and Y_2 are geometrically joinable.

Thus, $\lambda = \lambda'$ and, $x = x'$ and $w = w'$. Therefore, X is geometrically joinable to $Y_1 \cup Y_2$. □

Notation 3.6 Now let $n = (k+1)m$, and for $i \in [m]$, consider the following notations:

- $v_i^\ell = e_{(k+1)(i-1)+\ell}$, for any $\ell = 1, \cdots, k+1$,
- $a_i = \dfrac{1}{k+1} \sum_{\ell=1}^{k+1} v_i = \mathrm{bar}\{v_i^\ell\}_{\ell=1}^{k+1}$, that is, a_i is the barycenter of $\{v_i^\ell\}_{\ell=1}^{k+1}$,
- $\Delta_i = \mathrm{cx}\left\{v_i^\ell\right\}_{\ell=1}^{k+1}$,
- $S_i = \partial \Delta_i$.

Lemma 3.7 *For any subset $\sigma \subseteq [m]$, the collections $\{\Delta_i\}_{i\in\sigma}$, $\{S_i\}_{i\in\sigma}$, and $\{a_i\}_{i\in\sigma}$ are, respectively, families of geometrically joinable compact subspaces of $\mathbb{R}^n$.*

Proof Consider the following identity of convex combinations

$$\sum_{i=1}^{|\sigma|} \lambda_i x_i = \sum_{i=1}^{|\sigma|} \lambda_i' x_i', \tag{11}$$

for some $x_i, x_i' \in \Delta_i$ and $\lambda_i, \lambda_i' \in I$ with $\sum_{i=1}^{k} \lambda_i = 1 = \sum_{i=1}^{k} \lambda_i'$. The Eq. (11) is equivalent to

$$\sum_{i=1}^{|\sigma|} \sum_{\ell=1}^{k+1} \lambda_i t_i^\ell v_i^\ell = \sum_{i=1}^{|\sigma|} \sum_{\ell=1}^{k+1} \lambda_i' s_i^\ell v_i^\ell,$$

where $x_i = \sum_{\ell=1}^{k+1} t_i^\ell v_i^\ell$ and $x_i' = \sum_{\ell=1}^{k+1} s_i^\ell v_i^\ell$ are convex combinations. Since $\{v_i^\ell : i \in \sigma, \ell = 1, \cdots, k+1\}$ is affinely independent, then $\lambda_i t_i^\ell = \lambda_i' s_i^\ell$, for all $i \in \sigma, \ell = 1, \cdots, k+1$. Without loss of generality if $\lambda_i \neq 0$ (the proof is similar if we assume $\lambda_i' \neq 0$), then $t_i^\ell = \dfrac{\lambda_i'}{\lambda_i} s_i^\ell$. Hence, we have

$$\begin{aligned}\sum_{\ell=1}^{k+1} t_i^\ell = 1 &\Rightarrow \sum_{\ell=1}^{k+1} \frac{\lambda_i'}{\lambda_i} s_i^\ell = 1\\ &\Rightarrow \frac{\lambda_i'}{\lambda_i} \sum_{\ell=1}^{k+1} s_i^\ell = 1\end{aligned}$$

$$\begin{aligned}\sum_{\ell=1}^{k+1} t_i^\ell = 1 &\Rightarrow \frac{\lambda_i'}{\lambda_i} = 1, \text{ since } \sum_{\ell=1}^{k+1} s_i^\ell = 1\\ &\Rightarrow \lambda_i' = \lambda_i\\ &\Rightarrow t_i^\ell = s_i^\ell, \text{ for all } i \in \sigma, \ell = 1, \cdots, k+1\\ &\Rightarrow x_i = \sum_{\ell=1}^{k+1} t_i^\ell v_i^\ell = \sum_{\ell=1}^{k+1} s_i^\ell v_i^\ell = x_i'\\ &\Rightarrow x_i = x_i'.\end{aligned}$$

Therefore, the collection $\{\Delta_i\}_{i\in\sigma}$ is geometrically joinable. As the collection of boundaries of disjoint k-simplices Δ_i in $\mathbb{R}^n$, respectively, $\{S_i\}_{i\in\sigma}$ is a family of geometrically joinable compact subspaces of $\mathbb{R}^n$. Likewise, the collection of barycenters $\{a_i\}_{i\in\sigma}$ of the k-simplices Δ_i is geometrically joinable. □

Notation 3.8 For any $\sigma \subseteq [m]$ and by Lemma 3.7, consider the setting

- $J(\sigma) = \cup_{i\in\sigma}\{(k+1)(i-1)+\ell\}_{\ell=1}^{k+1} \subseteq [n]$,
- $\Delta_\sigma = \Delta(J(\sigma)) = \overline{*}_{i\in\sigma}\Delta_i$,
- $S_\sigma = \overline{*}_{i\in\sigma} S_i$,
- $S_\sigma^* = \overline{*}_{j\notin\sigma} S_j$,
- $a_\sigma = \text{cx}\{a_i : i \in \sigma\}$.

An example of this setup is illustrated in Fig. 3.

Lemma 3.9 *For any $\sigma \subseteq [m]$, the compact spaces a_σ and S_σ are geometrically joinable, and we have*

$$\Delta_\sigma = a_\sigma \overline{*} S_\sigma.$$

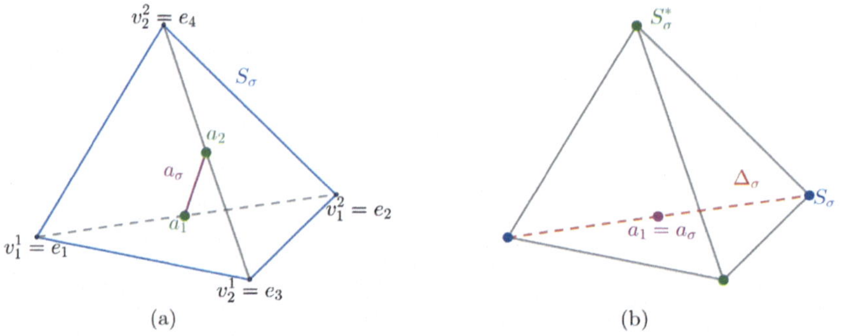

Fig. 3 Examples of a_i, S_i, a_σ, S_σ, and S_σ^*, for $m = 2$ and $k = 1$, that is $n = 4$. (**a**) For $\sigma = \{1, 2\}$. (**b**) For $\sigma = \{1\}$

Proof Let us prove a_σ and S_σ are geometrically joinable for $k = 1$ and for $\sigma = \{1, 2\}$; the general case can be proven similarly. Hence, S_σ can be split as follows:

$$S_\sigma = \underbrace{[e_1, e_3]}_{F_1} \cup \underbrace{[e_1, e_4]}_{F_2} \cup \underbrace{[e_2, e_3]}_{F_3} \cup \underbrace{[e_2, e_4]}_{F_4} \text{ and}$$

$$a_\sigma = \left[\frac{1}{2}e_1 + \frac{1}{2}e_2, \frac{1}{2}e_3 + \frac{1}{2}e_4\right].$$

The complexes a_σ and F_i, for each $i = 1, 2, 3, 4$, are both 1-simplices, and all their four vertices are not coplanar. So a_σ and F_i, for each $i = 1, 2, 3, 4$, are geometrically joinable, and their join $a_\sigma \overline{*} F_i$ is a 3-simplex. Also we have

$$a_\sigma \overline{*} (F_1 \cap F_2) = a_\sigma \overline{*} \{e_1\} = (a_\sigma \overline{*} F_1) \cap (a_\sigma \overline{*} F_2).$$

Hence, by Lemma 3.5, a_σ is geometrically joinable to $F_1 \cup F_2$. Similarly, we have

$$\begin{aligned} a_\sigma \overline{*} ((F_1 \cup F_2) \cap F_3) &= a_\sigma \overline{*} \{e_3\} \\ &= (a_\sigma \overline{*} (F_1 \cup F_2)) \cap (a_\sigma \overline{*} F_3). \end{aligned}$$

Hence, a_σ is geometrically joinable to $F_1 \cup F_2 \cup F_3$. Similarly, we have

$$\begin{aligned} a_\sigma \overline{*} ((F_1 \cup F_2 \cup F_3) \cap F_4) &= a_\sigma \overline{*} \{e_2, e_4\} \\ &= (a_\sigma \overline{*} (F_1 \cup F_2 \cup F_3)) \cap (a_\sigma \overline{*} F_4). \end{aligned}$$

Hence, a_σ is geometrically joinable to $S_\sigma = F_1 \cup F_2 \cup F_3 \cup F_4$. For any $i \in [m]$, $S_i = \partial \Delta_i$ and $a_i = \text{bar}\left\{v_i^\ell\right\}_{\ell=1}^{k+1}$. Then a_i and S_i are geometrically joinable, and we have $\Delta_i = a_i \overline{*} S_i$. We deduce

$$\begin{aligned}\Delta_\sigma &= \overline{\circledast}_{i\in\sigma}\Delta_i\\ &= \overline{\circledast}_{i\in\sigma}(a_i\overline{*}S_i)\\ &= (\overline{\circledast}_{i\in\sigma}a_i)\overline{*}(\overline{\circledast}_{i\in\sigma}S_i)\\ &= a_\sigma\overline{*}S_\sigma.\end{aligned}$$

□

4 Proof of Theorem 1.5

Now we have all the tools to write down the proof of Theorem 1.5 for $k \geq 1$. The case of $k = 0$ will be treated in the next section. In the following, we consider the notations introduced in Sects. 2 and 3.

Proof Setting $W_\sigma = \Delta_\sigma\overline{*}S^*_\sigma$ for each $\sigma \in K$, we have

$$\bigcup_{\sigma\in K} W_\sigma = \bigcup_{\sigma\in K}\left((\overline{\circledast}_{i\in\sigma}\Delta_i)\,\overline{*}\,(\overline{\circledast}_{j\notin\sigma}S_j)\right). \tag{12}$$

Consider the collection of simplices $A(K) = \{a_\sigma\}_{\sigma\in K}$, which is a geometric realization of the simplicial complex K and

$$|A(K)| = \bigcup_{\sigma\in K} a_\sigma \tag{13}$$

to be the underlying subspace of $\mathbb{R}^n = \mathbb{R}^{(k+1)m}$. The complexes $a_{[m]}$ and $S_{[m]}$ are geometrically joinable by Lemma 3.9, and also $A(K)$ is a subcomplex of $a_{[m]}$. Therefore, $|A(K)|$ and $S_{[m]}$ are geometrically joinable by Remark 3.4, and we have $S_{[m]}\overline{*}|A(K)| \cong S^{km-1} * |K| \cong \Sigma^{km}|K|$, since $S_{[m]} \cong S^{km-1}$. Then,

$$\begin{aligned}W_\sigma &= \Delta_\sigma\overline{*}S^*_\sigma\\ &= a_\sigma\overline{*}(S_\sigma\overline{*}S^*_\sigma) \text{ by Lemma 3.9}\\ &= a_\sigma\overline{*}S_{[m]}.\end{aligned}$$

This is illustrated by Fig. 4, where W_σ is the union of the two blue triangular surfaces. In this example, we have $S_{[m]} \cong \partial W_\sigma$.

Hence, from Eq. (13), we also have

$$\bigcup_{\sigma\in K} W_\sigma \cong \Sigma^{km}|K|. \tag{14}$$

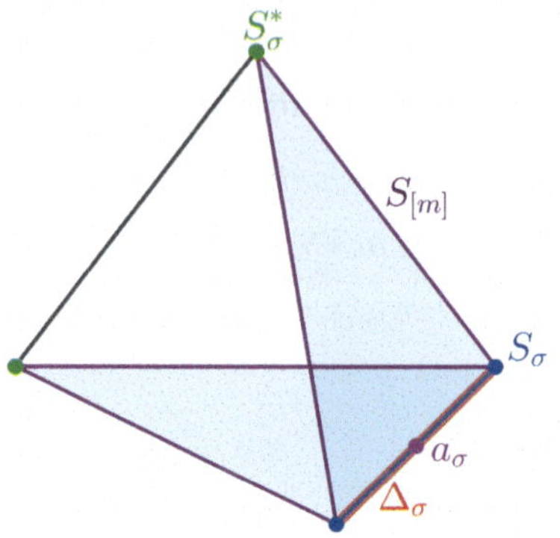

Fig. 4 Examples of W_σ, for $m = 2$, $\sigma = \{1\}$, and $k = 1$, that is, $n = 4$

Therefore, we have

$$\begin{aligned}\Psi\left(C\left(\bigcup_{\sigma\in K} W_\sigma\right)\right) &\cong \bigcup_{\sigma\in K}\Psi\left(CW_\sigma\right)\\ &= \bigcup_{\sigma\in K}\Psi\left(C\left(\left(\overline{\ast}_{i\in\sigma}\Delta_i\right)\overline{\ast}\left(\overline{\ast}_{j\notin\sigma}S_j\right)\right)\right) \text{ by (12)}\end{aligned}$$

$$\begin{aligned}\Psi\left(C\left(\bigcup_{\sigma\in K} W_\sigma\right)\right) &\cong \bigcup_{\sigma\in K}\Psi\left(\prod_{i\in\sigma}(C\Delta_i)\times\prod_{j\notin\sigma}(CS_j)\right) \text{ by Lemma 2.9}\\ &\cong \bigcup_{\sigma\in K}\Psi\left(\prod_{i\in\sigma}C_i^{k+1}\times\prod_{j\notin\sigma}\partial_- C_j^{k+1}\right) \text{ by Lemma 2.7,} \qquad (15)\\ &\quad\text{where } C_i^{k+1} \text{ and } C_j^{k+1} \text{ are copies of the } (k+1)\text{-cube.}\end{aligned}$$

Then we get

$$\begin{aligned}\overline{\Psi}\left(\Sigma\bigcup_{\sigma\in K} W_\sigma\right) &\cong \bigcup_{\sigma\in K}\overline{\Psi}\left(\Sigma W_\sigma\right)\\ &\cong \bigcup_{\sigma\in K}\omega\left(\prod_{i\in\sigma}C_i^{k+1}\times\prod_{j\notin\sigma}\partial_- C_j^{k+1}\right) \text{ by (15) and Lemma 2.7}\\ &\cong \bigcup_{\sigma\in K}\left(\bigwedge_{i\in\sigma}\widetilde{D}_i^{k+1}\wedge\bigwedge_{j\notin\sigma}\partial\widetilde{D}_i^{k+1}\right) \text{ by Remark 2.8, where } \widetilde{D}_i^{k+1}\\ &\quad\text{are copies of the nonstandard } (k+1)\text{-disk } \widetilde{D}^{k+1}.\end{aligned}$$

$$\overline{\Psi}\left(\Sigma \bigcup_{\sigma\in K} W_\sigma\right) \cong \bigcup_{\sigma\in K}\left(\bigwedge_{i\in\sigma} D_i^{k+1} \wedge \bigwedge_{j\notin\sigma} S_j^k\right)$$ by Lemma 2.3, where D_i^{k+1} and S_j^k are (resp.) copies of the $(k+1)$-disk D^{k+1} and the k-sphere S^k.

$$\cong \widehat{Z}\left(K; \left(D^{k+1}, S^k\right)\right). \tag{16}$$

On the other hand, we obtain

$$\begin{aligned}\overline{\Psi}\left(\Sigma \bigcup_{\sigma\in K} W_\sigma\right) &\cong \overline{\Psi}\left(\Sigma\Sigma^{km}|K|\right) \text{ by (14)}\\ &\cong \overline{\Psi}\left(\Sigma^{km+1}|K|\right)\\ &\cong \Sigma^{km+1}|K| \text{ by Lemma 2.7.}\end{aligned} \tag{17}$$

Therefore, by (16) and (17), we have

$$\widehat{Z}\left(K; \left(D^{k+1}, S^k\right)\right) \cong \Sigma^{km+1}|K|.$$

Hence, we have the result. □

5 The Case $k = 0$

In the previous section, we have proven Theorem 1.5 for $k \geq 1$. Here we prove the remaining case, namely, $k = 0$, given by as follows.

Theorem 5.1 *There is a homeomorphism*

$$\widehat{Z}\left(K; \left(D^1, S^0\right)\right) \cong \Sigma|K|.$$

For the purpose of the proof, we will consider the categorical definition of the polyhedral smash product given by (4). As in the third bullet of Notation 2.4, consider the functor

$$\begin{aligned}\Delta : \text{CAT}(K) &\to \text{Top}\\ \sigma &\mapsto \Delta(\sigma) \cong \Delta^{|\sigma|-1}.\end{aligned}$$

The geometric realization of K is given by

$$|K| = \operatorname*{colim}_{\sigma\in K} \Delta(\sigma). \tag{18}$$

Proof Consider the composite functor $\Sigma\Delta$ given by

$$\begin{aligned}\Sigma\Delta : \text{CAT}(K) &\to \text{Top}\\ \sigma &\mapsto \Sigma\Delta(\sigma) \cong \Sigma\Delta^{|\sigma|-1}.\end{aligned}$$

Let $\tau \subseteq \sigma$ be a face inclusion in K, with $|\tau| = p \leq \ell = |\sigma|$. We will look at the case $\tau = \{1, \cdots, p\} \subseteq \{1, \cdots, \ell\} = \sigma$ for simplicity; the same argument works for the general case. Consider the two following maps:

$$\begin{aligned}\phi_1 : C\Delta^{p-1} &\to C\Delta^{\ell-1}\\ \left[c, (t_1, \cdots, t_p), \lambda\right] &\mapsto \left[c, (t_1, \cdots, t_p, \underbrace{0, \cdots, 0}_{\ell-p \text{ times}}), \lambda\right]\end{aligned}$$

and

$$\begin{aligned}\phi_2 : C^p &\to C^\ell\\ (t_1, \cdots, t_p) &\mapsto (t_1, \cdots, t_p, \underbrace{0, \cdots, 0}_{\ell-p \text{ times}}).\end{aligned}$$

The transformation $\overline{\Psi} : \Sigma\Delta \Rightarrow \widehat{D}$, where the functor $\widehat{D}$ is defined by (3) for the pair $\left(D^1, S^0\right)$, is a natural isomorphism if the diagram (19) commutes since it induces the commutative diagram (20) below:

$$\begin{array}{ccc} C\Delta^{p-1} & \xrightarrow[\cong]{\Psi_p} & C^p \\ \phi_1 \downarrow & & \downarrow \phi_2 \\ C\Delta^{\ell-1} & \xrightarrow[\Psi_\ell]{\cong} & C^\ell \end{array} \tag{19}$$

$$\begin{array}{ccc} \Sigma\Delta^{p-1} & \xrightarrow[\cong]{\overline{\Psi}_p} & \widetilde{D}^p \\ \overline{\phi}_1 \downarrow & & \downarrow \overline{\phi}_2 \\ \Sigma\Delta^{\ell-1} & \xrightarrow[\overline{\Psi}_\ell]{\cong} & \widetilde{D}^\ell, \end{array} \tag{20}$$

where $\overline{\phi}_1([s_i, x, \lambda]) = [\phi_1([c, x, \lambda])]$, for all $[s_i, x, \lambda] \in \Sigma\Delta^{p-1}$, $\widetilde{D}^p = \underbrace{C^1 \wedge \cdots \wedge C^1}_{p \text{ times}}$ and so $\overline{\phi}_2([y]) = [\phi_2(y)]$, for all $y \in C^p$. For $[c, (t_1, \cdots, t_p), \lambda] \in C\Delta^{p-1}$, we have

$$
\begin{aligned}
&\Psi_\ell\phi_1\ ([c,(t_1,\cdots,t_p),\lambda]) \\
&= \Psi_\ell([c,(t_1,\cdots,t_p,\underbrace{0,\cdots,0}_{\ell-p\text{ times}}),\lambda]) \\
&= \begin{cases} 2\lambda(t_1,\cdots,t_p,\underbrace{0,\cdots,0}_{\ell-p\text{ times}}), \text{ if } 0\le\lambda\le\frac{1}{2} \\ \left((2-2\lambda)+(2\lambda-1)\frac{2}{\max_{1\le i\le p}\{t_i\}}\right)(t_1,\cdots,t_p,\underbrace{0,\cdots,0}_{\ell-p\text{ times}}), \text{ if } \frac{1}{2}\le\lambda\le 1 \end{cases} \quad \text{by (6)}
\end{aligned}
$$

$$
\begin{aligned}
&\phi_2\Psi_p\ ([c,(t_1,\cdots,t_p),\lambda]) \\
&= \begin{cases} \phi_2(2\lambda(t_1,\cdots,t_p)), \text{ if } 0\le\lambda\le\frac{1}{2} \\ \phi_2\left(\left((2-2\lambda)+(2\lambda-1)\frac{2}{\max_{1\le i\le p}\{t_i\}}\right)(t_1,\cdots,t_p)\right), \text{ if } \frac{1}{2}\le\lambda\le 1 \end{cases} \quad \text{by (6)} \\
&= \begin{cases} 2\lambda(t_1,\cdots,t_p,\underbrace{0,\cdots,0}_{\ell-p\text{ times}}), \text{ if } 0\le\lambda\le\frac{1}{2} \\ \left((2-2\lambda)+(2\lambda-1)\frac{2}{\max_{1\le i\le p}\{t_i\}}\right)(t_1,\cdots,t_p,\underbrace{0,\cdots,0}_{\ell-p\text{ times}}), \text{ if } \frac{1}{2}\le\lambda\le 1. \end{cases}
\end{aligned}
$$

Then the two diagrams commute, and therefore $\overline{\Psi}$ is a natural isomorphism. Passing to the colimit, we have

$$
\operatorname{colim}\overline{\Psi} : \operatorname*{colim}_{\sigma\in K}\Sigma\Delta(\sigma) \xrightarrow{\cong} \operatorname*{colim}_{\sigma\in K}\widehat{D}(\sigma). \tag{21}
$$

But we have

$$
\operatorname*{colim}_{\sigma\in K}\Sigma\Delta(\sigma) \cong \Sigma\operatorname*{colim}_{\sigma\in K}\Delta(\sigma) = \Sigma|K| \text{ by (18)}
$$

and also considering identity (4), the homeomorphism (21) yields a homeomorphism

$$
\Sigma|K| \cong \widehat{Z}\left(K;\left(D^1,S^0\right)\right).
$$

Hence, we have the result. □

6 Generalization of Theorem 1.5

In this section, we generalize Theorem 1.5 further, using an argument kindly provided by the referee. Instead of doubling all the vertices of K simultaneously as in David Stone's original construction, we double one vertex at a time and argue inductively, starting from the case $k = 0$ in Theorem 5.1. Let us start by setting some notation and stating intermediate results.

Notation 6.1 In this section, we consider the following setup:

- For $J = (j_1, \cdots, j_m)$ an m-tuple from $(\mathbb{N} \cup \{0\})^m$, denote the family of CW-pairs

$$\left(\underline{D^{J+1}}, \underline{S^J}\right) = \left\{\left(D^{j_1+1}, S^{j_1}\right), \cdots, \left(D^{j_m+1}, S^{j_m}\right)\right\}.$$

- Set $J_i = (0, \cdots, 1, \cdots, 0)$ to be the m-tuple having 1 only at the i-th position and 0 elsewhere. For a simplicial complex K over $[m]$ and $i \in [m]$, consider the new simplicial complex $K(J_i)$ with $m + 1$ vertices labeled $\{1, \cdots, i-1, i_a, i_b, i+1, \cdots, m\}$ and defined by

$$\begin{aligned} K(J_i) := \; & \{(\sigma \backslash \{i\}) \cup \{i_a, i_b\} \mid \sigma \in K \text{ and } i \in \sigma\} \\ & \cup \{\sigma \cup \{i_a\} \mid \sigma \in K \text{ and } i \notin \sigma\} \\ & \cup \{\sigma \cup \{i_b\} \mid \sigma \in K \text{ and } i \notin \sigma\} \cup \{\text{all their subsets}\}. \end{aligned}$$

The meaning behind the introduction of $K(J_i)$ is illustrated in the following example.

Example 6.2 For $m = 2$, consider $K = \{\emptyset, \{1\}, \{2\}\}$ and $J_1 = (1, 0)$. We have

$$\begin{aligned} \widehat{Z}\left(K; \left(\underline{D^{J_1+1}}, \underline{S^{J_1}}\right)\right) &= \widehat{Z}\left(K; \left\{\left(D^2, S^1\right), \left(D^1, S^0\right)\right\}\right) \\ &= \widehat{D}(\{1\}) \cup \widehat{D}(\{2\}) \quad \text{as a subspace of } D^2 \wedge D^1 \\ &= D^2 \wedge S^0 \cup S^1 \wedge D^1 \\ &\cong \left(D^1 \wedge D^1\right) \wedge S^0 \cup \left(D^1 \wedge S^0 \cup S^0 \wedge D^1\right) \wedge D^1 \\ &= \left(D^1 \wedge D^1 \wedge S^0\right) \cup \left(D^1 \wedge S^0 \wedge D^1\right) \cup \left(S^0 \wedge D^1 \wedge D^1\right) \\ &= \widehat{D}\left(\{1_a, 1_b\}\right) \cup \widehat{D}\left(\{2\} \cup \{1_a\}\right) \cup \widehat{D}\left(\{2\} \cup \{1_b\}\right) \\ &= \widehat{Z}\left(K(J_1); \left\{\left(D^1, S^0\right), \left(D^1, S^0\right), \left(D^1, S^0\right)\right\}\right) \\ &= \widehat{Z}\left(K(J_1); \left(D^1, S^0\right)\right), \text{ see Fig. 5.} \end{aligned}$$

Fig. 5 $K(J_1) = \{\{1_a, 1_b\}, \{1_a, 2\}, \{1_b, 2\}, \text{their subsets}\}$

The next lemma suggests that the polyhedral smash product $\widehat{Z}\left(K; (\underline{D^{J+1}}, \underline{S^J})\right)$ can be computed iteratively with steps involving $K(J_i)$ for some $i \in [m]$.

Lemma 6.3 *Let $J = (j_1, \cdots, j_m)$ to be an m-tuple and $i \in [m]$ such that $j_i \neq 0$. There is a homeomorphism*

$$\widehat{Z}\left(K; \left(\underline{D^{J+1}}, \underline{S^J}\right)\right) \cong \widehat{Z}\left(K(J_i); \left(\underline{D^{J'+1}}, \underline{S^{J'}}\right)\right),$$

where J' is the $(m+1)$-tuple $J' = (j_1, \cdots, j_i - 1, 0, \cdots, j_m)$.

Proof The polyhedral smash product $\widehat{Z}\left(K; (\underline{D^{J+1}}, \underline{S^J})\right)$ is defined as follows:

$$\begin{aligned}
\widehat{Z}\left(K; \left(\underline{D^{J+1}}, \underline{S^J}\right)\right) &= \widehat{Z}\left(K; \left\{\left(D^{j_1+1}, S^{j_1}\right), \cdots, \left(D^{j_i+1}, S^{j_i}\right), \cdots, \left(D^{j_m+1}, S^{j_m}\right)\right\}\right) \\
&= \bigcup_{\sigma \in K} \left(\widehat{D}(\sigma)\right) \quad \text{as a subspace of } \bigwedge_{\ell=1}^{m} D^{j_\ell+1} \cong \widetilde{D}^{j_1+\cdots+j_m+m} \\
&= \bigcup_{\sigma \in K} \left(\bigwedge_{\ell=1}^{m} Y_\ell\right), \quad \text{where } Y_\ell \text{ is given by (2)} \\
&= \bigcup_{\sigma \in K, i \in \sigma} \left(\bigwedge_{\ell=1}^{m} Y_\ell\right) \cup \bigcup_{\sigma \in K, i \notin \sigma} \left(\bigwedge_{\ell=1}^{m} Y_\ell\right) \\
&= \bigcup_{\sigma \in K, i \in \sigma} \left(Y_1 \wedge \cdots \wedge D^{j_i+1} \wedge \cdots \wedge Y_m\right) \\
&\quad \cup \bigcup_{\sigma \in K, i \notin \sigma} \left(Y_1 \wedge \cdots \wedge S^{j_i} \wedge \cdots \wedge Y_m\right) \\
&\cong \bigcup_{\sigma \in K, i \in \sigma} \left(Y_1 \wedge \cdots \wedge \left(D^{j_i} \wedge D^1\right) \wedge \cdots Y_m\right) \\
&\quad \cup \bigcup_{\sigma \in K, i \notin \sigma} \left(Y_1 \wedge \cdots \wedge \left(D^{j_i} \wedge S^0 \cup S^{j_i-1} \wedge D^1\right) \wedge \cdots Y_m\right)
\end{aligned}$$

$$
\begin{aligned}
&= \bigcup_{\sigma\in K, i\in\sigma} \left(Y_1 \wedge \cdots \wedge D^{j_i} \wedge D^1 \wedge \cdots Y_m\right) \\
&\cup \bigcup_{\sigma\in K, i\notin\sigma} \left(Y_1 \wedge \cdots \wedge D^{j_i} \wedge S^0 \wedge \cdots Y_m\right) \\
&\cup \bigcup_{\sigma\in K, i\notin\sigma} \left(Y_1 \wedge \cdots \wedge S^{j_i-1} \wedge D^1 \wedge \cdots Y_m\right)
\end{aligned}
$$

$$
\begin{aligned}
\widehat{Z}\left(K; \left(\underline{D^{J+1}}, \underline{S^J}\right)\right) &= \bigcup_{\sigma\in K, i\in\sigma} \widehat{D}\left((\sigma\backslash\{i\}) \cup \{i_a, i_b\}\right) \cup \bigcup_{\sigma\in K, i\notin\sigma} \widehat{D}\left(\sigma \cup \{i_a\}\right) \\
&\cup \bigcup_{\sigma\in K, i\notin\sigma} \widehat{D}\left(\sigma \cup \{i_b\}\right) \\
&= \widehat{Z}\left(K(J_i); \left(\underline{D^{J'+1}}, \underline{S^{J'}}\right)\right),
\end{aligned}
$$

where J' is the $(m+1)$-tuple $J' = (j_1, \cdots, j_i - 1, 0, \cdots, j_m)$. □

Example 6.4

1. For $m = 3$, consider $K = \{\{1, 2\}, \{3\}, \text{their subsets}\}$ and $J = (1, 1, 0)$. We have

$$
\begin{aligned}
\widehat{Z}\left(K; \left(\underline{D^{J+1}}, \underline{S^J}\right)\right) &= \widehat{Z}\left(K; \left\{\left(D^2, S^1\right), \left(D^2, S^1\right), \left(D^1, S^0\right)\right\}\right) \\
&= \widehat{D}(\{1, 2\}) \cup \widehat{D}(\{3\}) \text{ as a subspace of } D^2 \wedge D^2 \wedge D^1 \\
&= D^2 \wedge D^2 \wedge S^0 \cup S^1 \wedge S^1 \wedge D^1 \\
&\cong D^2 \wedge \left(D^1 \wedge D^1\right) \wedge S^0 \cup S^1 \wedge \left(D^1 \wedge S^0 \cup S^0 \wedge D^1\right) \\
&\quad \wedge D^1 \\
&= \left(D^2 \wedge D^1 \wedge D^1 \wedge S^0\right) \cup \left(S^1 \wedge D^1 \wedge S^0 \wedge D^1\right) \\
&\quad \cup \left(S^1 \wedge S^0 \wedge D^1 \wedge D^1\right) \\
&= \widehat{D}((\{1, 2_a, 2_b\}) \cup \widehat{D}(\{2_a, 3\}) \cup \widehat{D}(\{2_b, 3\}) \\
&= \widehat{Z}\left(K(J_2); \left\{\left(D^2, S^1\right), \left(D^1, S^0\right), \left(D^1, S^0\right), \left(D^1, S^0\right)\right\}\right) \\
&= \widehat{Z}\left(K(J_2); \left(\underline{D^{J_1+1}}, \underline{S^{J_1}}\right)\right), \text{ with } J_1 = (1, 0, 0, 0) \\
&= \widehat{Z}\left(K(J_2)(J_1); \left(D^1, S^0\right)\right) \\
&= \widehat{Z}\left(K(J); \left(D^1, S^0\right)\right), \text{ where } K(J) = K(J_2)(J_1).
\end{aligned}
$$

Fig. 6 $K(J_2) = \{\{1, 2_a, 2_b\}, \{2_a, 3\}, \{2_b, 3\}, \text{their subsets}\}$

See Fig. 6.

2. For $m = 3$, consider $K = \{\{1, 2\}, \{1, 3\}, \{2, 3\}, \text{their subsets}\}$ and $J = (2, 1, 1)$. We have

$$\widehat{Z}\left(K; \left(\underline{D^{J+1}}, \underline{S^J}\right)\right) = \widehat{Z}\left(K; \left\{\left(D^3, S^2\right), \left(D^2, S^1\right), \left(D^2, S^1\right)\right\}\right)$$
$$\cong \widehat{Z}\left(K(J_1); \left(\underline{D^{J'+1}}, \underline{S^{J'}}\right)\right), \text{ where } J' = (1, 0, 1, 1).$$

With

$$K(J_1) = \{\{1_a, 1_b, 2\}, \{1_a, 1_b, 3\}, \{1_a, 2, 3\}, \{1_b, 2, 3\}, \text{their subsets}\},$$

which is a simplicial complex with $m + 1 = 4$ vertices.

The second intermediate result is given by the following lemma, which states that the geometric realization of the simplicial complex $K(J_i)$ can be obtained just by considering a single suspension of the geometric realization of the simplicial complex K.

Lemma 6.5 *For any $i \in [m]$, we have*

$$|K(J_i)| \cong \Sigma|K|.$$

Proof Set $S^0 = \{s_1, s_2\}$ to be the 0-sphere. We have

$$\begin{aligned}
\Sigma|K| &= S^0 * |K| \\
&= \{s_1, s_2\} * \left(\bigcup_{\sigma \in K} |\sigma|\right) \\
&= \bigcup_{\sigma \in K} (\{s_1, s_2\} * |\sigma|) \\
&= \left(\bigcup_{\sigma \in K, i \in \sigma} (\{s_1, s_2\} * |\sigma|)\right) \cup \left(\bigcup_{\sigma \in K, i \notin \sigma} (\{s_1, s_2\} * |\sigma|)\right)
\end{aligned}$$

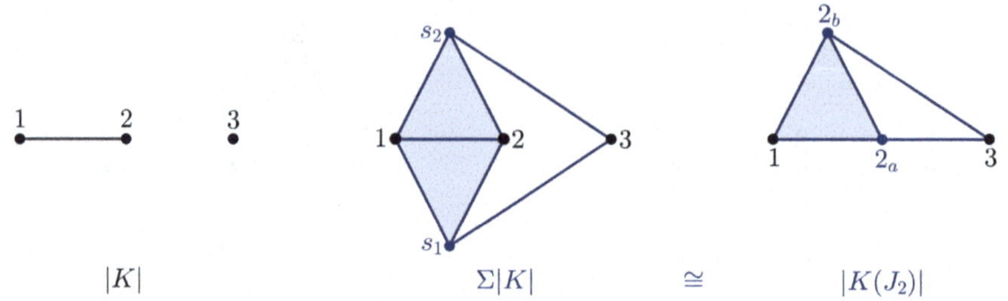

Fig. 7 $|K(J_2)| \cong \Sigma|K|$

$$
\begin{aligned}
&\cong \left(\bigcup_{\sigma\in K, i\in\sigma} (|\sigma\backslash\{i\}\cup\{s_1, s_2\}|)\right) \cup \left(\bigcup_{\sigma\in K, i\notin\sigma} (|\sigma\cup\{s_1\}|\cup|\sigma\cup\{s_2\}|)\right)\\
&= \left(\bigcup_{\sigma\in K, i\in\sigma} (|\sigma\backslash\{i\}\cup\{s_1, s_2\}|)\right) \cup \left(\bigcup_{\sigma\in K, i\notin\sigma} (|\sigma\cup\{s_1\}|)\right)\\
&\quad \cup \left(\bigcup_{\sigma\in K, i\notin\sigma} (|\sigma\cup\{s_2\}|)\right)\\
&\cong \left(\bigcup_{\sigma\in K, i\in\sigma} (|\sigma\backslash\{i\}\cup\{i_a, i_b\}|)\right) \cup \left(\bigcup_{\sigma\in K, i\notin\sigma} (|\sigma\cup\{i_a\}|)\right)\\
&\quad \cup \left(\bigcup_{\sigma\in K, i\notin\sigma} (|\sigma\cup\{i_b\}|)\right)\\
&= \bigcup_{\tau\in K(J_i)} |\tau|\\
&= |K(J_i)|.
\end{aligned}
$$

□

The lemma is illustrated in Fig. 7 for the simplicial complex K from Example 6.4(1). Now we can state and prove the main result.

Theorem 6.6 *For any m-tuple $J = (j_1, \cdots, j_m)$ in $(\mathbb{N}\cup\{0\})^m$, there is a homeomorphism*

$$\widehat{Z}\left(K; \left(\underline{D^{J+1}}, \underline{S^J}\right)\right) \cong \Sigma^{j_1+\cdots+j_m+1}|K|.$$

Proof Applying Lemma 6.3 $\sum_{i=1}^{m} j_i$ times, we get

$$\widehat{Z}\left(K; \left(\underline{D^{J+1}}, \underline{S^{J}}\right)\right) \cong \widehat{Z}\left(K(J); \left(\underline{D^{1}}, \underline{S^{0}}\right)\right), \tag{22}$$

where $K(J)$ is a simplicial complex obtained by applying the basic move (doubling a single vertex) $\sum_{i=1}^{m} j_i$ times. By the base case $k = 0$ in Theorem 5.1, we have

$$\widehat{Z}\left(K(J); \left(\underline{D^{1}}, \underline{S^{0}}\right)\right) \cong \Sigma|K(J)|. \tag{23}$$

Also, applying Lemma 6.5 $\sum_{i=1}^{m} j_i$ times, we have

$$\begin{aligned} \Sigma|K(J)| &\cong \Sigma\Sigma^{j_1+\cdots+j_m}|K| \\ &= \Sigma^{j_1+\cdots+j_m+1}|K|. \end{aligned} \tag{24}$$

Therefore, by putting Eqs. (22), (23), and (24) together, we obtain

$$\widehat{Z}\left(K; \left(\underline{D^{J+1}}, \underline{S^{J}}\right)\right) \cong \Sigma^{j_1+\cdots+j_m+1}|K|.$$

Hence, we have the result. □

Acknowledgments I would like to thank the Fields Institute at the University of Toronto where I started this work during the Thematic Program on Toric Topology and Polyhedral Products, for the inspiring work environment and the valuable financial support. I am thankful to Anthony Bahri for sharing the private correspondence from David Stone with me. Also I would like to express my deep gratitude to both Donald Stanley and Martin Frankland for their scientific and financial contribution to the realization of this work. Finally, I am deeply grateful to the referee for the concrete idea and guideline regarding the methodology that helped me to build the entire Sect. 6.

References

1. A. Bahri, M. Bendersky, F.R. Cohen, S. Gitler, The polyhedral product functor: a method of decomposition for moment-angle complexes, arrangements and related spaces. Adv. Math. **225**(3), 1634–1668 (2010)
2. A. Bahri, M. Bendersky, F.R. Cohen, S. Gitler, Operations on polyhedral products and a new topological construction of infinite families of toric manifolds. Homology Homotopy Appl. **17**(2), 137–160 (2015)
3. V.M. Buchstaber, T.E. Panov, Toric topology, in *Mathematical Surveys and Monographs*, vol. 204 (American Mathematical Society, Providence, RI, 2015)
4. A. Hatcher, *Algebraic Topology* (Cambridge University, Cambridge, 2002)
5. D. Stone, Private communication with A. Bahri (2006)

Invariance of Polarization Induced by Symplectomorphisms

Ethan Ross

1 Introduction

Geometric quantization roughly amounts to associating complex vector spaces (preferably Hilbert spaces) to symplectic manifolds. The terminology arises from the realization of a symplectic manifold as a classical phase space and a Hilbert space as the space of quantum wave functions. One fruitful direction in quantization has been the Kirillov-Kostant-Souriau picture in which one associates to a given symplectic manifold (M, ω) a complex Hermitian line bundle with compatible covariant derivative $(L, \nabla) \to M$ so that the curvature $\mathrm{curv}(\nabla)$ is given by the symplectic form ω. The Hilbert space of L^2 sections then satisfies many of the naive axioms of quantization given by Dirac [11, axioms Q1-Q3, page 155], but it can be "too large" in some sense. In order to correct this issue with size, one possible approach is to introduce an object called a polarization. In this paper I shall only be considering real polarizations, which are possibly singular Lagrangian foliations P of the symplectic manifold (M, ω). Kähler polarizations, given by complex structures compatible with the symplectic form ω, are also widely used in the literature, but shall not be considered in this paper.

From here, one has at least two choices for what a quantization could be. One option uses the covariant derivative ∇, to define a sheaf of polarized sections $\mathcal{S}_{(P,\nabla)}$ of L. I then define the sheaf quantization to be

$$Q_{shf}(M, P, \nabla) = \bigoplus_n \check{H}^n(M, \mathcal{S}_{(P,\nabla)}), \tag{1}$$

E. Ross (✉)
Department of Mathematics, University of Toronto, Toronto, ON, Canada
e-mail: ethan.ross@mail.utoronto.ca

A. Bahri et al. (eds.), *Toric Topology and Polyhedral Products*, Fields Institute Communications 89, https://doi.org/10.1007/978-3-031-57204-3_14

where $\check{H}^n(M, \mathcal{S}_{(P,\nabla)})$ are the Čech cohomology groups associated to the sheaf. The other option makes use of the fact that P is a foliation to define distinguished leaves $\iota : B \to M$, called Bohr-Sommerfeld leaves, which admit nontrivial covariantly constant sections, that is, sections s of the pull back bundle $(\iota^* L, \iota^* \nabla)$ satisfying $\iota^* \nabla s = 0$ along TB. Writing $BS(P)$ for the Bohr-Sommerfeld leaves of P, I then define the other quantization, Bohr-Sommerfeld quantization, by

$$Q_{BS}(M, P, \nabla) = \bigoplus_{B \in BS(P)} \check{H}^0(B, \mathcal{S}_{(P,\nabla)}|_B), \tag{2}$$

where $\mathcal{S}_{(P,\nabla)}|_B$ denotes the sheaf of covariantly constant sections on B. Sometimes these two quantizations agree as shown by Sniatycki [8, Theorems 1.1 and 1.2], and other times they do not as shown by Hamilton [6, Theorem 8.10]. Since both are of interest, I will consider both in this paper.

A famous body of results in geometric quantization are the "invariance of polarization" results, where it can be shown that two naturally arising polarizations induce the same quantization. Standard examples include the Gelfand-Zeitlin system [5, Theorem 6.1] and the moduli space of flat $SU(2)$ connections [7, Theorem 8.3]. In this paper, I show a new kind of invariance theorem arising from the action of a symplectomorphism on the polarizations. The reason this invariance is "new" is due to the fact that I am comparing quantizations coming from two real polarizations. The more classic results cited above arose from comparing the quantizations of a real polarization with a Kähler polarization.

In particular, given a real polarization P and a symplectomorphism Φ on (M, ω), I define a new real polarization $\Phi^* P$ via the pushforward of the inverse of Φ. It then appears to be a tautology that the quantizations of M with respect to P and $\Phi^* P$ should agree; however, I could only show this holds if the symplectomorphism Φ lifts to a connection-preserving isomorphism on the prequantum line bundle $(L, \nabla) \to M$. It is not known to the author if this condition is necessary.

To set up the statements of the main results, let $\underline{\mathbb{C}}^\times_M$ denote the sheaf of locally constant, nonvanishing complex functions on M. I then obtain the following two theorems.

Theorem 1.1 *Let (M, ω) be a symplectic manifold, P a real polarization, $\Phi : M \to M$ a symplectomorphism, and $(L, \nabla) \to M$ a prequantum line bundle. If $\check{H}^1(M, \underline{\mathbb{C}}^\times_M) = 0$, then Φ induces an isomorphism between the sheaf quantizations*

$$Q_{shf}(M, P, \nabla) \to Q_{shf}(M, \Phi^* P, \nabla),$$

where $Q_{shf}(M, P, \nabla)$ is defined in Eq. (1).

Theorem 1.2 *Let (M, ω) be a symplectic manifold, P a real polarization, $\Phi : M \to M$ a symplectomorphism, and $(L, \nabla) \to M$ a prequantum line bundle. If $\check{H}^1(M, \underline{\mathbb{C}}^\times_M) = 0$, then Φ induces an isomorphism between the Bohr-Sommerfeld quantizations*

$$Q_{BS}(M, P, \nabla) \rightarrow Q_{BS}(M, \Phi^* P, \nabla),$$

where $Q_{BS}(M, P, \nabla)$ *is defined in Eq.* (2).

The upshot of the proofs of these theorems is that the isomorphism can explicitly be defined via lifts of the symplectomorphism Φ to the prequantum line bundle. Historically, most invariance results arise from counting the dimensions of the quantizations with respect to two choices of polarization, then showing these agree. So, even though the isomorphisms of Theorems 1.1 and 1.2 depend on choices, they could be considered more "canonical" in some sense.

Once again, it should be noted here that although Theorems 1.1 and 1.2 are invariance of polarization results, they are of a different nature than the classical results cited above. In particular, I am only comparing the quantizations arising from two real polarizations. There is a more general notion of polarization where one consider subbundles P of the complexified tangent bundle $T_{\mathbb{C}}M$. This gives rise to three classes of polarizations: real (which I am considering in this paper), mixed, and complex. In the cases of the Gelfand-Zeitlin system and the moduli space of flat $SU(2)$ connections, the main point of interest is that quantization by naturally arising real and complex polarizations gives the same quantization. The ideas in this paper do not apply to those results since symplectomorphisms do not change the "type" of a polarization. An interesting future direction of research could involve finding larger classes of symmetries than just symplectomorphisms which interpolate between the various kinds of polarizations and investigating if similar invariance results can be obtained. I am also unaware of any example of a symplectic manifold (M, ω) with $\check{H}^1(M, \underline{\mathbb{C}}^{\times}_M) \neq 0$ such that quantization is not preserved under the action of symplectomorphisms. This could also be an interesting area of future research.

I would like to thank Dr. Lisa Jeffrey, Dan Hudson, Carrie Clark, and the referee for reading through the drafts of this paper and pointing out the numerous grammatical mistakes.

2 Quantization

Let (M, ω) be a symplectic manifold.

Definition 2.1 A prequantum line bundle over M is a complex Hermitian line bundle with compatible connection $(L, \nabla) \rightarrow M$ such that

$$curv(\nabla) = \omega.$$

Naively, a quantization should attach to a symplectic manifold (thought of as a classical phase space) a Hilbert space and a map taking the classical observables $C^{\infty}(M)$ to self-adjoint operators on the Hilbert space. Prequantum line bundles achieve both of these objectives. The Hermitian structure of the complex line bundle together with the canonical volume form $\omega^n/n!$ allows us to equip the space $\Gamma_c(L)$ of compactly supported sections with the structure of a pre-Hilbert space. Then, using the compatible covariant derivative ∇, one can construct the desired map between classical and quantum observables. See Woodhouse [11, Chapter 8] for more details.

The reason why these are called "prequantum" line bundles and not "quantum" line bundles is due to the fact that even in the simplest cases, the resulting Hilbert space is too large in some sense. For example, equip $\mathbb{R}^{2n}$ with standard coordinates $(x_1, \ldots, x_n, y_1, \ldots, y_n)$ and the standard symplectic structure

$$\omega_0 = \sum_j dx_j \wedge dy_j.$$

Also define the prequantum line bundle (L, ∇) over $(\mathbb{R}^{2n}, \omega_0)$ by

$$L = \mathbb{R}^{2n} \times \mathbb{C}, \quad \nabla = d + i \sum_{j=1}^{n} x_j dy_j,$$

where I identify the sections of L with smooth complex-valued functions on $\mathbb{R}^{2n}$. It's then an easy exercise to show that the resulting Hilbert space is $L^2(\mathbb{R}^{2n})$. This is unsatisfactory since the quantization of symplectic $\mathbb{R}^{2n}$ should be $L^2(\mathbb{R}^n)$. Furthermore, important properties like the Heisenberg uncertainty relations between the coordinates will not hold in this Hilbert space.

An easy remedy for these problems would be to only consider functions f which only depend on half the variables, say $f = f(y_1, \ldots, y_n)$. Observe that these are precisely the functions which obey

$$\nabla_{\frac{\partial}{\partial x_j}} f = 0$$

for each j, that is, functions which are covariantly constant along the Lagrangian subbundle

$$P = \text{span}\{\frac{\partial}{\partial x_1}, \ldots, \frac{\partial}{\partial x_n}\}.$$

Generalizing this idea is where the concept of a polarization arises.

Definition 2.2 A (singular, real) polarization of (M, ω) is a singular subbundle $P \subset TM$ such that its sheaf of sections $\mathfrak{F}_P$ satisfies the following axioms.

(i) (Involutivity) If $X, Y \in \mathfrak{F}_P$, then so is $[X, Y]$.
(ii) (Locally Finitely Generated) For any $x \in M$, there exists an open neighborhood $U \subset M$ of x and sections $X_1, \ldots, X_k \in \mathfrak{F}_P(U)$ such that

$$\mathfrak{F}_P(U) = \text{span}_{C^\infty(U)}\{X_1, \ldots, X_k\}.$$

(iii) (Lagrangian) There exists open dense subset $U \subset M$ such that for each $x \in U$, $P_x \subset T_x M$ is Lagrangian.

Remark 2.3

(i) Involutivity and the locally finitely generated property of P mean that it defines a singular foliation in the sense of [2]. It then automatically follows that there exists an open dense subset $U \subset M$ such that $P|_U$ is an involutive smooth subbundle of TU.

(ii) As was noted in the introduction, there are other kinds of polarizations. In the non-singular case, a polarization is an involutive Lagrangian subbundle $P \subset T_{\mathbb{C}}M$, where $T_{\mathbb{C}}M$ is the complexified tangent bundle. Usually it is also demanded that $E = (P+\overline{P}) \cap TM$ and $D = P \cap \overline{P} \cap TM$ are also subbundles. P is called real if $D = E$, complex if $E = TM$ and $D = 0$, and mixed otherwise. See Andersen [1] for a more in-depth discussion.

Since Lagrangian subspaces have half the dimension of the symplectic manifold, demanding sections be covariantly constant along the directions associated to P is one mechanism for cutting down the variables on which the Hilbert space depends. With this motivation, define the sheaf of covariantly constant sections $\mathcal{S}_{(P,\nabla)}$ by

$$\mathcal{S}_{(P,\nabla)}(U) := \{s \in \Gamma(U, L) \mid \nabla_X s = 0 \; \forall \; X \in \mathfrak{F}_P(U)\}, \tag{3}$$

for open $U \subset M$.

This enables us to formally define the sheaf quantization for the purposes of this paper.

Definition 2.4 Define the sheaf quantization of (M, ω) with respect to the prequantum line bundle $(L, \nabla) \to M$ and the polarization $P \subset TM$ by

$$Q_{shf}(M, P, \nabla) := \bigoplus_n \check{H}^n(M, \mathcal{S}_{(P,\nabla)}), \tag{4}$$

where $\check{H}^n(M, \mathcal{S}_{(P,\nabla)})$ denotes the n-th Čech cohomology group of M with respect to the sheaf $\mathcal{S}_{(P,\nabla)}$.

As alluded to in the introduction, another common way of quantizing a symplectic manifold is by a Bohr-Sommerfeld quantization. To do this, observe that if $P \subset TM$ is a polarization, then P induces a decomposition of M into disjoint immersed submanifolds called leaves [9, Theorem 4.2]. If M is connected, the leaves of maximal dimension are immersed Lagrangian submanifolds.

Now, if $\iota : B \to M$ is a leaf of P, one can pull back the covariant derivative ∇ on L to the pull back bundle $\iota^* L \to B$ and write $\iota^* \nabla$ for the resulting covariant derivative. Thus, define a sheaf $\mathcal{S}_{(P,\nabla)}|_B$ on B by

$$\mathcal{S}_{(P,\nabla)}|_B(U) = \{s \in \Gamma(U, \iota^* L) \mid \iota^* \nabla_X s = 0 \; \forall X \in \mathfrak{X}(U)\}, \tag{5}$$

for open $U \subset B$.

Definition 2.5 A leaf B of P is called Bohr-Sommerfeld if

$$H^0(B, \mathcal{S}_{(P,\nabla)}|_B) \neq 0.$$

Writing $BS(P)$ for the set of Bohr-Sommerfeld leaves of P, define the Bohr-Sommerfeld quantization of (M, ω) with respect to the prequantum line bundle $(L, \nabla) \to M$ and the polarization P by

$$Q_{BS}(M, P, \nabla) := \bigoplus_{B \in BS(P)} H^0(B, \mathcal{S}_{(P,\nabla)}|_B). \tag{6}$$

Remark 2.6

(i) An equivalent definition of a Bohr-Sommerfeld leaf used in the literature is as follows. A Bohr-Sommerfeld leaf is a leaf $\iota : B \to M$ that admits a nontrivial covariantly constant section $s : B \to \iota^* L$, i.e., a covariantly constant section which is not identically the zero section.

(ii) If B is a Bohr-Sommerfeld leaf and is connected, then it follows that

$$H^0(B, \mathcal{S}_{(P,\nabla)}|_B) \cong \mathbb{C}.$$

3 Lifts of Symplectomorphisms

The main mechanism for comparing the quantizations by various polarizations is by lifts of symplectomorphisms to automorphisms of the line bundle. For the following discussion, fix a prequantum line bundle $(L, \nabla) \xrightarrow{\pi} (M, \omega)$, with M connected, and a symplectomorphism $\Phi : M \to M$.

Definition 3.1 A lift of the symplectomorphism Φ is a diffeomorphism $F : L \to L$ such that

(i) $\pi \circ F = \Phi \circ \pi$.

(ii) For each $x \in M$, the map

$$F_x : L_x \to L_{\Phi(x)}$$

is a linear isomorphism.

Example 3.2 If $L = M \times \mathbb{C}$ is a trivial line bundle, then a lift of $\Phi : M \to M$ is given by

$$F : L \to L; \quad (x, z) \mapsto (\Phi(x), f(x)z),$$

where $f \in C^\infty(M, \mathbb{C}^\times)$ is a nonvanishing smooth function.

Remark 3.3 Let X and Y be topological spaces and $f : X \to Y$ a continuous map. f then defines a functor between the category of sheaves on X to the category of sheaves on Y. Indeed, let $\mathcal{S}$ be a sheaf of X and define sheaf $f_*\mathcal{S}$ on Y by

$$f_*\mathcal{S}(U) := \mathcal{S}(f^{-1}(U)) \tag{7}$$

for open $U \subset Y$. Similarly, given a natural transformation

$$\eta : \mathcal{S} \to \mathcal{S}'$$

between sheaves $\mathcal{S}$ and $\mathcal{S}'$ on X, we can define natural transformation

$$f_*\eta : f_*\mathcal{S} \to f_*\mathcal{S}'$$

by

$$f_*\eta(U) = \eta(f^{-1}(U)) : \mathcal{S}(f^{-1}(U)) \to \mathcal{S}'(f^{-1}(U)) \tag{8}$$

for open $U \subset Y$.

Write Γ_L for the sheaf of sections of L. A lift F of Φ then defines a natural isomorphism

$$F_* : \Phi_*\Gamma_L \to \Gamma_L, \tag{9}$$

where for each open $U \subset M$ and each $s \in \Gamma_L(\Phi^{-1}(U))$, define $F_*s \in \Gamma_L(U)$ by

$$F_*s(x) := F_{\Phi^{-1}(x)}(s(\Phi^{-1}(x))), \quad x \in U.$$

It's easy to see that F_* respects restrictions and hence is indeed a natural transformation. The inverse to F thus also induces a natural transformation

$$(F^{-1})_* : (\Phi^{-1})_*\Gamma_L \to \Gamma_L.$$

Pushing this forward by Φ,

$$\Phi_*(F^{-1})_* : \Gamma_L \to \Phi_*\Gamma_L,$$

we thus get our inverse to F_*.

Now, let $\mathfrak{X}_M$ denote the sheaf of vector fields on M. Since Φ is a diffeomorphism, its derivative $d\Phi$ also defines a natural isomorphism

$$\Phi_* : \Phi_*\mathfrak{X}_M \to \mathfrak{X}_M.$$

Thus, since the covariant derivative ∇ is a natural transformation $\mathfrak{X}_M \otimes \Gamma_L \to \Gamma_L$, the natural isomorphisms Φ_* and F_* are then used to define a pullback covariant derivative.

Definition 3.4 Given a lift $F : L \to L$ of Φ, define the pullback covariant derivative $F^*\nabla$ to be the unique natural transformation

$$F^*\nabla : \mathfrak{X}_M \otimes \Gamma_L \to \Gamma_L$$

so that the diagram commutes

$$\begin{array}{ccc} \Phi_*(\mathfrak{X}_M \otimes \Gamma_L) & \xrightarrow{\Phi_*(F^*\nabla)} & \Phi_*\Gamma_L \\ \downarrow{\scriptstyle \Phi_* \times F_*} & & \downarrow{\scriptstyle F_*} \\ \mathfrak{X}_M \otimes \Gamma_L & \xrightarrow{\nabla} & \Gamma_L \end{array}$$

Remark 3.5 For any lift F of Φ, $F^*\nabla$ is a prequantum covariant derivative for $L \to (M, \omega)$, that is, the curvature of F^∇ is the symplectic form ω,

$$\text{curv}(F^*\nabla) = \omega.$$

This can be shown as follows. Let $U \subset M$ be open and $s \in \Gamma_L(U)$ nonvanishing, and then there exists unique $\alpha \in \Omega^1(U)$ such that

$$\nabla s = i\alpha \otimes s. \tag{10}$$

Since $\text{curv}(\nabla) = \omega$, it follows that $d\alpha = \omega|_U$. Now, define $t \in \Gamma_L(\Phi^{-1}(U))$ uniquely by $F_*t = s$. It's then a matter of unfolding the above definitions to see that t is nonvanishing and that

$$F^*\nabla t = i\Phi^*\alpha \otimes t. \tag{11}$$

Thus, since Φ is a symplectomorphism,

$$d\Phi^*\alpha = \omega|_{\Phi^{-1}(U)}.$$

Hence, $\text{curv}(F^*\nabla) = \omega$.

It will be important later to determine when a symplectomorphism Φ admits a covariant derivative preserving lift, that is, a lift $F : L \to L$ satisfying

$$F^*\nabla = \nabla.$$

If Φ does admit such a lift, then Φ will always preserve the quantization with respect to any polarization.

Lemma 3.6 *Let $\underline{\mathbb{C}}_M^\times$ be the sheaf of locally constant $\mathbb{C}^\times$-valued functions on M. If $\check{H}^1(M, \underline{\mathbb{C}}_M^\times) = 0$, then Φ admits a lift $F : L \to L$ satisfying $F^*\nabla = \nabla$.*

Proof Choose a good cover $\{U_j\}$[3], that is, for any finite collection of indices $j_1, \ldots, j_k$, the intersection

$$U_{j_1 \cdots j_k} := U_{j_1} \cap \cdots \cap U_{j_k} \tag{12}$$

is either empty or contractible. One can always choose such a cover since contractible covers are cofinal among all covers. Furthermore, since U_j is contractible for each j, there exists a nonvanishing section $s_j \in \Gamma_L(U_j)$.

Since each of the s_j is nonvanishing, we obtain useful local data.

- (Local primitives of ω) As was noted above, on each U_j, there exists unique $\alpha_j \in \Omega^1(U_j)$ such that

$$\nabla s_j = i\alpha_j \otimes s_j. \tag{13}$$

  These α_j all satisfy

$$d\alpha_j = \omega|_{U_j}. \tag{14}$$

- (Transition Functions) If $U_{jk} \neq \emptyset$, then there exists unique $\lambda_{jk} \in C^\infty(U_{jk}, \mathbb{C}^\times)$ satisfying

$$\lambda_{jk} s_k|_{U_{jk}} = s_j|_{U_{jk}}. \tag{15}$$

It's a straightforward calculation to show that on overlaps U_{jk}, the local primitives α_j and the transition functions λ_{jk} are related by

$$\alpha_j - \alpha_k = -id \log(\lambda_{jk}) \tag{16}$$

for any choice of branch of log.

Now, using the inverse image of Φ, we obtain another good cover $\{\Phi^{-1}(U_j)\}$, and so we can choose a new collection of nonvanishing local sections where $t_j \in \Gamma_L(\Phi^{-1}(U_j))$. Write β_j for the associated local primitives of ω, and write μ_{jk} for the associated transition functions. The sections t_j will now be suitably rescaled to define the lift F.

First, we may assume that $\beta_j = \Phi^*\alpha_j$. Otherwise, since U_j is contractible and since β_j and $\Phi^*\alpha_j$ are local primitives of ω, there exists $f_j \in C^\infty(\Phi^{-1}(U_j))$ such that

$$\beta_j - \Phi^*\alpha_j = df_j.$$

Now redefine $t'_j := e^{-if_j}t_j$. It then follows that

$$\nabla t'_j = i\Phi^*\alpha_j \otimes t'_j.$$

Next, we may assume that the transition functions μ_{jk} for the t_j are related to the transition functions of the s_j by

$$\mu_{jk} = \lambda_{jk} \circ \Phi.$$

Indeed, otherwise using Eq. (16), we have

$$-id\log(\lambda_{jk}) = \alpha_j - \alpha_k$$

and, by assumption, we also have

$$-id\log(\mu_{jk}) = \Phi^*\alpha_j - \Phi^*\alpha_k.$$

Thus,

$$d\log(\mu_{jk}) = d\log(\Phi^*\lambda_{jk}).$$

Thus, since U_{jk} is contractible, there exists $C_{jk} \in \mathbb{C}^\times$ such that

$$\mu_{jk} = C_{jk}\Phi^*\lambda_{jk}.$$

Observe that on triple overlaps $U_{jk\ell}$, the following cocycle condition holds

$$\lambda_{k\ell}\lambda_{j\ell}^{-1}\lambda_{jk} = 1.$$

Thus, the constants C_{jk} satisfy an analogous cocycle condition

$$C_{k\ell}C_{j\ell}^{-1}C_{jk} = 1.$$

Hence, the constants C_{jk} define a closed two-cocycle $\{C_{jk}\} \in \check{Z}^1(\{U_j\}, \underline{\mathbb{C}}_M^\times)$. Since $\check{H}^1(\{U_j\}, \underline{\mathbb{C}}_M^\times) = 0$, there exists a collection of constants $\{e_j\} \subset \mathbb{C}^\times$ such that

$$C_{jk} = e_k e_j^{-1}.$$

Now, redefine $t'_j := e_j t_j$. It's then straightforward to check that

$$\Phi^*\lambda_{jk}t'_k = t'_j$$

and the local primitives are still $\Phi^*\alpha_j$

$$\nabla t_j' = i\Phi^*\alpha_j \otimes t_j'.$$

Thus, we may locally define lifts F_j of Φ. For each j, set

$$F_j : L|_{\Phi^{-1}(U_j)} \to L|_{U_j}$$

uniquely by $(F_j)_* t_j = s_j$. Since the transition functions for t_j are given by the pullbacks of the transition functions of the s_j, it follows that $F_j = F_k$ on overlaps U_{jk}. Thus, we obtain a global lift $F : L \to L$ of Φ. By construction, we have

$$\nabla t_j = i\Phi^*\alpha_j \otimes t_j.$$

Further,

$$F^*\nabla t_j = i\Phi^*\alpha_j \otimes t_j.$$

Therefore, $F^*\nabla = \nabla$. □

4 Action by Symplectomorphisms on Polarizations

Consider the motivating example for polarizations, $\mathbb{R}^{2n}$ with the standard prequantum line bundle introduced in the beginning of Section 2. Physicists will (implicitly) use one of two polarizations to cut down on the number of variables. Recall that $\mathbb{R}^{2n}$ is given coordinates $x_1, \ldots, x_n, y_1, \ldots, y_n$, and then there are two naturally arising polarizations:

$$P = \text{span}\{\frac{\partial}{\partial x_1}, \ldots, \frac{\partial}{\partial x_n}\}$$

and

$$Q = \text{span}\{\frac{\partial}{\partial y_1}, \ldots, \frac{\partial}{\partial y_n}\}.$$

Using P as the polarization gives the "momentum representation" $Q_{shf}(\mathbb{R}^{2n}, P)$, and Q returns the "position representation" $Q_{shf}(\mathbb{R}^{2n}, Q)$. If one were to equip these vector spaces with a Hilbert space structure, this can be done using half-forms [1], and then both would be isomorphic to $L^2(\mathbb{R}^n)$. In particular, they could be viewed as the same quantization.

Another, perhaps more geometric approach would be to realize that

$$\Phi : \mathbb{R}^{2n} \to \mathbb{R}^{2n}; \quad (x_1, \ldots, x_n, y_1, \ldots, y_n) \mapsto (y_1, \ldots, y_n, -x_1, \ldots, -x_n)$$

is a symplectomorphism which fiberwise swaps the polarizations

$$\Phi_* P = Q, \quad \Phi_* Q = P.$$

Furthermore, if $U \subset \mathbb{R}^{2n}$ is open, then it's easy to see that precomposition by Φ defines an isomorphism of sheaves

$$\Phi^* : \mathcal{S}_{(P,\nabla)} \to \Phi_* \mathcal{S}_{(Q,\nabla)}.$$

Hence, Φ induces an isomorphism

$$Q_{shf}(\mathbb{R}^{2n}, P) \to Q_{shf}(\mathbb{R}^{2n}, Q)$$

between the momentum and position representations as desired.

I now want to generalize the above construction to more general prequantum line bundles and polarizations.

Definition 4.1 Fix a symplectic manifold (M, ω), a symplectomorphism $\Phi : M \to M$, and a polarization $P \subset TM$. Define a new polarization $\Phi^* P \subset TM$ pointwise by

$$\Phi^* P_x := d_{\Phi(x)} \Phi^{-1}(P_{\Phi(x)}). \tag{17}$$

Since Φ is a diffeomorphism, it easily follows that $\Phi^* P$ is an involutive, locally finitely generated, smooth singular subbundle of TM. Further, Φ being a symplectomorphism gives us that $\Phi^* P$ is generically Lagrangian in the sense of Definition 2.2.

4.1 Symplectomorphisms and Sheaf Quantization

For this section, fix a prequantum line bundle $(L, \nabla) \to (M, \omega)$, a polarization $P \subset TM$, and a symplectomorphism $\Phi : M \to M$. Recall that $\mathfrak{F}_P$ denotes the sheaf of sections of P.

Proposition 4.2 *Let $F : L \to L$ be a lift of Φ. Then the inverse of F defines a natural isomorphism*

$$\Phi_*(F^{-1})_* : \mathcal{S}_{(P,\nabla)} \to \Phi_* \mathcal{S}_{(\Phi^* P, F^* \nabla)},$$

where $\Phi_*(F^{-1})_*$ is the pushforward of the natural isomorphism $(F^{-1})_*$ by Φ as in Eq. (8).

Proof Let $U \subset M$ be open. Recall that the lift F defines an isomorphism

$$F_* : \Gamma_L(\Phi^{-1}(U)) \to \Gamma_L(U)$$

from Eq. (9). Fix $s \in \mathcal{S}_{(P,\nabla)}(U)$. I will now show that $(F^{-1})_* s \in \mathcal{S}_{(\Phi^* P, F^*\nabla)}(\Phi^{-1}(U))$, that is, for any $X \in \mathfrak{F}_{\Phi^* P}(\Phi^{-1}(U))$, I want to show

$$(F^*\nabla)_X F_*^{-1} s = 0.$$

This is quite straightforward since by definition of $F^*\nabla$, we have

$$(F^*\nabla)_X F_*^{-1} s = F_*^{-1}(\nabla_{\Phi_* X} s).$$

By construction of $\Phi^* P$, we have $\Phi_* X \in \mathfrak{F}_P(U)$. Hence, $\nabla_X s = 0$, and thus $(F^*\nabla)_X F_*^{-1} s = 0$. Clearly $F_* : \Phi_* \mathcal{S}_{(\Phi^* P, F^*\nabla)} \to \mathcal{S}_{(P,\nabla)}$ is the inverse. □

Corollary 4.3 *For any lift $F : L \to L$ of Φ, there exists a canonical isomorphism*

$$Q_{shf}(M, P, \nabla) \to Q_{shf}(M, \Phi^* P, F^*\nabla).$$

Proof The natural isomorphisms of Proposition 4.2 give an isomorphism

$$\check{H}^\bullet(M, \mathcal{S}_{(P,\nabla)}) \to \check{H}^\bullet(M, \Phi_* \mathcal{S}_{(\Phi^* P, F^*\nabla)})$$

All that's left to show is that there is an isomorphism between cohomology groups

$$\check{H}^\bullet(M, \Phi_* \mathcal{S}_{(\Phi^* P, F^*\nabla)}) \cong \check{H}^\bullet(M, \mathcal{S}_{(\Phi^* P, F^*\nabla)}).$$

For sake of convenience, set $\mathcal{S} = \mathcal{S}_{(\Phi^* P, F^*\nabla)}$. For any open cover $\{U_j\}$, we have equality of chain complexes

$$\check{C}^\bullet(\{U_j\}, \mathcal{S}) = \check{C}^\bullet(\{\Phi^{-1}(U_j), \Phi_* \mathcal{S}).$$

Thus, we get an isomorphism of cohomologies

$$\check{H}^\bullet(\{U_j\}, \mathcal{S}) \to \check{H}^\bullet(\{\Phi^{-1}(U_j)\}, \Phi_* \mathcal{S}).$$

It's easy to check that this isomorphism is compatible with refinement and hence induces a map

$$\check{H}^\bullet(M, \mathcal{S}) \to \check{H}^\bullet(M, \Phi_* \mathcal{S}).$$

It follows since Φ is a diffeomorphism that it is a bijection on open covers; hence, the above map is in fact an isomorphism. See Chapter 10 of Wedhorn [10] for more details. □

Proof of Theorem 1.1 Applying Lemma 3.6, there exists a lift $F : L \to L$ of Φ such that $F^*\nabla = \nabla$. Corollary 4.3 then gives the desired isomorphism

$$Q_{shf}(M, P, \nabla) \to Q_{shf}(M, \Phi^* P, \nabla).$$

□

4.2 *Symplectomorphisms and Bohr-Sommerfeld Leaves*

As before, fix a polarization $P \subset TM$ and a symplectomorphism $\Phi : M \to M$.

Lemma 4.4 *If $\iota : B \to M$ is a leaf of P, then $\Phi^{-1} \circ \iota : B \to M$ is a leaf of $\Phi^* P$.*

Proof By construction of $\Phi^* P$, it is clear that if $\iota : B \to M$ is an immersed integral submanifold of P, then $\Phi^{-1} \circ \iota : B \to M$ is an immersed integral submanifold of $\Phi^* P$. Indeed, for any $x \in B$, by definition

$$\iota_*(T_x B) = P_{\iota(x)}.$$

Hence,

$$(\Phi^{-1} \circ \iota)_*(T_x B) = \Phi_*^{-1} P_{\iota(x)} = \Phi^* P_{\Phi^{-1} \circ \iota(x)}.$$

The only issue now is maximality of $\Phi^{-1} \circ \iota : B \to M$. Suppose $\iota' : B' \to M$ is another integral submanifold of $\Phi^* P$ through $\Phi^{-1} \circ \iota(x)$ for some $x \in B$. Then, by the same argument, $\Phi \circ \iota' : B' \to M$ is an integral submanifold of P through $\iota(x)$. Thus, by the maximality of $\iota : B \to M$, there exists an open embedding

$$H : B' \to B$$

making the diagram commute

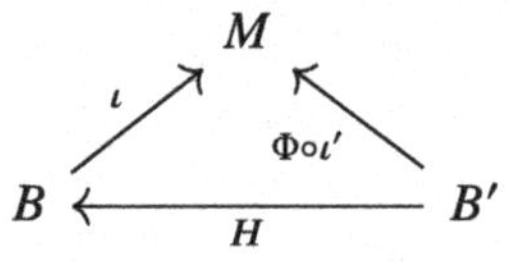

Since Φ is a diffeomorphism, it then follows that the below diagram commutes

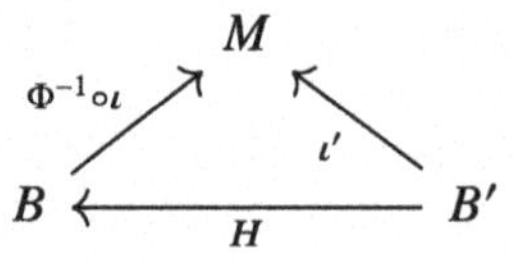

Hence, $\Phi^{-1} \circ \iota : B \to M$ is a maximal integral submanifold of $\Phi^* P$. □

For the sake of convenience, simply write B for the data of a leaf $\iota : B \to M$ of P and $\Phi^* B$ for the data of $\Phi^{-1} \circ \iota : B \to M$. It then becomes clear that Φ induces a bijection between the leaves of P and $\Phi^* P$.

Corollary 4.5 *If $Lf(P)$ denotes the set of leaves of P, then the map*

$$Lf(P) \to Lf(\Phi^* P); \quad B \mapsto \Phi^* B.$$

is a bijection.

Lemma 4.6 *Let Φ be a symplectomorphism and $F : L \to L$ a lift. Then, for any leaf $\iota : B \hookrightarrow M$, F induces an isomorphism*

$$(F_B)_* : H^0(B, \mathcal{S}_{(\Phi^* P, F^* \nabla)}|_{\Phi^* B}) \to H^0(B, \mathcal{S}_{(P, \nabla)}|_B). \tag{18}$$

Proof Fix a leaf $\iota : B \to M$ and a lift $F : L \to L$ of Φ. Define an isomorphism of line bundles

$$\begin{array}{ccc} (\Phi^{-1} \circ \iota)^* L & \xrightarrow{\quad F_B \quad} & \iota^* L \\ & \searrow \quad \swarrow & \\ & B & \end{array}$$

by

$$F_B(x, z) = (x, F_{\iota(x)}(z)).$$

One then checks that the following identity holds

$$F_B^*(\iota^* \nabla) = (\Phi^{-1} \circ \iota)^* (F^* \nabla).$$

Thus, if $s : B \to (\Phi^{-1} \circ \iota)^* L$ is a section satisfying

$$(\Phi^{-1} \circ \iota)^* (F^* \nabla)_X s = 0$$

for all vector fields X on B, then

$$(\iota^* \nabla)_X (F_B)_* s = (F_B)_* ((\Phi^{-1} \circ \iota)^* (F^* \nabla)_X s) = (F_B)_*(0) = 0.$$

Hence, one obtains a map

$$(F_B)_* : H^0(B, \mathcal{S}_{(\Phi^* P, F^* \nabla)}|_{\Phi^* B}) \to H^0(B, \mathcal{S}_{(P, \nabla)}|_B).$$

It's clear that $(F_B^{-1})_*$ is the inverse. □

Proof of Theorem 1.2 By Lemma 3.6, Φ admits a lift $F : L \to L$ such that $F^*\nabla = \nabla$. Due to Lemma 4.6, the bijection in Corollary 4.5 restricts to a bijection

$$BS(P) \to BS(\Phi^* P).$$

Furthermore, taking the direct sum of the isomorphisms in Eq. (18), the desired isomorphism is given by

$$\bigoplus_{B \in BS(P)} (F_B)_* : Q_{BS}(M, \Phi^* P, \nabla) \to Q_{BS}(M, P, \nabla)$$

□

5 Application to Toric Geometry

To finish off, let's discuss an application to toric varieties. There are easier proofs that don't require the machinery developed above, but it at least illustrates how the vanishing of the first cohomology with coefficients $\mathbb{C}^\times$ can appear.

Theorem 5.1 *Let (M, ω) be a smooth compact symplectic toric variety together with a prequantum line bundle $(L, \nabla) \to M$. Then, for any choice of symplecto-morphism $\Phi : M \to M$ and any choice of real polarization $P \subset TM$, there exists isomorphisms*

$$Q_{shf}(M, P, \nabla) \cong Q_{shf}(M, \Phi^* P, \nabla)$$
$$Q_{BS}(M, P, \nabla) \cong Q_{BS}(M, \Phi^* P, \nabla)$$

Proof Since M is a compact toric variety, it follows that $\pi_1(M) = 0$ [4, First Proposition, Section 3.2]. Since

$$H^1(M, \mathbb{C}^\times) \cong \mathrm{Hom}(\pi_1(M), \mathbb{C}^\times),$$

it then follows that $H^1(M, \mathbb{C}^\times) = 0$. Thus, the hypotheses of Theorems 1.1 and 1.2 hold. □

As an application of this result, I obtained a kind of universality for quantization of toric manifolds under twisting. In more detail, let $T \cong (S^1)^n$ be an n-torus with $\mathfrak{t} = \mathrm{Lie}(T)$ and $\mu : (M, \omega) \to \mathfrak{t}^*$ be a compact T-toric manifold. The momentum map μ naturally defines a polarization $P(\mu) \subset TM$ on (M, ω), where for each $x \in M$ we define

$$P(\mu)_x := T_x \mu^{-1}(\mu(x)). \tag{19}$$

Given a symplectomorphism $\Phi : M \to M$, one can twist the action of T on M by

$$T \times M \to M; \quad (t, x) \mapsto \Phi(t \cdot \Phi^{-1}(x)).$$

This is clearly a symplectic action. Furthermore, define

$$\mu^{\Phi} := \mu \circ \Phi^{-1}.$$

Then, $\mu^{\Phi} : (M, \omega) \to \mathfrak{t}^*$ is a T-toric manifold once again. Call this the twisting of $\mu : (M, \omega) \to \mathfrak{t}^*$ by Φ.

Thus, each symplectomorphism Φ of a toric manifold generates a new polarization $P(\mu^{\Phi})$ defined as in Eq. (19), but with μ^{Φ} replacing μ. It is a triviality to unwind the definitions to show that

$$P(\mu^{\Phi}) = \Phi^* P(\mu).$$

This computation together with the previous theorem provides us with the following result.

Corollary 5.2 *Quantization of a compact toric manifold is invariant under twisting by symplectomorphisms.*

Of course, this is nothing new as Hamilton in [6, Theorem 8.10] showed for any compact toric manifold $\mu : (M, \omega) \to \mathfrak{t}^*$ that

$$Q_{shf}(M, P(\mu), \nabla) \cong \bigoplus_{\Lambda \cap \mu(M)^{\circ}} \mathbb{C},$$

where Λ is the dual of the lattice $\ker(\exp : \mathfrak{t} \to T)$ and $\mu(M)^{\circ}$ is the interior of the associated Delzant polytope to M. Since the cardinality of $\Lambda \cap \mu(M)^{\circ}$ is a isomorphism invariant of a toric manifold, we immediately arrive at Corollary 5.2.

References

1. J.E. Andersen, Geometric quantization of symplectic manifolds with respect to reducible non-negative polarizations. Commun. Math. Phys. **183**, 401–421 (1997)
2. I. Androulidakis, G. Skandalis, The Holonomy groupoid of a singular foliation. J. für die Reine und Angewandte Mathematik **2009**, 1–37 (2009)
3. R. Bott, L.W. Tu, *Differential Forms in Algebraic Topology* (Springer, New York, 1982)
4. W. Fulton, *Introduction to Toric Varieties* (Princeton University Press, Princeton, 1993)
5. V. Guillemin, S. Sternberg, The Gelfand-Cetlin system and quantization of the complex flag manifolds. J. Funct. Analy. **52**, 106–128 (1983)
6. M. Hamilton, *Locally Toric Manifolds and Singular Bohr-Sommerfeld leaves*. Memoirs of the American Mathematical Society, vol. 207 (American Mathematical Society, Providence, 2010)

7. L.C. Jeffrey, J. Weitsman, Bohr-Sommerfeld orbits in the moduli space of flat connections and the verlinde dimension formula. Commun. Math. Phys. **150**, 593–630 (1992)
8. J. Sniatycki, On cohomology groups appearing in geometric quantization, in *Differential Geometrical Methods in Mathematical Physics*. Lectures Notes in Mathematics, vol. 570 (Springer, Berlin, 1977), pp. 46–66
9. H.J. Sussmann, Orbits of families of vector fields and integrability of distributions. Trans. Amer. Math. Soc. **180**, 171–188 (1973)
10. T. Wedhorn, *Manifolds, Sheaves, and Cohomology*. (Springer Fachmedien Wiesbaden, Wiesbaden, 2016)
11. N.M.J. Woodhouse, *Geometric Quantization* (Oxford University Press, Oxford, 1991)

Polyhedral Products for Wheel Graphs and Their Generalizations

Stephen Theriault

1 Introduction

Polyhedral products have emerged as an important class of topological spaces, and a key problem is identifying their homotopy type. While there has been a certain amount of success in doing this in special cases, it is a very difficult problem in general. In this paper, we make the case that it is sometimes, perhaps paradoxically, easier to determine the homotopy type of the loop space. This may be a new way forward in the analysis of the homotopy theory of polyhedral products.

Let K be an abstract simplicial complex on the vertex set $[m] = \{1, 2, \ldots, m\}$. In other words, K is a collection of subsets $\sigma \subseteq [m]$ such that for any $\sigma \in K$ all subsets of σ also belong to K. We refer to K as a simplicial complex rather than an abstract simplicial complex. A subset $\sigma \in K$ is a *simplex* or *face* of K. The emptyset $\emptyset$ is assumed to belong to K.

Given a simplicial complex K on the vertex set $[m]$, for $1 \leq i \leq m$ let (X_i, A_i) be a pair of pointed CW-complexes, where A_i is a pointed subspace of X_i. Let $(\underline{X}, \underline{A}) = \{(X_i, A_i)\}_{i=1}^{m}$ be the sequence of CW-pairs. For each simplex (face) $\sigma \in K$, let $(\underline{X}, \underline{A})^{\sigma}$ be the subspace of $\prod_{i=1}^{m} X_i$ defined by

$$(\underline{X}, \underline{A})^{\sigma} = \prod_{i=1}^{m} Y_i \qquad \text{where} \qquad Y_i = \begin{cases} X_i & \text{if } i \in \sigma \\ A_i & \text{if } i \notin \sigma. \end{cases}$$

The *polyhedral product* determined by $(\underline{X},\underline{A})$ and K is

S. Theriault (✉)
School of Mathematical Sciences, University of Southampton, Southampton, UK
e-mail: s.d.theriault@soton.ac.uk

A. Bahri et al. (eds.), *Toric Topology and Polyhedral Products*, Fields Institute Communications 89, https://doi.org/10.1007/978-3-031-57204-3_15

$$(\underline{X}, \underline{A})^K = \bigcup_{\sigma \in K} (\underline{X}, \underline{A})^\sigma \subseteq \prod_{i=1}^{m} X_i.$$

A fundamental case is the *moment-angle complex* $\mathcal{Z}_K$ that is central to toric topology, which occurs when each pair of spaces (X_i, A_i) is (D^2, S^1).

We first give a general decomposition for the loops on certain polyhedral products that generalizes work of Félix and Tanré [6]. Let K, L, and M be simplicial complexes with L a subcomplex of K, and let $\overline{K}$ be the pushout of the simplicial maps $L \longrightarrow K$ and $L \longrightarrow L * M$, where $L * M$ is the join of L and M. In Theorem 2.6, a homotopy equivalence for $\Omega(\underline{X}, \underline{A})^{\overline{K}}$ is given in terms of $\Omega(\underline{X}, \underline{A})^K$, $\Omega(\underline{X}, \underline{A})^L$ and two related spaces. This is then specialized considerably in order to get concrete, explicit homotopy equivalences for a family of moment-angle complexes. Taking $K = P_m$ as the boundary of the m-gon, $L = V_m$ as its vertex set, and M as a single vertex, the simplicial complex $\overline{K} = W_m$ is known as a *wheel graph*. With the same K and L but taking M to be any simplicial complex, we obtain what will be called a *wheel complex* $W_m(M)$ (as it need no longer be a graph). In Theorem 5.9, explicit homotopy equivalences for $\Omega\mathcal{Z}_{P_m}$ and the homotopy fiber of $\mathcal{Z}_{V_m} \longrightarrow \mathcal{Z}_{P_m}$ are used to give an explicit homotopy equivalence for $\Omega\mathcal{Z}_{W_m(M)}$ in terms of spheres, loops on spheres, and $\Omega\mathcal{Z}_M$. In particular, $\Omega\mathcal{Z}_{W_m}$ is homotopy equivalent to a product of spheres and loops on spheres.

This suggests there may be a wide class of simplicial complexes K with the property that $\Omega\mathcal{Z}_K$ is homotopy equivalent to a product of spheres and loops on spheres. It would be interesting to investigate this problem further.

The author would like to thank the referee for several comments that have improved the paper.

2 A General Decomposition for the Loops on Certain Polyhedral Products

This section generalizes work of Félix and Tanré [6] on the homotopy theory of certain polyhedral products. The main result is Theorem 2.6; it requires two tools, presented in Lemmas 2.1 and 2.3. The first is Mather's cube Lemma [15].

Lemma 2.1 *Suppose that there is a homotopy commutative diagram of spaces and maps*

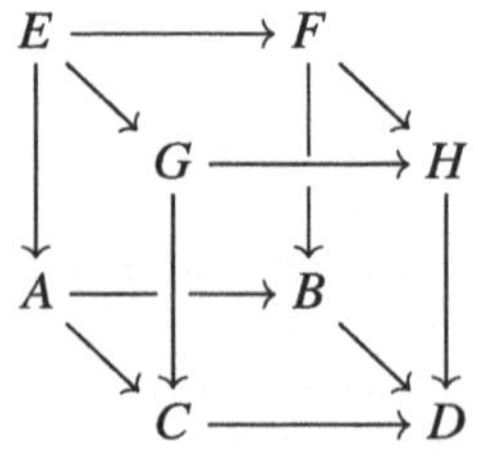

where the bottom face is a homotopy pushout and the four sides are homotopy pullbacks. Then the top face is a homotopy pushout. □

A typical construction of such a cube is to start with a homotopy pushout A-B-C-D and a map $f : D \longrightarrow Z$. Define the space H as the homotopy fiber of f, and define F, G, and E by pulling back with the map $H \longrightarrow D$. This gives a homotopy commutative cube with the bottom face a homotopy pushout and all four sides being homotopy pullbacks, so Lemma 2.1 implies that the top face is also a homotopy pushout.

The second tool requires some setup. In general, if L and M are simplicial complexes, the *join* of L and M is the simplicial complex

$$L * M = \{\sigma \cup \tau \mid \sigma \in L \text{ and } \tau \in M\}.$$

By the definitions of the join and the polyhedral product, there is a homeomorphism

$$(\underline{X}, \underline{A})^{L*M} = (\underline{X}, \underline{A})^{L} \times (\underline{X}, \underline{A})^{M}.$$

Let K be a simplicial complex on the vertex set $\{1, \ldots, m\}$, and let M be a simplicial complex on the vertex set $\{m+1, \ldots, n\}$. Let L be a subcomplex of K, and define the simplicial complex $\overline{K}$ by the pushout

$$\begin{array}{ccc} L & \longrightarrow & L * M \\ \downarrow & & \downarrow \\ K & \longrightarrow & \overline{K}. \end{array} \qquad (1)$$

Note that $\overline{K}$ has vertex set $\{1, \ldots, n\}$.

Example 2.2 If $M = \{v\}$ has only a single vertex, then (1) attaches a cone to the subcomplex L of K. In terms of the standard star-link-restriction pushout with respect to the vertex v, we have $K = \overline{K} \backslash v$, $L * M = \text{star}_{\overline{K}}(v)$ and $L = \text{link}_{\overline{K}}(v)$.

The pushout (1) induces a commutative diagram of polyhedral products, but this can be strengthened to a pushout of polyhedral products if correctly interpreted. Regard K, L, and M as simplicial complexes on the vertex set $\{1, \ldots, n\}$. In particular, K has *ghost vertices* $m+1, \ldots, n$, and the simplicial map $K \longrightarrow \overline{K}$ induces a map of polyhedral products $(\underline{X}, \underline{A})^K \times \prod_{i=1}^{m+1} A_i \longrightarrow (\underline{X}, \underline{A})^{\overline{K}}$. While L, as a subcomplex of K, may have fewer vertices, it will be convenient to distinguish the ghost vertices $m+1, \ldots, n$ and regard the simplicial map $L \longrightarrow K$ as inducing a map of polyhedral products $(\underline{X}, \underline{A})^L \times \prod_{i=m+1}^{n} A_i \xrightarrow{g \times 1} (\underline{X}, \underline{A})^K \times \prod_{i=m+1}^{n} A_i$ where g is the map of polyhedral products induced when restricted to the vertex set $\{1, \ldots, m\}$ and 1 is the identity map on $\prod_{i=m+1}^{n} A_i$. The simplicial map $L * M \longrightarrow \overline{K}$ induces the map $(\underline{X}, \underline{A})^L \times (\underline{X}, \underline{A})^M \longrightarrow (\underline{X}, \underline{A})^{\overline{K}}$ and the simplicial inclusion $L \longrightarrow L * M$ induces the map $(\underline{X}, \underline{A})^L \times \prod_{i=m+1}^{n} A_i \xrightarrow{1 \times h} (\underline{X}, \underline{A})^L \times (\underline{X}, \underline{A})^M$

where 1 is the identity map on $(\underline{X},\underline{A})^L$ and h is induced by the simplicial map $\overline{\emptyset} \longrightarrow M$ (where $\overline{\emptyset}$ is the simplicial complex on ghost vertices $\{m+1,\ldots,n\}$). By Grbić and Theriault [9, Proposition 3.1], all this combines to give the following.

Lemma 2.3 *There is a (point-set) pushout*

$$\begin{array}{ccc}(\underline{X},\underline{A})^L \times \prod_{i=m+1}^n A_i & \xrightarrow{1\times h} & (\underline{X},\underline{A})^L \times (\underline{X},\underline{A})^M \\ \downarrow{\scriptstyle g\times 1} & & \downarrow \\ (\underline{X},\underline{A})^K \times \prod_{i=m+1}^n A_i & \longrightarrow & (\underline{X},\underline{A})^{\overline{K}}.\end{array}$$

□

The pushout in Lemma 2.3 will serve as the starting point for a cube that lets us apply Lemma 2.1. To produce the four sides of the cube, we will construct a map from $(\underline{X},\underline{A})^{\overline{K}}$ to an appropriate polyhedral product and take fibers.

Including K and $L*M$ into $K*M$, by (1), there is a pushout map

$$\overline{K} \longrightarrow K*M.$$

Since the composite $K \longrightarrow \overline{K} \longrightarrow K*M$ is the inclusion of the first factor the composite and $L*M \longrightarrow \overline{K} \longrightarrow K*M$ is the join of the inclusion of L into K and the identity map on M, there are induced maps of polyhedral products

$$(\underline{X},\underline{A})^K \times \prod_{i=m+1}^n A_i \xrightarrow{1\times h} (\underline{X},\underline{A})^K \times (\underline{X},\underline{A})^M$$

$$(\underline{X},\underline{A})^L \times (\underline{X},\underline{A})^M \xrightarrow{g\times 1} (\underline{X},\underline{A})^K \times (\underline{X},\underline{A})^M.$$

Therefore, if the four corners of the diagram in Lemma 2.3 are composed with the map $(\underline{X},\underline{A})^{\overline{K}} \longrightarrow (\underline{X},\underline{A})^K \times (\underline{X},\underline{A})^M$ we obtain homotopy fibrations

$$F \longrightarrow (\underline{X},\underline{A})^{\overline{K}} \longrightarrow (\underline{X},\underline{A})^K \times (\underline{X},\underline{A})^M$$

$$H \longrightarrow (\underline{X},\underline{A})^K \times \prod_{i=m+1}^n A_i \xrightarrow{1\times h} (\underline{X},\underline{A})^K \times (\underline{X},\underline{A})^M$$

$$G \longrightarrow (\underline{X},\underline{A})^L \times (\underline{X},\underline{A})^M \xrightarrow{g\times 1} (\underline{X},\underline{A})^K \times (\underline{X},\underline{A})^M$$

$$G\times H \longrightarrow (\underline{X},\underline{A})^L \times \prod_{i=m+1}^n A_i \xrightarrow{g\times h} (\underline{X},\underline{A})^K \times (\underline{X},\underline{A})^M$$

where the first fibration defines F, and G and H are defined as the homotopy fibers of the maps g and h, respectively. Since all homotopy fibers are given by composing into a common base, we obtain a homotopy commutative cube

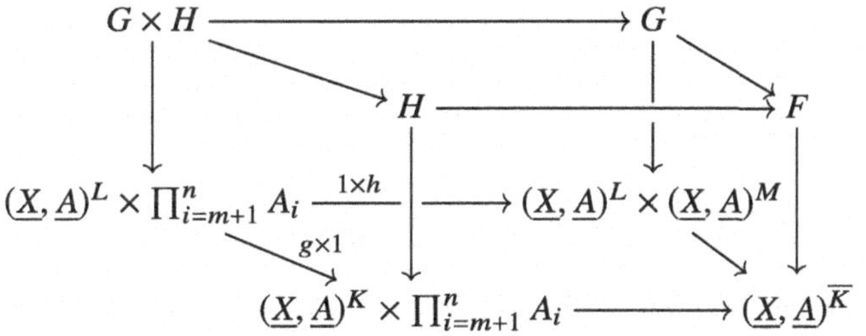

(2)

where the bottom face is a pushout and the four sides are homotopy pullbacks. Therefore, by Lemma 2.1, the top face is also a homotopy pushout.

The (reduced) *join* of two spaces A and B is the quotient space

$$A * B = (A \times [0, 1] \times B)/\sim$$

where $(a, 0, b) \sim (a', 0, b)$, $(a, 1, b) \sim (a, 1, b')$ and $(*, t, *) \sim (*, 0, *)$ for all $a, a' \in A$, $b, b' \in B$ and $t \in [0, 1]$. It is well known that there is a homotopy equivalence $A * B \simeq \Sigma A \wedge B$.

Lemma 2.4 *The maps $G \times H \longrightarrow G$ and $G \times H \longrightarrow H$ in (2) can be chosen to be the projections. Consequently, there is a homotopy equivalence $F \simeq G * H$.*

Proof Consider the homotopy fibration diagram

$$\begin{array}{ccccc}
G \times H & \longrightarrow & (\underline{X}, \underline{A})^L \times \prod_{i=m+1}^n A_i & \xrightarrow{g \times h} & (\underline{X}, \underline{A})^K \times (\underline{X}, \underline{A})^M \\
\downarrow & & \downarrow{\scriptstyle g \times 1} & & \| \\
H & \longrightarrow & (\underline{X}, \underline{A})^K \times \prod_{i=m+1}^n A_i & \xrightarrow{1 \times h} & (\underline{X}, \underline{A})^K \times (\underline{X}, \underline{A})^M .
\end{array}$$

Regarding H as $* \times H$, this fibration diagram is the product of the fibration diagrams for the lefthand and righthand factors. Thus, one choice of the map between fibers is $G \times H \longrightarrow * \times H$ is the product $* \times 1$. That is, this is the projection $G \times H \longrightarrow H$. The argument that $G \times H \longrightarrow G$ can be chosen to be a projection is similar.

In general, it is well known that the homotopy pushout of projections $S \times T \longrightarrow S$ and $S \times T \longrightarrow T$ are the joint $S * T$. So in our case, we obtain $F \simeq G * H$. □

Remark 2.5 Félix and Tanré [6] considered the special case of (2) when M is a single vertex.

We now identify a homotopy decomposition for $\Omega(\underline{X}, \underline{A})^{\overline{K}}$.

Theorem 2.6 *Let $\overline{K}$ be a pushout as in (1). Then, there is a homotopy fibration*

$$G * H \longrightarrow (\underline{X}, \underline{A})^{\overline{K}} \longrightarrow (\underline{X}, \underline{A})^{K} \times (\underline{X}, \underline{A})^{M}$$

where G is the homotopy fiber of $(\underline{X}, \underline{A})^{L} \longrightarrow (\underline{X}, \underline{A})^{K}$ and H is the homotopy fiber of $\prod_{i=m+1}^{n} A_i \longrightarrow (\underline{X}, \underline{A})^{M}$. Further, this fibration splits after looping, giving a homotopy equivalence

$$\Omega(\underline{X}, \underline{A})^{\overline{K}} \simeq \Omega(\underline{X}, \underline{A})^{K} \times \Omega(\underline{X}, \underline{A})^{M} \times \Omega(G * H).$$

Proof Consider the homotopy fibration $F \longrightarrow (\underline{X}, \underline{A})^{\overline{K}} \longrightarrow (\underline{X}, \underline{A})^{K} \times (\underline{X}, \underline{A})^{M}$ that defines F. By Lemma 2.4, $F \simeq G * H$, proving the first statement.

For the splitting, we have seen that the composite $(\underline{X}, \underline{A})^{K} \longrightarrow (\underline{X}, \underline{A})^{\overline{K}} \longrightarrow (\underline{X}, \underline{A})^{K} \times (\underline{X}, \underline{A})^{M}$ is the inclusion of the left factor while $(\underline{X}, \underline{A})^{L} \times (\underline{X}, \underline{A})^{M} \longrightarrow (\underline{X}, \underline{A})^{\overline{K}} \longrightarrow (\underline{X}, \underline{A})^{K} \times (\underline{X}, \underline{A})^{M}$ is $g \times 1$. Restricting the latter case to $(\underline{X}, \underline{A})^{M}$ is the inclusion of the right factor. Taking the wedge sum therefore gives a composite

$$(\underline{X}, \underline{A})^{K} \vee (\underline{X}, \underline{A})^{M} \longrightarrow (\underline{X}, \underline{A})^{\overline{K}} \longrightarrow (\underline{X}, \underline{A})^{K} \times (\underline{X}, \underline{A})^{M}$$

which is the inclusion of the wedge into the product. It is well known that the inclusion of the wedge into a product has a right homotopy inverse after looping. Hence, the fibration $G * H \longrightarrow (\underline{X}, \underline{A})^{\overline{K}} \longrightarrow (\underline{X}, \underline{A})^{K} \times (\underline{X}, \underline{A})^{M}$ splits after looping, and the asserted homotopy equivalence for $\Omega(\underline{X}, \underline{A})^{\overline{K}}$ follows. □

3 An Initial Analysis of Theorem 2.6

The decomposition for $\Omega(\underline{X}, \underline{A})^{\overline{K}}$ in Theorem 2.6 has four ingredients: (i) $\Omega(\underline{X}, \underline{A})^{K}$, (ii) $\Omega(\underline{X}, \underline{A})^{M}$, (iii) the homotopy fiber of $(\underline{X}, \underline{A})^{L} \longrightarrow (\underline{X}, \underline{A})^{K}$, and (iv) the homotopy fiber of $\prod_{i=m+1}^{n} A_i \longrightarrow (\underline{X}, \underline{A})^{M}$. To go further, we would like to identify some or all of these components.

It will be helpful to reduce to analyzing a special case of polyhedral products. In general, for $1 \leq i \leq m$, let Y_i be the homotopy fiber of the inclusion $A_i \longrightarrow X_i$. In [10], the following was proven when each pair (X_i, A_i) has both X_i and A_i path-connected, but the same argument works in the more general case when only X_i is path-connected.

Theorem 3.1 *Let K be a simplicial complex on the vertex set $[m]$, and let $(\underline{X}, \underline{A})$ be any sequence of pointed pairs (X_i, A_i) where each X_i is path-connected. Then there is a homotopy fibration*

$$(\underline{CY}, \underline{Y})^{K} \longrightarrow (\underline{X}, \underline{A})^{K} \longrightarrow \prod_{i=1}^{m} X_i.$$

Further, this fibration splits after looping to give a homotopy equivalence

$$\Omega(\underline{X}, \underline{A})^K \simeq \left(\prod_{i=1}^{m} \Omega X_i\right) \times \Omega(\underline{CY}, \underline{Y})^K.$$

□

In our case, we obtain $\Omega(\underline{X}, \underline{A})^{\overline{K}} \simeq \left(\prod_{i=1}^{n} \Omega X_i\right) \times \Omega(\underline{CY}, \underline{Y})^{\overline{K}}$, so to determine the homotopy type of $\Omega(\underline{X}, \underline{A})^{\overline{K}}$ it is equivalent to determine the homotopy type of $\Omega(\underline{CY}, \underline{Y})^{\overline{K}}$. Put another way, in the context of Theorem 2.6, it suffices to consider the case when $(\underline{X}, \underline{A})$ is of the form $(\underline{CA}, \underline{A})$.

Now, from the point of view of $(\underline{CA}, \underline{A})$, the decomposition for $\Omega(\underline{CA}, \underline{A})^{\overline{K}}$ in Theorem 2.6 has four ingredients: (i)$'$ $\Omega(\underline{CA}, \underline{A})^K$, (ii)$'$ $\Omega(\underline{CA}, \underline{A})^M$, (iii)$'$ the homotopy fiber of $(\underline{CA}, \underline{A})^L \longrightarrow (\underline{CA}, \underline{A})^K$, and (iv)$'$ the homotopy fiber of $\prod_{i=m+1}^{n} A_i \longrightarrow (\underline{CA}, \underline{A})^M$. Component (iv)$'$ can be handled generically. In general, in [9, Corollary 3.4], the following was proven.

Proposition 3.2 *If K is a simplicial complex on the vertex set $[m]$, then the inclusion $\prod_{i=1}^{m} A_i \longrightarrow (\underline{CA}, \underline{A})^K$ is null homotopic.* □

In our case, we immediately obtain the following.

Corollary 3.3 *The homotopy fiber of the map $\prod_{i=m+1}^{n} A_i \longrightarrow (\underline{CA}, \underline{A})^M$ is homotopy equivalent to $\left(\prod_{i=m+1}^{n} A_i\right) \times \Omega(\underline{CA}, \underline{A})^M$.* □

Consequently, (iv)$'$ has been rewritten in terms of (ii)$'$. Components (i)$'$ to (iii)$'$ cannot be handled generically, so special cases need to be identified. In particular, (iii)$'$ is particularly contentious.

In what follows we will specialize considerably. The polyhedral products will be taken to be moment-angle complexes. The simplicial complex K will be the boundary of an m-gon and L will be the vertex set of the m-gon. The point in specializing so much is that then the homotopy types of $\mathcal{Z}_K$ and $\mathcal{Z}_L$ are known, and with a nontrivial amount of work, we will be able to identify the homotopy types of $\Omega\mathcal{Z}_K$ and the homotopy fiber of the map $\mathcal{Z}_L \longrightarrow \mathcal{Z}_K$.

4 Spaces Having the Homotopy Type of a Wedge of Spheres

This section establishes some preliminary properties for spaces having the homotopy type of a wedge of spheres.

Lemma 4.1 *If X is homotopy equivalent to a finite type product of path-connected spheres and loops on simply connected spheres, then ΣX is homotopy equivalent to a wedge of simply connected spheres.*

Proof By hypothesis,

$$X \simeq \Big(\prod_{\alpha \in \mathcal{I}} S^{n_\alpha}\Big) \times \Big(\prod_{\beta \in \mathcal{J}} \Omega S^{n_\beta}\Big)$$

for some index sets $\mathcal{I}$ and $\mathcal{J}$, with each $n_\alpha \geq 1$ and each $n_\beta \geq 2$. By the James construction [13], $\Sigma\Omega S^n$ is homotopy equivalent to a wedge of spheres if $n \geq 2$, and it is well known that $\Sigma(S \times T) \simeq \Sigma S \vee \Sigma T \vee (\Sigma S \wedge T)$. Iteratively using these two properties implies that ΣX is homotopy equivalent to a wedge of simply connected spheres. □

The *right half-smash* of pointed spaces A and B is the quotient space

$$A \rtimes B = (A \times B)/\sim$$

where $(a, *) \sim (*, *)$ for all $a \in A$. It is well known that if A is a co-H-space, then there is a homotopy equivalence $A \rtimes B \simeq A \vee (A \wedge B)$. A modest variation on Lemma 4.1 is the following.

Lemma 4.2 *If X is homotopy equivalent to a finite type product of path-connected spheres and loops on simply connected spheres, and Y is homotopy equivalent to a wedge of simply connected spheres, then $Y \rtimes X$ is homotopy equivalent to a wedge of simply connected spheres.*

Proof Since Y is homotopy equivalent to a wedge of simply connected spheres, we have $Y \simeq \Sigma Y'$ for some wedge of path-connected spheres Y'. Therefore, Y is a co-H-space so there is a homotopy equivalence $Y \rtimes X \simeq Y \vee (Y \wedge X)$. Further, $Y \wedge X \simeq Y' \wedge (\Sigma X)$ and by Lemma 4.1 ΣX is homotopy equivalent to a wedge of simply connected spheres. Hence, as Y' is a wedge of spheres, so is $Y' \wedge (\Sigma X)$, and the spheres are all simply connected because ΣX is. Thus, $Y \rtimes X$ is homotopy equivalent to a wedge of simply connected spheres. □

Lemma 4.3 *Suppose that R and S are wedges of simply connected spheres and $R \xrightarrow{f} S$ induces an epimorphism in homology. Then f has a right homotopy inverse.*

Proof Take homology with integral coefficients. Since f_* is an epimorphism, for each generator $x_\alpha \in H_*(S)$, there is an element $y_\alpha \in H_*(R)$ such that $f_*(y_\alpha) = x_\alpha$. Since R is a wedge of spheres, the basis for $H_*(R)$ induced by including each sphere into the wedge implies that each basis generator is in the image of the Hurewicz homomorphism. As the Hurewicz homomorphism is a homomorphism, any linear combination of basis elements in $H_*(R)$ is also in the image of the Hurewicz homomorphism. In particular, y_α is in the image of the Hurewicz homomorphism, and so there is a map $s_\alpha \colon S^{n_\alpha} \longrightarrow R$ whose Hurewicz image is y_α. Let

$$s \colon \bigvee_\alpha S^{n_\alpha} \longrightarrow R$$

be the wedge sum of the maps s_α as α runs over a basis for $H_*(S)$. Then the composite $\bigvee_\alpha S^{n_\alpha} \xrightarrow{s} R \xrightarrow{f} S$ induces an isomorphism in homology. As all spaces are simply connected, this isomorphism in homology implies that $f \circ s$ is a homotopy equivalence by Whitehead's theorem. Thus, f has a right homotopy inverse. □

Improving on Lemma 4.3, the next lemma shows that the map $R \xrightarrow{f} S$ is obtained by taking the cofiber of some map $T \xrightarrow{u} R$.

Lemma 4.4 *Suppose that R and S are wedges of simply connected spheres and $R \xrightarrow{f} S$ induces an epimorphism in homology. Then there is a wedge of simply connected spheres T and a map $u\colon T \longrightarrow R$ such that there is a homotopy cofibration $T \xrightarrow{u} R \xrightarrow{f} S$.*

Proof Let $s\colon S \longrightarrow R$ be the right homotopy inverse of f in Lemma 4.3. Define the space T and the map t by the homotopy cofibration

$$S \xrightarrow{s} R \xrightarrow{t} T. \tag{3}$$

As R is a wedge of spheres, it is a co-H-space so it has a comultiplication σ. The right homotopy inverse for f implies that the composite

$$e\colon R \xrightarrow{\sigma} R \vee R \xrightarrow{f \vee t} S \vee T$$

is a homotopy equivalence. Note that as T retracts off a simply connected space, it is simply connected, and as it retracts off a wedge of spheres, it is homotopy equivalent to a wedge of spheres.

Define the map u by the composite

$$u\colon T \xrightarrow{i_2} S \vee T \xrightarrow{e^{-1}} R$$

where i_2 is the inclusion of the second wedge summand. Let C be the homotopy cofiber of u. By definition of u, the composite $e \circ u \simeq i_2$. Therefore, there is a homotopy pushout diagram

$$\begin{array}{ccccc} T & \xrightarrow{u} & R & \longrightarrow & C \\ \| & & \big\downarrow{\scriptstyle e} & & \big\downarrow{\scriptstyle e'} \\ T & \xrightarrow{i_2} & S \vee T & \xrightarrow{p_1} & S \end{array}$$

where p_1 is the pinch map to the first wedge summand and e' is an induced map of cofibers. Since e is a homotopy equivalence, the five lemma implies that e' induces an isomorphism in homology, so as all spaces are simply connected, e' is a homotopy equivalence by Whitehead's theorem. Thus, there is a homotopy cofibration $T \xrightarrow{u} R \xrightarrow{p_1 \circ e} S$. It remains to show that $p_1 \circ e \simeq f$. But this follows from the definition of e and the naturality of the pinch map p_1. □

5 Wheel Graphs

Let P_m be the boundary of an m-gon, and let V_m be its vertex set, so V_m consists of m disjoint points. Define the simplicial complex W_m by the pushout

$$\begin{array}{ccc} V_m & \longrightarrow & V_m * \{v\} \\ \downarrow & & \downarrow \\ P_m & \longrightarrow & W_m \end{array}$$

where $\{v\}$ is a vertex disjoint from those in V_m. The simplicial complex W_m is called a *wheel graph*, where v is regarded as a hub with spokes (edges) connecting it to each vertex in the n-gon. Pictorially, representations of P_5 and W_5 are as follows:

P_5: 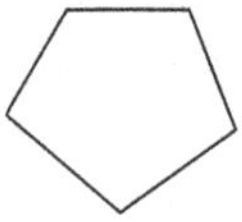 W_5: 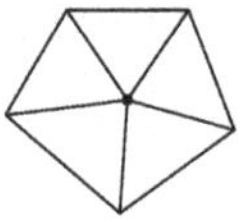

The homotopy type of $\mathcal{Z}_{P_m}$ is known. In fact, a much stronger identification was proven by MacGavran [14] in work that predated moment-angle complexes. Since P_m is a triangulation of a sphere, it is known [4] that the corresponding moment-angle complex $\mathcal{Z}_{P_m}$ is a manifold. Reformulating MacGavran's result in terms of moment-angle complexes, he showed that for $m \geq 4$, there is a diffeomorphism

$$\mathcal{Z}_{P_m} \cong \#_{k=3}^{m-1}(S^k \times S^{m+2-k})^{\#(k-2)\binom{m-2}{k-1}} \tag{4}$$

where the right side is an iterated connected sum of products of two spheres.

It would be ideal to identify the homotopy type of $\mathcal{Z}_{W_m}$ as well. However, this seems to be difficult, but it is possible to determine the homotopy type of $\Omega\mathcal{Z}_{W_m}$. In fact, we will do more. Let M be any simplicial complex. Define the simplicial complex $W_m(M)$ by the pushout

$$\begin{array}{ccc} V_m & \longrightarrow & V_m * M \\ \downarrow & & \downarrow \\ P_m & \longrightarrow & W_m(M). \end{array} \tag{5}$$

The wheel graph W_m is the special case when M is a single point. We will determine the homotopy type of $\mathcal{Z}_{W_m(M)}$, provided the homotopy type of $\Omega\mathcal{Z}_M$ is known.

It is worth mentioning two facts that give some context to the homotopy type of $W_m(M)$.

- Observe that P_m is the full subcomplex of $W_m(M)$ on the vertex set V_m. Therefore, by Denham and Suciu [5], $\mathcal{Z}_{P_m}$ is a retract of $\mathcal{Z}_{W_m(M)}$. That is, $\mathcal{Z}_{W_m(M)}$ has the connected sum of products of two spheres in (4) retracting off it.
- Removing an edge from P_m and also the corresponding edge from W_m gives simplicial complexes that pictorially look like

P'_5: 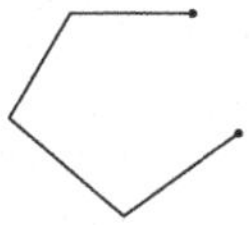 W'_5: 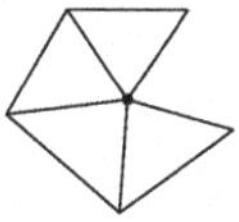

Observe that P'_m can be formed by iteratively gluing an edge to the previous one along a common endpoint, and W'_m can be formed by iteratively gluing a 2-simplex to the previous one along a common edge. By Theriault [18], both $\mathcal{Z}_{P'_m}$ and $\mathcal{Z}_{W'_m}$ are homotopy equivalent to wedges of spheres. So inserting the final edge to form P_m from P'_m and W_m from W'_m dramatically changes the homotopy type and also significantly changes cohomology by introducing nontrivial cup products.

To get started, suppose that M is on the vertex set $\{m+1, \ldots, n\}$. By Theorem 2.6 applied to (5), there is a homotopy equivalence

$$\Omega(\underline{X}, \underline{A})^{W_m(M)} \simeq \Omega(\underline{X}, \underline{A})^{P_m} \times \Omega(\underline{X}, \underline{A})^{M} \times \Omega(G * H) \tag{6}$$

where G is the homotopy fiber of the map $(\underline{X}, \underline{A})^{V_m} \longrightarrow (\underline{X}, \underline{A})^{P_m}$ and H is the homotopy fiber of the map $\prod_{i=m+1}^{n} A_i \longrightarrow (\underline{X}, \underline{A})^{M}$. Specialize to the case when each pair (X_i, A_i) is (D^2, S^1). Then the polyhedral products in (6) are moment-angle complexes, Corollary 3.3 applies to identify H, and we obtain the following:

Lemma 5.1 *Let V_m, P_m and M be as in (5), and suppose that M is on the vertex set $\{m+1, \ldots, n\}$. Then there is a homotopy equivalence*

$$\Omega\mathcal{Z}_{W_m(M)} \simeq \Omega\mathcal{Z}_{P_m} \times \Omega\mathcal{Z}_M \times \Omega(G * H)$$

where G is the homotopy fiber of $\mathcal{Z}_{V_m} \longrightarrow \mathcal{Z}_{P_m}$ and $H \simeq \left(\prod_{i=m+1}^{n} S^1\right) \times \Omega\mathcal{Z}_M$.

□

Lemma 5.1 implies that to understand the homotopy type of $\mathcal{Z}_{W_m(M)}$, we need to understand the homotopy type of (i) $\Omega\mathcal{Z}_{P_m}$, (ii) $\Omega\mathcal{Z}_M$, and (iii) G. Both (i) and (iii) depend only on V_m and P_m, so the next few lemmas will focus solely on these cases.

While $\mathcal{Z}_{P_m}$ is a connected sum of products of two spheres, the homotopy type of the loops on a connected sum is not easy to explicitly identify. However, in this case, by Beben and Theriault [2, Example 3.1], we have the following.

Lemma 5.2 *For $m \geq 4$, there is a homotopy equivalence*

$$\Omega\mathcal{Z}_{P_m} \simeq \Omega S^3 \times \Omega S^{m-1} \times \Omega S(P_m)$$

where $S(P_m)$ is a wedge of simply connected spheres. □

The construction in [2] describes the wedge $S(P_m)$ explicitly. In general, let $A^{\vee t}$ be the wedge sum of t copies of A. Let

$$R_m = \left(\bigvee_{k=3}^{m-1} (S^k \vee S^{m+2-k})^{\vee(k-2)\binom{m-2}{k-1}}\right).$$

Observe that R_m is the $(m+1)$-skeleton of $\#_{k=3}^{m-1}(S^k \times S^{m+2-k})^{\#(k-2)\binom{m-2}{k-1}}$. Equivalently, R_m is homotopy equivalent to the connected sum with a puncture. Define R'_m by the cofibration

$$S^3 \vee S^{m-1} \longrightarrow R_m \longrightarrow R'_m$$

where the left map is the inclusion of one copy of $S^3 \vee S^{m-1}$ into R_m. Then R'_m is a wedge of simply connected spheres and

$$S(P_m) = R'_m \rtimes (\Omega S^3 \times \Omega S^{m-1}).$$

By Lemma 4.2, $S(P_m)$ is homotopy equivalent to a wedge of simply connected spheres.

Next, we aim toward Lemma 5.7, which identifies the space G in Lemma 5.1. Since V_m is m disjoint points, by Grbić and Theriault [8], Porter [17], the following holds.

Lemma 5.3 *For $m \geq 4$, there is a homotopy equivalence*

$$\mathcal{Z}_{V_m} \simeq \bigvee_{k=3}^{m+1} (S^k)^{\vee(k-2)\binom{m}{k-1}}.$$

□

Observe that the dimensions of $\mathcal{Z}_{V_m}$ and $\mathcal{Z}_{P_m}$ are $m+1$ and $m+2$, respectively, so the map $\mathcal{Z}_{V_m} \longrightarrow \mathcal{Z}_{P_m}$ factors through the $(m+1)$-skeleton R_m of $\mathcal{Z}_{P_m}$, giving a homotopy commutative diagram

$$\begin{array}{ccc} \mathcal{Z}_{V_m} & \longrightarrow & \mathcal{Z}_{P_m} \\ \Big\downarrow \theta & & \Big\| \\ R_m & \longrightarrow & \mathcal{Z}_{P_m} \end{array} \tag{7}$$

for some map θ.

Lemma 5.4 *The map θ induces a surjection in homology.*

Proof In general, by Bahri et al. [1, Corollary 2.23], there is a homotopy equivalence

$$\Sigma\mathcal{Z}_K \simeq \bigvee_{I\notin K} \Sigma^{2+|I|}|K_I|$$

where $I = \{i_1, \ldots, i_k\}$ is a subsequence of $[m]$ with $1 \le i_1 < \cdots < i_k \le m$, K_I is the full subcomplex of K on the vertex set I, $|K_I|$ is the geometric realization of K_I, and $|I|$ is the number of vertices in I. This homotopy equivalence is natural for simplicial maps $L \longrightarrow K$ and induces a $\mathbb{Z}$-module decomposition in integral homology,

$$H_*(\mathcal{Z}_K) \cong \bigoplus_{I\notin K} H_*(\Sigma^{1+|I|}|K_I|). \tag{8}$$

In our case, if $I = [m]$, then $|(P_m)_I| = |P_m| \simeq S^1$, and this case accounts for the generator of $H_{m+2}(\mathcal{Z}_{P_m})$. Therefore, as R_m is the $(m+1)$-skeleton of $\mathcal{Z}_{P_m}$, there is an isomorphism

$$H_*(R_m) \cong \bigoplus_{\substack{I\notin P_m \\ I\neq[m]}} H_*(\Sigma^{1+|I|}|(P_m)_I|).$$

In general, the inclusion of the vertex set V into a simplicial complex K induces an epimorphism $H_0(|V|) \longrightarrow H_0(|K|)$ since H_0 counts the number of connected components, $|K|$ has at most m components where m is the number of vertices in V, and each connected component of $|K|$ contains at least one of the vertices of V. Consequently, if $|K|$ is homotopy equivalent to some number of disjoint points, then the inclusion $V \longrightarrow K$ induces an epimorphism $H_*(|V|) \longrightarrow H_*(|K|)$.

In our case, consider (8) applied to the simplicial map $V_m \longrightarrow K_m$. Assume that $I \notin P_m$ and $I \neq [m]$. Observe that $I \notin V_m$ as well. Since I is a proper subset of $[m]$, we have $|(P_m)_I|$ homotopy equivalent to some number of disjoint points. Therefore, as the vertex set of $(P_m)_I$ is $(V_m)_I$, the simplicial map $(V_m)_I \longrightarrow (P_m)_I$ induces an epimorphism $H_*(|(V_m)_I|) \longrightarrow H_*(|(P_m)_I|)$. Hence there is an epimorphism

$$\bigoplus_{\substack{I \notin V_m \\ I \neq [m]}} H_*(\Sigma^{1+|I|}|(V_m)_I|) \bigoplus_{\substack{I \notin P_m \\ I \neq [m]}} H_*(\Sigma^{1+|I|}|(P_m)_I|) \cong H_*(R_m).$$

Observe that the left side is a submodule of $H_*(\mathcal{Z}_{V_m})$ by (8), and therefore the homotopy commutativity of (7) implies that θ_* is an epimorphism. □

Observe that $\mathcal{Z}_{V_m}$ is a wedge of simply connected spheres by Lemma 5.3, R_m is a wedge of simply connected spheres by definition, and $\mathcal{Z}_{V_m} \xrightarrow{\theta} R_m$ induces an epimorphism in homology by Lemma 5.4. Therefore, Lemmas 4.3 and 4.4 imply that θ has a right homotopy inverse and there is a homotopy cofibration

$$T_m \xrightarrow{u} \mathcal{Z}_{V_m} \xrightarrow{\theta} R_m$$

where T_m is a wedge of simply connected spheres.

By definition, R_m is the $(m+1)$-skeleton of $\#_{k=3}^{m-1}(S^k \times S^{m+2-k})^{\#(k-2)\binom{m-2}{k-1}}$. Writing the connected sum as $\mathcal{Z}_{P_m}$, there is a homotopy cofibration

$$S^{m+1} \xrightarrow{g} R_m \longrightarrow \mathcal{Z}_{P_m} \qquad (9)$$

where g attaches the top cell. Since θ has a right homotopy inverse, g lifts to a map $g' \colon S^{m+1} \longrightarrow \mathcal{Z}_{V_m}$.

Lemma 5.5 *There is a homotopy cofibration* $S^{m+1} \vee T_m \xrightarrow{g' \vee u} \mathcal{Z}_{V_n} \longrightarrow \mathcal{Z}_{P_m}$.

Proof Consider the diagram

$$\begin{array}{ccccc}
T_m & = & T_m & & \\
\downarrow{\scriptstyle i_2} & & \downarrow{\scriptstyle u} & & \\
S^{m+1} \vee T & \xrightarrow{g' \vee u} & \mathcal{Z}_{V_m} & \longrightarrow & \mathcal{Z}_{P_m} \\
\downarrow{\scriptstyle p_1} & & \downarrow{\scriptstyle \theta} & & \| \\
S^{m+1} & \xrightarrow{g} & R_m & \longrightarrow & \mathcal{Z}_{P_m}.
\end{array}$$

The upper left square clearly commutes. Taking cofibers vertically gives the lower left square. The lower left square is therefore a homotopy pushout, so taking cofibers horizontally we obtain the lower right square (which matches (7)). □

We will use the following result proven in [3].

Proposition 5.6 *Suppose that* $A \longrightarrow X \xrightarrow{h} Z$ *is a homotopy cofibration and* Ωh *has a right homotopy inverse. Then there is a homotopy fibration*

$$A \rtimes \Omega Z \longrightarrow X \xrightarrow{h} Z. \qquad \square$$

In our case, consider the homotopy cofibration $S^{m+1} \vee T_m \longrightarrow \mathcal{Z}_{V_m} \xrightarrow{h} \mathcal{Z}_{P_m}$ from Lemma 5.5, where h is simply a label for the right map. Since P_m is a flag simplicial complex and V_m is its vertex set, by Panov and Theriault [16] the map Ωh has a right homotopy inverse. Therefore, the hypotheses of Proposition 5.6 are satisfied, implying that the homotopy fiber G of h can be identified.

Lemma 5.7 *There is a homotopy equivalence: $G \simeq (S^{m+1} \vee T_m) \rtimes \Omega\mathcal{Z}_{P_m}$.* $\square$

Remark 5.8 By definition, T_m is a wedge of simply connected spheres, and by Lemma 5.2, $\Omega\mathcal{Z}_{P_m}$ is homotopy equivalent to a product of loops on simply connected spheres. Therefore, Lemma 4.2 implies that $G \simeq (S^{m+1} \vee T_m) \rtimes \Omega\mathcal{Z}_{P_m}$ is homotopy equivalent to a wedge of simply connected spheres.

The homotopy equivalence for $\Omega\mathcal{Z}_{W_m(M)}$ in Lemma 5.1 can now be refined by substituting in the homotopy equivalences for $\Omega\mathcal{Z}_{P_m}$ and G in Lemmas 5.2 and 5.7, respectively.

Theorem 5.9 *For $m \geq 4$, there is a homotopy equivalence*

$$\Omega\mathcal{Z}_{W_m(M)} \simeq \Omega S^3 \times \Omega S^{m+1} \times \Omega S(P_m) \times \Omega\mathcal{Z}_M \times \Omega(G * H)$$

where $G \simeq (S^{m+1} \vee T_m) \rtimes \Omega\mathcal{Z}_{P_m}$ and $H = (\prod_{i=m+1}^{n} S^1) \times \Omega\mathcal{Z}_M$. $\square$

Corollary 5.10 *If $\Omega\mathcal{Z}_M$ is homotopy equivalent to a product of path-connected spheres and loops on simply connected spheres, then so is $\Omega\mathcal{Z}_{W_M}$.*

Proof Since $\Omega\mathcal{Z}_M$ is homotopy equivalent to a product of path-connected spheres and loops on simply connected spheres, so is $H = (\prod_{i=m+1}^{n} S^1) \times \Omega\mathcal{Z}_M$. Therefore, by Lemma 4.1, ΣH homotopy equivalent to a wedge of simply connected spheres. Since G is also homotopy equivalent to a wedge of simply connected spheres by Remark 5.8, so is $G * H$. The Hilton-Milnor theorem then implies that $\Omega(G * H)$ is homotopy equivalent to a product of loops on simply connected spheres. Thus, in the homotopy equivalence for $\Omega\mathcal{Z}_{W_m(M)}$ in Theorem 5.9, each of the factors is homotopy equivalent to a product of path-connected spheres and loops on simply connected spheres, and hence so is $\Omega\mathcal{Z}_{W_m(M)}$. $\square$

Example 5.11 Return to the wheel graph W_m itself. This is $W_m(M)$ with $M = \{v\}$ being a single vertex. By definition of the polyhedral product, $\mathcal{Z}_{\{v\}} = D^2$, which is contractible, so Theorem 5.9 implies that there is a homotopy equivalence

$$\Omega\mathcal{Z}_{W_m} \simeq \Omega S^3 \times \Omega S^{m+1} \times \Omega S(P_m) \times \Omega(G * H)$$

where $G \simeq (S^{m+1} \vee T_m) \rtimes \Omega\mathcal{Z}_{P_m}$ and $H = S^1$. In particular, $G * H \simeq \Sigma^2 G$.

Example 5.12 Take $M = \{u, v\}$ be two disjoint points. A pictorial representation of $W_5(\{u, v\})$ is

$W_5(\{u, v\})$:

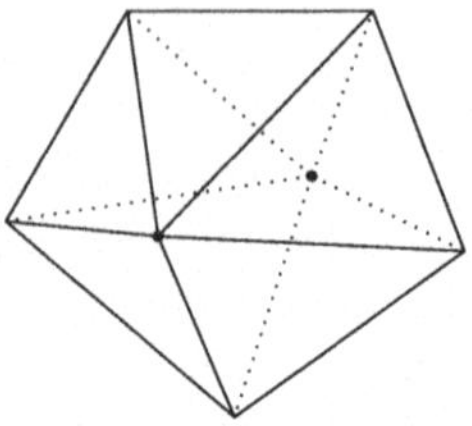

By Lemma 5.3, $\mathcal{Z}_M \simeq S^3$. Theorem 5.9 therefore implies that there is a homotopy equivalence

$$\Omega \mathcal{Z}_{W_m(\{u,v\})} \simeq \Omega S^3 \times \Omega S^{m+1} \times \Omega S(P_m) \times \Omega S^3 \times \Omega(G * H)$$

where $G \simeq (S^{m+1} \vee T_m) \rtimes \Omega \mathcal{Z}_{P_m}$ and $H = S^1 \times S^1 \times \Omega S^3$.

More generally, there is a wide class of simplicial complexes M with the property that $\mathcal{Z}_M$ is homotopy equivalent to a wedge of simply connected spheres, implying by the Hilton-Milnor theorem that $\Omega \mathcal{Z}_M$ is homotopy equivalent to a product of loops on simply connected spheres, and hence Corollary 5.10 can be applied to decompose $\Omega \mathcal{Z}_{W_m(M)}$. This class of simplicial complexes includes shifted complexes [9, 11], or more generally extractible simplicial complexes [12], and flag simplicial complexes whose 1-skeleton is a chordal graph [7].

References

1. A. Bahri, M. Bendersky, F.R. Cohen, S. Gitler, The polyhedral product functor: a method of decomposition for moment-angle complexes, arrangements and related spaces. Adv. Math. **225**, 1634–1668 (2010)
2. P. Beben, S. Theriault, The loop space homotopy type of simply-connected four-manifolds and their generalizations. Adv. Math. **262**, 213–238 (2014)
3. P. Beben, S. Theriault, Homotopy groups of highly connected Poincaré duality complexes. Doc. Math. **27**, 183–211 (2022)
4. V.M. Buchstaber, T.E. Panov, Toric Topology. Mathematical Surveys and Monographs, vol. 204 (American Mathematical Society, Providence, 2015)
5. G. Denham, A.I. Suciu, Moment-angle complexex, monomial ideals, and massey products. Pure Appl. Math. Q. **3**, 25–60 (2007)
6. Y. Félix, D. Tanré, Rational homotopy of the polyhedral product functor. Proc. Amer. Math. Soc. **137**, 891–898 (2009)
7. J. Grbić, T. Panov, S. Theriault, J. Wu, The homotopy types of moment-angle complexes for flag complexes. Trans. Amer. Math. Soc. **368**, 6663–6682 (2016)
8. J. Grbić, S. Theriault, The homotopy type of the complement of the codimension two coordinate subspace arrangement. Russ. Math. Surv. **59**, 203–204 (2005)
9. J. Grbić, S. Theriault, The homotopy type of the polyhedral product for shifted complexes. Adv. Math. **245**, 690–715 (2013)
10. Y. Hao, Q. Sun, S. Theriault, Moore's conjecture for polyhedral products. Math. Proc. Camb. Philos. Soc. **167**, 23–33 (2019)

11. K. Iriye, D. Kishimoto, Decompositions of polyhedral products for shifted complexes. Adv. Math. **245**, 716–736 (2013)
12. K. Iriye, D. Kishimoto, Fat-wedge filtration and decomposition of polyhedral products. Kyoto J. Math. **59**, 1–51 (2019)
13. I.M. James, Reduced product spaces. Ann. Math. **62**, 170–197 (1955)
14. D. MacGavran, Adjacent connected sums and torus actions. Trans. Amer. Math. Soc. **251**, 235–254 (1979)
15. M. Mather, Pull-backs in homotopy theory. Canad. J. Math. **28**, 225–263 (1976)
16. T. Panov, S. Theriault, The homotopy theory of polyhedral products associated with flag complexes. Compos. Math. **155**, 206–228 (2019)
17. G.J. Porter, The homotopy groups of wedges of suspensions. Amer. J. Math. **88**, 655–663 (1966)
18. S. Theriault, Moment-angle manifolds and Panov's problem. Int. Math. Res. Not. IMRN **20**, 10154–10175 (2015)

On the Cohomology Ring of Real Moment-Angle Complexes

Elizabeth Vidaurre

1 Introduction

Fixing a pair of topological spaces (X, A), polyhedral product spaces $Z_K(X, A)$ give a family of spaces where K is a simplicial complex (see Definition 2.1). Examples include moment-angle complexes, complements of complex coordinate subspace arrangements, and intersections of quadrics, among others. In certain cases, polyhedral products provide geometric realizations of right-angled Artin groups and the Stanley-Reisner ring (see Definition 2.5).

The real moment-angle complex, $Z_K(D^1, S^0)$, and its complex analog (arising from the pair of spaces, the unit disc D^2, and the circle S^1) feature in toric topology, as they have been key in showing applications in combinatorics and algebraic geometry, among others [5]. The cohomology ring of the moment-angle complex is shown to be isomorphic to the Tor-algebra $Tor_{\mathbb{Z}[v_1,\ldots,v_m]}(\mathbb{Z}[K], \mathbb{Z})$ in [4], where $\mathbb{Z}[K]$ is the Stanley-Reisner (or face ring) of K and the indeterminates v_i are of degree 2 (see Sect. 2). The generators correspond to certain subsets of integers, and the product of two generators corresponding to non-disjoint subsets is trivial, forming a symplectic basis.

On the other hand, the cohomology ring of the real moment-angle complex is not completely understood. The group structure is known to be given by $Tor_{\mathbb{Z}[v_1,\ldots,v_m]}(\mathbb{Z}[K], \mathbb{Z})$ with indeterminates v_i of degree 1. Theorem 3.5 gives the ring structure for real moment-angle complexes over certain simplicial complexes. A consequence of Theorem 3.5 is that the multiplicative structure does not have the same nice closed form as that of moment-angle complexes. In other words,

E. Vidaurre (✉)
Molloy College, Rockville Centre, NY, USA
e-mail: evidaurre@molloy.edu

A. Bahri et al. (eds.), *Toric Topology and Polyhedral Products*, Fields Institute Communications 89, https://doi.org/10.1007/978-3-031-57204-3_16

generators corresponding to non-disjoint subsets do not necessarily have trivial product.

This set of combinatorially defined generators can be identified using Bahri-Bendersky-Cohen-Gitler's splitting theorem [1] and Welker-Ziegler-Živaljević's wedge lemma [13]. In this paper, we consider the case when the simplicial complex K is the boundary of an n-gon and describe the ring structure in terms of the combinatorial generators in Theorem 3.5.

In full generality, for a simplicial complex K on m vertices, polyhedral product spaces $Z_K(\underline{X}, \underline{A})$ are defined in terms of a collection of pairs of spaces $(\underline{X}, \underline{A}) = \{X_i, A_i\}_{i=1}^m$. The ring structure for the real moment-angle complex $Z_K(D^1, S^0)$ is particularly useful in that the cohomology ring of the more general polyhedral product $Z_K(\underline{CA}, \underline{A})$ when CA_i is the cone on A_i can be described in terms of the ring structure of $H^*(Z_K(D^1, S^0))$ and $H^*(A)$ [3].

Moreover, this problem of understanding the cohomology ring of a real moment-angle complex has connections to studying the topology of intersections of quadrics associated to simple polytopes, as well as that of real coordinate subspace arrangements. In particular, the case when K is the pentagon is discussed in [10]. The cohomology of real moment-angle complexes and related spaces has also been studied in [7], in the case of rational coefficients.

In Sect. 3.2, we will illustrate the main theorem with some examples. As a corollary we will see that, even though real moment-angle complexes over an n-gon are orientable surfaces, the combinatorial generators do not form a symplectic basis.

2 Polyhedral Product Spaces

In this section, we will give a brief introduction to polyhedral products, moment-angle complexes, and real moment-angle complexes, with an emphasis on the multiplicative structure of their respective cohomology rings.

Let $[m] = \{1, 2, \ldots, m\}$ denote the set of integers from 1 to m. An *abstract simplicial complex*, K, on $[m]$ is a subset of the power set of $[m]$, such that

1. $\emptyset \in K$.
2. If $\sigma \in K$ with $\tau \subset \sigma$, then $\tau \in K$.

An n-simplex is the full power set of $[n+1]$ and is denoted Δ^n. Associated to an abstract simplicial complex is its *geometric realization*, denoted $\mathcal{K}$ or $|K|$ (also called a geometric simplicial complex). A (geometric) n-simplex, Δ^n, is the convex hull of $n+1$ points.

We do not assume m is minimal, i.e., there may exist $[n] \subsetneq [m]$ such that K is contained in the power set of $[n]$.

Let I be a subset of $[m]$. The *full subcomplex of* K *in* I is denoted K_I. It is a simplicial complex on the set I and defined

$$K_I := \{\sigma \in K | \sigma \subset I\}.$$

It is often called the restriction of K to I in the literature.

Given an abstract simplicial complex K, let $\mathcal{S}_K$ be the category with simplices of K as the objects and inclusions as the morphisms. In particular, for $\sigma, \tau \in ob(\mathcal{S}_K)$, there is a morphism $\sigma \rightarrow \tau$ whenever $\sigma \subset \tau$. Define $\mathcal{CW}$ to be the category of CW-complexes and continuous maps. Define $(\underline{X}, \underline{A})$ to be a collection of pairs of CW-complexes $\{(X_i, A_i)\}_{i=1}^m$, where A_i is a subspace of X_i for all i.

Definition 2.1 Given an abstract simplicial complex K on $[m]$, simplices σ, τ of K, and a collection of pairs of CW-complexes $(\underline{X}, \underline{A})$, define a diagram $D : \mathcal{S}_K \rightarrow \mathcal{CW}$ given by

$$D(\sigma) = \prod_{i \in [m]} Y_i \qquad \text{where } Y_i = \begin{cases} X_i & i \in \sigma \\ A_i & i \in [m] \backslash \sigma \end{cases}.$$

For a morphism $f : \sigma \rightarrow \tau$, the functor D maps f to $\iota : D(\sigma) \rightarrow D(\tau)$ where ι is the canonical injection.

The *polyhedral product space* is defined as

$$Z_K(\underline{X}, \underline{A}) := colim_{\sigma \in K} D(\sigma) = \bigcup_{\sigma \in K} D(\sigma)$$

and is topologized as a subspace of $\prod_{i \in [m]} X_i$.

Notice that it suffices to take the colimit over the maximal simplices of K. In fact, simplicial complexes can be defined by their maximal simplices, and this description will be used throughout. In the case where $(X_i, A_i) = (X, A)$ for all i, we write $Z_K(X, A)$.

Some examples of polyhedral products are moment-angle complexes $Z_K(D^2, S^1)$, which have the homotopy type of the complement of a complex coordinate subspace arrangement, and Davis-Januszkiewicz spaces $Z_K(\mathbb{C}P^\infty, *)$, which have the Stanley-Reisner ring as cohomology ring. For a simple example, consider the following. Let K be the boundary of a 2-simplex with vertices labeled $1, 2, 3$.

$$\begin{aligned} Z_K(D^1, S^0) &= D(\{1, 2\}) \cup D(\{1, 3\}) \cup D(\{2, 3\}) \\ &= D^1 \times D^1 \times S^0 \cup D^1 \times S^0 \times D^1 \cup S^0 \times D^1 \times D^1 \\ &= \partial(D^1 \times D^1 \times D^1) \\ &\cong S^2 \end{aligned}$$

In general, $Z_{\partial\Delta^m}(D^1, S^0) \cong S^m$ (see examples in [1]). Next we will define the polyhedral smash product, a space analogous to the polyhedral product with the smash product operation in place of the Cartesian product. Define $\mathcal{CW}_*$ to be the category of based CW-complexes and based continuous maps.

Definition 2.2 Let the CW-pairs $(\underline{X}, \underline{A})$ be pointed. Likewise, define a functor $\widehat{D}(\sigma) : \mathcal{S}_K \to \mathcal{CW}_*$ by

$$\widehat{D}(\sigma) = \wedge Y_i \qquad \text{where } Y_i = \begin{cases} X_i & i \in \sigma \\ A_i & i \notin \sigma \end{cases}.$$

Then the *polyhedral smash product* is

$$\widehat{Z}_K(\underline{X}, \underline{A}) = \bigcup \widehat{D}(\sigma).$$

For the remainder of the paper, we will assume that (X_i, A_i) is a pair of pointed CW-complexes where A_i is a subspace of X_i.

The following theorem of Bahri, Bendersky, Cohen, and Gitler (BBCG) gives a stable decomposition of a polyhedral product.

Theorem 2.3 (Splitting Theorem, [1]) *Let* $(\underline{X_I}, \underline{A_I}) = \{(X_i, A_i)\}_{i \in I}$. *Then*

$$\Sigma Z_K(\underline{X}, \underline{A}) \simeq \Sigma \bigvee_{I \subset [m]} \widehat{Z}_{K_I}(\underline{X_I}, \underline{A_I}))$$

where Σ *denotes the reduced suspension.*

In [1], the authors apply the wedge lemma from [13] to polyhedral smash products and obtain the following:

Theorem 2.4 (Wedge Lemma, [13]) *If* X_i *is contractible for all* i, *then*

$$\widehat{Z}_K(\underline{X}, \underline{A}) \simeq \Sigma |K| \wedge A^{\wedge [m]} \simeq |K| * A^{\wedge [m]}$$

where $A^{\wedge [m]} = A_1 \wedge \ldots \wedge A_m$.

Since S^0 serves as an identity for the smash product operation, computing the cohomology groups of real moment-angle complexes becomes a combinatorial process that involves examining only the simplicial complex. This follows from the previous two theorems.

$$H^*(Z_K(D^1, S^0)) = \bigoplus_{I \subset [m]} H^*(\Sigma \mathcal{K}_I) \tag{1}$$

The generators of the cohomology ring are given by the subsets of $[m]$ that yield a noncontractible full subcomplex of K after suspension, which we call the *combinatorial generators*.

To describe the ring structure of the cohomology of the moment-angle complex, we will introduce some notation. The graded ring $\mathbb{Z}[m]$ is the polynomial ring on m variables $\mathbb{Z}[v_1, v_2, \ldots, v_m]$ with $|(|v_i) = 2$.

Definition 2.5 The *Stanley-Reisner* ring (or face ring) of the simplicial complex K is the quotient of $\mathbb{Z}[m]$ by the ideal generated by square-free monomials associated to nonfaces of K:

$$\mathbb{Z}[K] := \mathbb{Z}[m]/\langle v_{i_1} v_{i_2} \ldots v_{i_k} \mid \{i_1, i_2, \ldots, i_k\} \notin K\rangle.$$

The following was first proven by Franz in [9] and stated in terms of smooth toric varieties. Another proof was later given by Baskakov, Buchstaber, and Panov in [4].

Theorem 2.6 (Franz, [9]) *The cohomology ring of the moment-angle complex* $Z_K(D^2, S^1)$ *is given by*

$$H^*(Z_K(D^2, S^1)) \cong Tor_{\mathbb{Z}[m]}(\mathbb{Z}[K], \mathbb{Z}).$$

A description of the multiplicative structure in terms of full subcomplexes comes from Hochster's theorem in commutative algebra on the Tor-module [11]. We obtain the following analogous formula:

$$H^k(Z_K(D^2, S^1)) \cong \bigoplus_{J \subset [m]} \widetilde{H}^{k-|J|-1}(K_J).$$

For the multiplicative structure, take classes $\alpha \in H^i(Z_K(D^2, S^1))$ and $\beta \in \widetilde{H}^k(Z_K(D^2, S^1))$. Then α corresponds to some class in $\widetilde{H}^{i-|J|-1}(K_J)$ for some subset $J \subset [m]$ and similarly β to some class in $\widetilde{H}^{k-|L|-1}(K_L)$ for some $L \subset [m]$. Their product is

$$\alpha \smile \beta = \begin{cases} \gamma & \text{if } J \cap L = \emptyset \\ 0 & \text{if } J \cap L \neq \emptyset \end{cases}$$

for some γ coming from $\widetilde{H}^{i+k-|L|-|J|-1}(K_{J \cup L})$. See [12] for more details.

2.1 The BBCG Spectral Sequence

We will use a spectral sequence developed by BBCG [3]. It gives a Künneth-like formula for the cohomology of a polyhedral product as long as the pairs $(\underline{X}, \underline{A})$ satisfy the following freeness condition.

Definition 2.7 Given the pair (X_i, A_i), the associated long exact sequence is given by

$$\ldots \xrightarrow{\delta} \widetilde{H}^*(X_i/A_i) \xrightarrow{g} H^*(X_i) \xrightarrow{f} H^*(A_i) \xrightarrow{\delta} \widetilde{H}^{*+1}(X_i/A_i) \xrightarrow{g} \ldots$$

The pair is said to satisfy the *strong* h^* *freeness* condition if there are free h^*−modules E_i, B_i, C_i and W_i satisfying

$$\begin{aligned} H^*(A_i) &= B_i \oplus E_i \\ H^*(X_i) &= B_i \oplus C_i \\ \widetilde{H}^*(X_i/A_i) &= C_i \oplus W_i \end{aligned}$$

where W_i is sE_i the suspension of E_i. Additionally, assume $1 \in B_i$, and for $b \in B_i, c \in C_i, e \in E_i, w \in W_i = sE_i$, we have

$$b \overset{f}{\mapsto} b \overset{\delta}{\mapsto} 0, \qquad c \overset{g}{\mapsto} c \overset{f}{\mapsto} 0, \qquad e \overset{\delta}{\mapsto} w \overset{g}{\mapsto} 0.$$

Before defining the spectral sequence, we will give some notation and recall the definition of a half smash product:

1. for $\sigma = \{i_1, \ldots, i_k\}$, define $\widehat{X}^\sigma := X_{i_1} \wedge \ldots \wedge X_{i_k}$ and $A^\sigma = A_{i_1} \times \ldots \times A_{i_k}$
2. the complement of a set $\sigma \subset [m]$ is $\sigma^c = [m] \backslash \sigma$
3. given a basepoint $x_0 \in X$, the right half smash product $X \rtimes Y = (X \times Y)/(x_0 \times Y)$
4. for a subset I and a simplex σ such that $\sigma \subset I$, define

$$Y^{I,\sigma} := \bigotimes_{i \in \sigma} C_i \otimes \bigotimes_{i \in I - \sigma} B_i.$$

Choosing a lexicographical ordering for the simplices of K gives a filtration of the associated polyhedral product space and polyhedral smash product, which in turn leads to a spectral sequence converging to the reduced cohomology of $Z_K(\underline{X}, \underline{A})$ and a spectral sequence converging to the reduced cohomology of $\widehat{Z}_K(\underline{X}, \underline{A})$. The $E_1^{s,t}$ term for $Z_K(\underline{X}, \underline{A})$ has the following description.

Theorem 2.8 (Bahri et al. [3]) *There exist spectral sequences*

$$E_r^{s,t} \to H^*(Z_K(\underline{X}, \underline{A}))$$

$$\widehat{E}_r^{s,t} \to H^*(\widehat{Z}_K(X, A))$$

with $E_1^{s,t} = \widetilde{H}^t((\widehat{X/A})^\sigma \rtimes A^{\sigma^c})$ *and* $\widehat{E}_1^{s,t} = \widetilde{H}^t((\widehat{X/A})^\sigma \wedge \widehat{A}^{\sigma^c})$ *where s is the index of σ in the lexicographical ordering and the differential* $d_r : E_r^{s,t} \to E_r^{s+r,t+1}$ *is induced by the coboundary map* $\delta : E \to W = sE$. *Moreover, the spectral sequence* *is natural for embeddings of simplicial maps with the same number of vertices and with respect to maps of pairs. The natural quotient map*

$$Z_K(\underline{X}, \underline{A}) \to \widehat{Z}_K(\underline{X}, \underline{A})$$

induces a morphism of spectral sequences, and the Splitting Theorem 2.3 induces a morphism of spectral sequences.

Following [3], Definition 2.7 and the Künneth Theorem imply that the entries $\widetilde{H}^t(\widehat{(X/A)}^\sigma \wedge \widehat{A}^{\sigma^c})$ in the first page of the spectral sequence for $\widehat{Z}_K(\underline{X}, \underline{A})$ decompose as a direct sum of spaces $W^N \otimes C^S \otimes B^T \otimes E^J$ such that $N \cup S = \sigma$, $T \cup J = \sigma^c$ and N, S, J, T are disjoint. We have that S is a simplex in K as $N \cup S$ is a simplex in K. Since the differential is induced by the coboundary $\delta : E \to W$, consider all the possible summands $W^N \otimes C^S \otimes B^T \otimes E^J$ for S and T fixed. It must be the case that N is a simplex in K and that N is a subset of $[m]\backslash(S \cup T)$. Therefore, all such N correspond to simplices in the link of S in K restricted to the vertex set $[m]\backslash(S \cup T)$.

Theorem 2.9 (Bahri et al. [3]) *Let* $(\underline{X}, \underline{A})$ *satisfy the decomposition described in Definition 2.7*

$$\begin{aligned} H^*(A_i) &= B_i \oplus E_i \\ H^*(X_i) &= B_i \oplus C_i \end{aligned}.$$

Then

$$H^*(Z_K(\underline{X}, \underline{A})) = \bigoplus_{I \subset [m], \sigma \subset I} E^{I^c} \otimes Y^{I,\sigma} \otimes \widetilde{H}^*(\Sigma|lk(\sigma)_{I^c}|)$$

where

1. *σ is a simplex in K,*
2. *$lk(\sigma)_{I^c} = \{\tau \subset [m]\backslash I \mid \tau \cup \sigma \in K\}$ is the link of σ in K restricted to the set $[m]\backslash I$,*
3. *$Y^{I,\sigma} = \bigotimes_{i\in\sigma} C_i \otimes \bigotimes_{i \in I-\sigma} B_i$, and*
4. *$\widetilde{H}^*(\Sigma\emptyset) = 1$.*

Theorem 2.10 (Bahri et al. [3]) *Let*

$$\begin{aligned} \widetilde{H}^*(A_i) &= \widetilde{B}_i \oplus E_i \\ \widetilde{H}^*(X_i) &= \widetilde{B}_i \oplus C_i \end{aligned}.$$

Then

$$H^*(\widehat{Z}_K(\underline{X}, \underline{A})) = \bigoplus_{I \subset [m], \sigma \subset I} E^{I^c} \otimes Y^{I,\sigma} \otimes \widetilde{H}^*(\Sigma|lk(\sigma)_{I^c}|)$$

where

1. *σ is a simplex in K,*
2. *$lk(\sigma)_{I^c} = \{\tau \subset [m]\backslash I \mid \tau \cup \sigma \in K\}$ is the link of σ in K restricted to the set $[m]\backslash I$,*
3. *$Y^{I,\sigma} = \bigotimes_{i\in\sigma} C_i \otimes \bigotimes_{i\in I-\sigma} \widetilde{B}_i$ where $\widetilde{B}_i = B_i\backslash\{1\}$,*
4. *$\widetilde{H}^*(\Sigma\emptyset) = 1$.*

A description of the ring structure in $H^*(Z_K(\underline{X}, \underline{A}))$ is given using the decomposition from Theorems 2.9 and 2.10. It is induced by a pairing involving links

$$\widetilde{H}^*(\Sigma|lk(\sigma_1)|_{I_1^c}) \otimes \widetilde{H}^*(\Sigma|lk(\sigma_2)|_{I_2^c}) \to \widetilde{H}^*(\Sigma|lk(\sigma_3)|_{I_3^c})$$

defined in terms of the $*$-product, introduced in [2], where I_3 and σ_3 are defined in terms of σ_1, σ_2, I_1 and I_2.

Theorem 2.11 (Theorem 6.1 in [3]) *Two classes*

$$\alpha, \beta \in H^*(Z_K(\underline{X}, \underline{A})) = \bigoplus_{I\subset[m],\sigma\subset I} E^{[m]-I} \otimes C^{\sigma} \otimes B^{I-\sigma} \otimes \widetilde{H}^*(\Sigma lk(\sigma)_{I^c}),$$

are of the form

$$\alpha = a_1 \otimes a_2 \otimes \ldots \otimes a_m \otimes n_\alpha$$
$$\beta = b_1 \otimes b_2 \otimes \ldots \otimes b_m \otimes n_\beta$$

where $n_\alpha \in \widetilde{H}^*(\Sigma lk(\sigma)_{I^c})$ and $n_\beta \in \widetilde{H}^*(\Sigma lk(\tau)_{J^c})$.

The cup product of α and β is given in terms of the $*$-product and a component-wise product induced by the multiplicative structure of $H^*(X_i)$ and $H^*(A_i)$.

For the pair of spaces (CA_i, A_i), where CA_i is the cone on A_i, the modules are given by $B_i = 1$, $C_i = 0$, and $E_i = \widetilde{H}^*(A)$. The links are all of the form K_I for $I \subset [m]$. Therefore, it can be seen from Theorem 2.9 that the product structure in $H^*(Z_K(\underline{CA}, \underline{A}))$ can be described in terms of the product structure in $H^*(A)$ and $\widetilde{H}^*(\Sigma \mathcal{K}_I)$.

Due to the decomposition in Eq. 1 and work in [2], the ring structure in $H^*(Z_K(\underline{CA}, \underline{A}))$ can be described in terms of the ring structure in $H^*(A)$ and $H^*(Z_K(D^1, S^0))$.

Theorem 2.12 (Theorem 1.9 in [2]) *Assume that any finite product of A_i with $Z_{K_I}(D^1, S^0)$ for all I satisfies the strong form of the Künneth Theorem. Then the cup product structure for the cohomology algebra $H^*(Z_K(\underline{CA}, \underline{A}))$ is a functor of the cohomology algebras of A_i, and $Z_{K_I}(D^1, S^0)$ for all I.*

3 Multiplicative Structure of $H^*(Z_K(D^1, S^0))$

Recall from Eq. 1 that each subset I of $[m]$ such that the full subcomplex $\mathcal{K}_I$ is not contractible corresponds to a generator of $H^*(Z_K(D^1, S^0))$.

To compute the cohomology of a real moment-angle complex, we will use a filtered chain complex induced by the long exact sequence of the pair (D^1, S^0), denoted C_K and constructed in [3]. For $(X_i, A_i) = (D^1, S^0)$, let $\widetilde{H}^*(A_k) = \widetilde{H}^*(S^0)$ be generated by t_k and $\widetilde{H}^*(X_k/A_k) = \widetilde{H}^*(S^1)$ be generated by s_k.

Definition 3.1 The chain complex $C(K_I)$ is generated by $y_\sigma := \otimes y_i$ where $\sigma \in K_I$ and

$$y_i = \begin{cases} s_i & i \in \sigma \\ t_i & i \in I - \sigma \\ 1 & k \notin I \end{cases}.$$

The differential is defined by

$$d_I(y_\sigma) = \sum_\tau (-1)^{n(\tau)} y_\tau$$

where $\sigma \subset \tau \in K_I$ and $\tau = \sigma \cup v$ for some vertex $v \in I$. The integer $n(\tau)$ is defined by the usual sign convention of a graded derivation. In particular, the coboundary δ acts on each factor of y_σ by $\delta(s_i) = 0$ and $\delta(t_i) = s_i$, and every time it passes an s_i, a factor of (-1) is introduced.

Then

$$C_K = \bigoplus_{I \subset [n]} C(K_I)$$

and $H^*(C_K) = H^*(Z_K(D^1, S^0))$.

It follows from work of Li Cai in [6] that the chain level cup product of two generators is induced by the following:

$$s_i \smile s_i = 0, \qquad t_i \smile t_i = t_i, \qquad s_i \smile t_i = s_i, \qquad t_i \smile s_i = 0.$$

3.1 *Boundary of a Polygon*

We will consider the case of K the boundary of a polygon. By Theorem 1, we need to consider all subsets of $[n]$ to find the cohomology groups. By convention, when I is the empty set, $H^*(\Sigma \mathcal{K}_I) = 1$. The suspension of the whole complex K is a degree 2 generator. The following lemma gives the generators of degree 1.

Lemma 3.2 *Suppose K is the boundary of an n-gon. Let $I = I_1 \sqcup I_2 \sqcup \ldots \sqcup I_p$ be a subset of $[n]$ such that $\mathcal{K}_I$ has exactly p maximal connected components, $\mathcal{K}_I \simeq \bigvee_{p-1} S^0$. Then*

$$H^1(\Sigma\mathcal{K}_I) = {\langle \sum_{i\in I_1} y_{\{i\}}, \sum_{i\in I_2} y_{\{i\}}, \ldots, \sum_{i\in I_p} y_{\{i\}} \rangle} \Big/ {\langle \sum_{i\in I} y_{\{i\}} \rangle}.$$

Proof Let $I_1 = \{i_1, \ldots, i_c\}$. If $i_1 = i_c$, then $d(y_{\{i_1\}}) = 0$, and $y_{\{i_1\}}$ is clearly a cocycle. If $i_1 \neq i_c$, then the differential will not be trivial. If $+y_{\{e\}}$ for some edge $e \in K_{I_1}$ appears as a summand in the image of $d(y_{\{v\}})$ for some vertex $v \in I_1$, then $e = \{v-1, v\}$ or $e = \{1, n\}$ (since $y_{\{e\}}$ was positive, we could not have passed an s). Additionally, since $y_{\{e\}}$ was in the image of $y_{\{v\}}$, it must be the case that $v - 1 \in I_1$, so $-y_{\{e\}}$ is a term in the image of $y_{\{v-1\}}$ under d. If it had been the case that $e = \{1, n\}$, then $\{1\} \in K_I$ and $-y_{\{e\}}$ would be in the image of $y_{\{1\}}$. Since it is only possible for $y_{\{e\}}$ to be in the image of $y_{\{v-1\}}$ or $y_{\{v\}}$, the terms $y_{\{e\}}$ cancel. This means that $d(y_{\{i_1\}} + \ldots + y_{\{i_c\}}) = 0$ since $i_1, \ldots, i_c$ are all the vertices in a connected component of $\mathcal{K}_I$. Without loss of generality, the same is true for the other connected components. Lastly, $d(\emptyset)$ is the sum of $y_{\{i\}}$ for $i \in I$. □

Corollary 3.3 *If $I = I_1 \sqcup I_2 \sqcup \ldots \sqcup I_p$ is a subset of $[n]$ such that $\mathcal{K}_I$ has exactly p maximal connected components with $p > 1$, then $H^1(\Sigma|\mathcal{K}_I|)$ has rank $p-1$, and a basis of generators can be chosen by picking any $p-1$ of the p disjoint subsets $I_1, I_2, \ldots, I_p$.*

Next, the following lemma will show how generators coming from different subsets of $[n]$ multiply. Consider subsets $I, J \subset [n]$ such that $I \cup J = [n]$. We will employ a slight change in notation: replacing y's associated to I with a's and y's associated to J with b's to differentiate between generators in $C(K_I)$ and generators in $C(K_J)$. Recall that $a_{\{i\}}$ is the generator associated to the vertex i, whereas a_i is the ith factor of a generator.

Lemma 3.4 *Suppose $a_{\{i\}} \in C(K_I)$ and $b_{\{j\}} \in C(K_J)$. Then $a_{\{i\}} = a_1 \otimes a_2 \otimes \ldots \otimes a_n$ where*

$$a_k = \begin{cases} s_i & k = i \\ t_k & k \in I \backslash \{i\} \\ 1 & k \notin I \end{cases}.$$

Define $b_{\{j\}}$ similarly. Then

$$a_{\{i\}} \smile b_{\{j\}} = \begin{cases} 0 & j \in I \text{ or } i - j \neq \pm 1 \pmod n \\ y_{\{i,j\}} & j > i \\ -y_{\{j,i\}} & j < i \end{cases}.$$

Proof If $|i - j| \neq 1$, then $\{i, j\}$ is not a simplex in K and $a_{\{i\}} \smile b_{\{j\}} = 0$. Therefore, we will now consider cases where $|i - j| = 1$.

Recall that $s \smile t = s \smile 1 = s$ and $t \smile s = 0$.

Suppose $j \in I$. Since $j \in I$, $a_j = t_j$. In particular, in the jth coordinate of $a_{\{i\}} \smile b_{\{j\}}$, we will have $a_j \smile b_j = t_j \smile s_j = 0$ so $a_{\{i\}} \smile b_{\{j\}} = 0$.

Next suppose $j \notin I$. Then $a_j = 1$ and $a_j \smile b_j = s_j$. If $j = i + 1$, then

$$\begin{aligned}
&a_{\{i\}} \smile b_{\{j\}} \\
&= (a_1 \smile b_1) \otimes \ldots \otimes (a_i \smile b_i) \otimes (a_j \smile b_j) \otimes \ldots \otimes (a_n \smile b_n) \\
&= \quad t_1 \quad \otimes \ldots \otimes \quad s_i \smile b_i \quad \otimes \quad 1 \smile s_j \quad \otimes \ldots \otimes \quad t_n \\
&= \quad t_1 \quad \otimes \ldots \otimes \quad s_i \quad \otimes \quad s_j \quad \otimes \ldots \otimes \quad t_n \\
&= \quad y_{\{i,j\}}
\end{aligned}$$

since the only coordinate of $b_{\{j\}}$ that is an s is b_j and all other coordinates are t or 1.

If $j = i - 1$, since $a_i \smile b_{i-1} = (-1)^{|b_{i-i}||a_i|}(b_{i-1} \smile a_i)$, we have

$$\begin{aligned}
&a_{\{i\}} \smile b_{\{j\}} \\
&= (a_1 \smile b_1) \otimes \ldots \otimes (-1)^{|b_j||a_i|}(a_j \smile b_j) \otimes (a_i \smile b_i) \otimes \ldots \otimes (a_n \smile b_n) \\
&= \quad t_1 \quad \otimes \ldots \otimes \quad (-1)(1 \smile s_j) \quad \otimes \quad s_i \smile b_i \quad \otimes \ldots \otimes \quad t_n \\
&= \quad t_1 \quad \otimes \ldots \otimes \quad (-1)s_j \quad \otimes \quad s_i \quad \otimes \ldots \otimes \quad t_n \\
&= \quad -y_{\{j,i\}}
\end{aligned}$$

□

The following theorem uses the previous lemmas to show what the only nontrivial products in $H^*(Z_K(D^1, S^0))$ are.

Theorem 3.5 *Let K be the boundary of an n-gon. Let $I = I_1 \sqcup I_2 \sqcup \ldots \sqcup I_p$ and $J = J_1 \sqcup J_2 \sqcup \ldots \sqcup J_q$ be nonempty subsets of $[n]$ such that $I \cup J = [n]$ and $K_I \simeq \bigvee_1^{p-1} S^0$ and $K_J \simeq \bigvee_1^{q-1} S^0$. Take generators α and β of $H^*(Z_K(D^1, S^0))$ such that one is associated to some I_g for $1 \leq g \leq p$ and the other is associated to some J_h for some $1 \leq h \leq q$. If γ is the second degree generator of $H^*(Z_K(D^1, S^0))$, then $\alpha \smile \beta = \pm\gamma$ if and only if the following conditions are met*

- $I_g \nsubseteq J_h$
- $J_h \nsubseteq I_g$
- $K_{I_g \cup J_h}$ *is contractible.*

Proof We will compute $H^*(Z_K(D^1, S^0))$ using the chain complex described previously.

$$\begin{aligned}
d(y_\emptyset) &= y_{\{1\}} + \ldots + y_{\{n\}} \\
d(y_{\{1\}}) &= -y_{\{1,2\}} - y_{\{1,n\}} \\
d(y_{\{2\}}) &= y_{\{1,2\}} - y_{\{2,3\}} \\
d(y_{\{3\}}) &= y_{\{2,3\}} - y_{\{3,4\}} \\
&\ldots \\
d(y_{\{n-1\}}) &= y_{\{n-2,n-1\}} - y_{\{n-1,n\}} \\
d(y_{\{n\}}) &= y_{\{n-1,n\}} + y_{\{1,n\}}
\end{aligned}$$

Therefore, all the classes in $H^*(Z(K;(D^1,S^0)))$ represented by an edge are cohomologous, except $y_{\{1,n\}}$, which is the negative. Note that if 1 and n are in I, then 1 and n are in I_i for some $1 \le i \le p$ (since $\mathcal{K}_I \simeq \bigvee_1^{p-1} S^0$, it cannot be that 1 and n are in different subsets of I). By Corollary 3.3, we only need to consider all but one of the disjoint subset of I and all but one of the disjoint subsets of J. Therefore, it suffices to only consider when $1, n \notin I_g \cup J_h$. As a consequence, the class $y_{\{1,n\}}$ cannot occur in the product $\alpha \smile \beta$.

Let $I_g = \{i_1, \ldots, i_c\}$ and $J_h = \{j_1, \ldots, j_d\}$. By Lemma 3.2, we have that

$$\alpha = \sum_{i \in I_g} a_{\{i\}}$$

Since $I \cup J = [n]$, we have that $i_1 \neq j_1$ and $i_c \neq j_d$.

First, suppose $I_g \cap J_h = \emptyset$. In the case that $i_c < j_1$, we must have $i_c = j_1 - 1$ so that there is at least one edge after expanding the product. Then there is only one nonzero term

$$\alpha \smile \beta = \sum_{i \in I_g, j \in J_h} a_{\{i\}} \smile b_{\{j\}} = y_{\{i_c, j_1\}}$$

by Lemma 3.4. Similarly, if $j_d = i_1 - 1$, then

$$\alpha \smile \beta = \sum_{i \in I_g, j \in J_h} a_{\{i\}} \smile b_{\{j\}} = a_{\{i_1\}} \smile b_{\{j_d\}} = -y_{\{j_d, i_1\}}$$

Secondly, suppose $I_g \cap J_h \neq \emptyset$ and that neither set is contained in the other. If $j_1 \le i_c$, then there exists $j \in J_h$ such that $j = i_c$. Note that $j+1 \in J_h$ (because $i_c \neq j_d$ and so $j \neq j_d$). Since $\{i_c, j+1\}$ is an edge and $j+1 = i_c + 1 \notin I$, by Lemma 3.4, we have only one nonzero term

$$\alpha \smile \beta = a_{\{i_c\}} \smile b_{\{j+1\}} = y_{\{i_c, i_c+1\}}$$

Similarly, if $i_1 \le j_d$ and $j = i_1$ for some $j \in J_h$, then $j-1 \notin I$ and $i_1 - (j-1) = 1$. Then

$$\alpha \smile \beta = a_{\{i_1\}} \smile b_{\{j-1\}} = -y_{\{i_1-1, i_1\}}$$

If $J_h \subset I_g$, then $\alpha \smile \beta = 0$ by Lemma 3.4. If $I_g \subset J_h$, then there are only two possible nonzero products between the summands of α and β. There exists $j \in J_h$ such that $j = i_1$. Then

$$\begin{aligned}
\alpha \smile \beta &= a_{\{i_1\}} b_{\{j-1\}} + a_{\{i_c\}} b_{\{j+c\}} \\
&= -y_{\{i_1-1, i_1\}} + y_{\{i_c, i_c+1\}} \\
&= -\gamma + \gamma \\
&= 0
\end{aligned}.$$

□

3.2 Example and Related Consequences

To illustrate an application of the theorem and some important consequences, we will consider the case when K is the boundary of the pentagon, denoted K_5. Let the 1-simplices of K_5 be labeled $\{1, 2\}, \{2, 3\}, \{3, 4\}, \{4, 5\}, \{1, 5\}$. It follows from [8] that the real moment-angle complex over the boundary of an n-gon is a closed orientable surface of genus $1 + (n - 4)2^{n-3}$, which means that in this example the associated real moment-angle complex has genus 5 (Fig. 1).

For the combinatorial generators, there are ten subsets I of [5] that yield a full subcomplex K_I equivalent to a wedge of 0-spheres. The cohomology of $Z_{K_I}(D^1, S^0)$ has an identity, ten degree one generators $x_0, \ldots, x_4, w_0, \ldots, w_4$, and a degree two generator z, subject to a graded commutative product. The identity corresponds to the empty set. The generators x_i correspond to the subsets that yield a full subcomplex of K of an edge and the opposite vertex, such as the subset $I = \{1, 2, 4\}$. The generators w_i correspond to the subsets that produce a full subcomplex of two disjoint vertices. Lastly, z corresponds to the full vertex set $[5] = \{1, 2, 3, 4, 5\}$.

$$H^*(Z_{K_5}(D^1, S^0)) = \langle 1, w_0, \ldots, w_4, x_0, \ldots, x_4, z \mid x_i x_j = z\delta_{j,i+1}, x_i w_j = z\delta_{i,j} \rangle$$

where δ is the Kronecker delta function and the subscripts i, j are integers modulo 5.

Notice that for subsets $J = \{2, 4, 5\}$ and $I = \{1, 3, 4\}$, $\alpha_I \smile \alpha_J = \gamma$. This is an example where generators coming from non-disjoint subsets have a nontrivial product, unlike the ring for moment-angle complexes. The cohomology ring of $Z_K(D^1, S^0)$ is not isomorphic to the Tor-module as rings. Moreover, this application of Theorem 3.5 also shows that the basis of combinatorial generators is not symplectic.

Lastly, recall that the multiplicative structure of the cohomology of real moment-angle complexes plays an important role in the product structure for more general

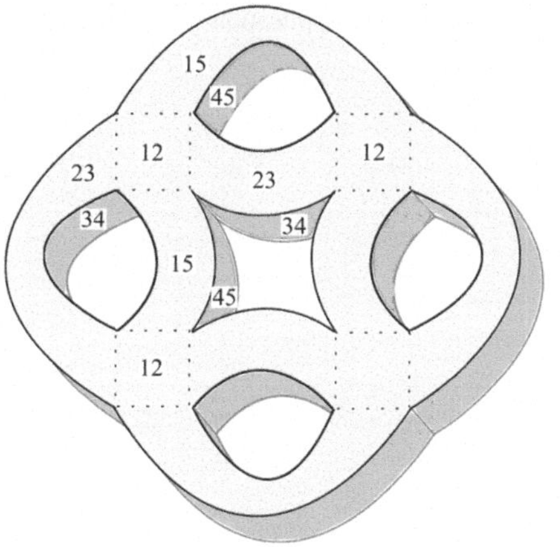

Fig. 1 Real moment-angle complex over the boundary of a pentagon. An edge $\{i, j\}$ of the pentagon contributes 2^3 copies of $D^1 \times D^1$. The copies of $D^1 \times D^1$ are labeled in the picture by which edge of the pentagon they come from

polyhedral product spaces [2]. Theorem 2.12 gives the algebra $H^*(Z_K(\underline{CA}, \underline{A}))$ in terms of the cohomology algebras of A_i and $H^*(Z_{K_I}(D^1, S^0))$.

Acknowledgments This work is part of the author's doctoral dissertation at the City University of New York Graduate Center. The author would like to thank Martin Bendersky for his guidance throughout this research.

References

1. A. Bahri, M. Bendersky, F.R. Cohen, S. Gitler, The polyhedral product functor: a method of decomposition for moment-angle complexes, arrangements and related spaces. Adv. Math. **225**(33), 1634–1668 (2010)
2. A. Bahri, M. Bendersky, F.R. Cohen, S. Gitler, Cup-products for the polyhedral product functor. Math. Proc. Cambridge Philosoph. Soc. **153**(3), 475–469 (2012)
3. A. Bahri, M. Bendersky, F.R. Cohen, S. Gitler, A spectral sequence for polyhedral products. Adv. Math. **308**, 767–814 (2017)
4. I.V. Baskakov, V.M. Buchshtaber, T.E. Panov, Algebras of cellular cochains and torus actions. Uspekhi Mat. Nauk **59**(3(357)), 159–160 (2004)
5. V.M. Buchstaber, T.E. Panov, *Toric Topology*. Mathematical Surveys and Monographs, vol. 204 (American Mathematical Society, Providence, 2015)
6. L. Cai, On products in a real moment-angle manifold. J. Math. Soc. Jpn **69**(2), 503–528 (2017)
7. S. Choi, H. Park, Multiplication structure of the cohomology ring of real toric spaces. arXiv:1711.04983
8. M.W. Davis, T. Januszkiewicz, Convex polytopes, Coxeter orbifolds and torus actions. Duke Math. J. **62**(2), 417–451 (1991)
9. M. Franz, The integral cohomology of toric manifolds, Tr. Mat. Inst. Steklova **252**, 53–62 (2006). Geom. Topol., Diskret. Geom. i Teor. Mnozh., 61–70
10. S. Gitler, S. López de Medrano, Intersections of quadrics, moment-angle manifolds and connected sums. Geom. Topol. **17**(3), 1497–1534 (2013)
11. M. Hochster, Cohen-Macaulay rings, combinatorics, and simplicial complexes. Lect. Notes Pure Appl. Math. **26**, 171–223 (1977)
12. T. Panov, Cohomology of face rings, and torus actions. Surveys in Contemporary Mathematics. London Math. Soc. Lecture Note Ser. **347**, 165–201 (2008)
13. V. Welker, G.M. Ziegler, R.T. Zivaljevic, Homotopy colimits – comparison lemmas for combinatorial applications. J. Reine Agnew Math. **509**, 117–149 (1999)

The manufacturer's authorised representative in the EU is Springer
Nature Customer Service Centre GmbH, Europaplatz 3, 69115 Heidelberg,
Germany. If you have any concerns regarding our products, please
contact ProductSafety@springernature.com

Printed and bound by CPI Group (UK) Ltd, Croydon, CR0 4YY
01/10/2026
02225904-0010